THE IMPORTATION INTO THE UNITED STATES OF THE PARASITES OF THE GIPSY MOTH AND THE BROWN-TAIL MOTH

This is a volume in the Arno Press collection

HISTORY OF ECOLOGY

Advisory Editor
Frank N. Egerton III

Editorial Board
John F. Lussenhop
Robert P. McIntosh

*See last pages of this volume for a
complete list of titles.*

THE IMPORTATION INTO THE UNITED STATES OF THE PARASITES OF THE GIPSY MOTH AND THE BROWN-TAIL MOTH

L[eland] O[ssian] Howard
and
W[illiam] F. Fiske

ARNO PRESS
A New York Times Company
New York / 1977

Editorial Supervision: LUCILLE MAIORCA

Reprint Edition 1977 by Arno Press Inc.

Reprinted from a copy in
The Pennsylvania State Library

HISTORY OF ECOLOGY
ISBN for complete set: 0-405-10369-7
See last pages of this volume for titles.

Manufactured in the United States of America

Publisher's Note: Plates I, IV, VII and all maps have been reproduced in black and white in this edition.

Library of Congress Cataloging in Publication Data

Howard, Leland Ossian, 1857-1950.
   The importation into the United States of the parasites of the gipsy moth and the brown-tail moth.

   (History of ecology)
   Reprint of the 1911 ed. published by the Govt. Print. Off., Washington, which was issued as no. 91 of the Bulletin of the U. S. Bureau of Entomology.
   1. Gipsy-moth--Biological control--United States.
2. Brown-tail moth--Biological control--United States.
3. Parasites--Gipsy-moth. 4. Parasites--Brown-tail moth. I. Fiske, William Fuller, joint author. II. Title. III. Series. IV. Series: United States. Bureau of Entomology. Bulletin ; no. 91.
SB945.G9H85 1977   632'.7'81   77-74230
ISBN 0-405-10400-6

THE IMPORTATION INTO THE UNITED STATES OF THE PARASITES OF THE GIPSY MOTH AND THE BROWN-TAIL MOTH

CALOSOMA SYCOPHANTA.

Adult eating gipsy-moth caterpillar, lower left; pupa, lower right; eggs, upper left; eaten chrysalides of gipsy moth, upper right; full-grown larvæ from above and from below. (Original.)

U. S. DEPARTMENT OF AGRICULTURE,
BUREAU OF ENTOMOLOGY—BULLETIN No. 91.
L. O. HOWARD, Entomologist and Chief of Bureau.

THE IMPORTATION INTO THE UNITED STATES OF THE PARASITES OF THE GIPSY MOTH AND THE BROWN-TAIL MOTH:

A REPORT OF PROGRESS, WITH SOME CONSIDERATION OF PREVIOUS AND CONCURRENT EFFORTS OF THIS KIND.

BY

L. O. HOWARD,
Chief, Bureau of Entomology,

AND

W. F. FISKE,
*In Charge, Gipsy Moth Parasite Laboratory,
Melrose Highlands, Mass.*

Issued July 29, 1911.

WASHINGTON:
GOVERNMENT PRINTING OFFICE.
1911.

BUREAU OF ENTOMOLOGY.

L. O. HOWARD, *Entomologist and Chief of Bureau.*
C. L. MARLATT, *Entomologist and Acting Chief in Absence of Chief.*
R. S. CLIFTON, *Executive Assistant.*
W. F. TASTET, *Chief Clerk.*

F. H. CHITTENDEN, *in charge of truck crop and stored product insect investigations.*
A. D. HOPKINS, *in charge of forest insect investigations.*
W. D. HUNTER, *in charge of southern field crop insect investigations.*
F. M. WEBSTER, *in charge of cereal and forage insect investigations.*
A. L. QUAINTANCE, *in charge of deciduous fruit insect investigations.*
E. F. PHILLIPS, *in charge of bee culture.*
D. M. ROGERS, *in charge of preventing spread of moths, field work.*
ROLLA P. CURRIE, *in charge of editorial work.*
MABEL COLCORD, *in charge of library.*

PREVENTING SPREAD OF MOTHS.

PARASITE LABORATORY.

W. F. FISKE, *in charge;* A. F. BURGESS, C. W. COLLINS, R. WOOLDRIDGE, J. D. TOTHILL, C. W. STOCKWELL, H. E. SMITH, W. N. DOVENER, F. H. MOSHER, *assistants.*

FIELD WORK.

D. M. ROGERS, *in charge;* H. B. DALTON, H. W. VINTON, D. G. MURPHY, I. L. BAILEY, H. L. MCINTYRE, *assistants.*

LETTER OF TRANSMITTAL.

U. S. DEPARTMENT OF AGRICULTURE,
BUREAU OF ENTOMOLOGY,
Washington, D. C., April 12, 1911.

SIR: I have the honor to transmit herewith the manuscript of a report of progress on the importation into the United States of the parasites of the gipsy moth and the brown-tail moth. To this has been added some consideration of previous and concurrent efforts to handle the parasites of destructive insects in a practical way. The work with the foreign parasites of the gipsy moth and the brown-tail moth has been going on now for rather more than five years. It promises excellent results, and the present seems the proper time to present to the people interested a somewhat detailed account of what has been done and of the present condition of the work. I recommend that this manuscript be published as Bulletin No. 91 of this bureau.

Respectfully,
L. O. HOWARD,
Entomologist and Chief of Bureau

Hon. JAMES WILSON,
Secretary of Agriculture.

CONTENTS.

	Page.
Introduction	13
Previous work in the practical handling of natural enemies of injurious insects	16
Early practical work	17
Permitting the parasites to escape	18
The transportation of parasites from one part of a given country to another part	20
The transfer of beneficial insects from one country to another	23
Early attempts	23
The Australian ladybird (*Novius cardinalis* Muls.) in the United States	24
Novius in Portugal	27
Icerya in Florida	28
Novius in Cape Colony	28
Novius in Egypt and the Hawaiian Islands	28
Icerya in Italy	29
Icerya in Syria	29
The reasons for the success of Novius	29
Introduction of *Entedon epigonus* Walk. into the United States	30
Other introductions by Koebele into California	31
International work with enemies of the black scale	31
The Hawaiian work	34
An importation of Clerus from Germany	36
Marlatt's journey for enemies of the San Jose scale	36
The parasites of *Diaspis pentagona* Targ	38
The work of Mr. George Compere	38
Work with the egg parasite of the elm leaf-beetle	39
Work with parasites of ticks	41
Mr. Froggatt's journey to various parts of the world in 1907–8	42
Other work of this kind (by Berlese; by Silvestri; in Algeria; in the Philippines; by De Bussy; in Peru)	44
Early ideas on introducing the natural enemies of the gipsy moth	47
Circumstances which brought about the actual beginning of the work	49
An investigation of the introduction work	50
Narrative of the progress of the work	54
Known and recorded parasites of the gipsy moth and of the brown-tail moth	84
Establishment and dispersion of the newly introduced parasites	94
Disease as a factor in the natural control of the gipsy moth and the brown-tail moth	97
Studies in the parasitism of native insects	102
Parasitism as a factor in insect control	105
The rate of increase of the gipsy moth in New England	109
Amount of additional control necessary to check the increase of the gipsy moth in America	114
The extent to which the gipsy moth is controlled through parasitism abroad	117
Parasitism of the gipsy moth in Japan	120
Parasitism of the gipsy moth in Russia	123
Parasitism of the gipsy moth in southern France	129

	Page.
Sequence of parasites of the gipsy moth in Europe	131
The brown-tail moth and its parasites in Europe	132
Sequence of parasites of the brown-tail moth in Europe	135
Parasitism of the gipsy moth in America	136
Summary of rearing work carried on at the laboratory in 1910	141
Parasitism of the brown-tail moth in America	143
Summary of rearing work in 1910	146
Importation and handling of parasite material	152
Egg masses of the gipsy moth	152
Gipsy-moth caterpillars, first stage	153
Gipsy-moth caterpillars, second to fifth stages	154
European importations	154
Japanese importations	155
Gipsy-moth caterpillars, full-fed and pupating	156
Gipsy-moth pupæ	159
Brown-tail moth egg masses	160
Hibernating nests of the brown-tail moth	161
Immature caterpillars of the brown-tail moth	161
Full-fed and pupating caterpillars of the brown-tail moth	162
Brown-tail moth pupæ	164
Cocoons of hymenopterous parasites	165
Tachinid puparia	166
Calosoma and other predaceous beetles	167
Quantity of parasite material imported	167
Localities from which the parasite material has been received	168
The egg parasites of the gipsy moth	168
Anastatus bifasciatus Fonsc	168
Schedius kuvanæ How	176
Life of Schedius and its relations to other egg parasites, primary and secondary	177
Rearing and colonization	184
The parasites of the gipsy-moth caterpillars	188
Apparently unimportant hymenopterous parasites	188
Apanteles solitarius Ratz	189
Meteorus versicolor Wesm	190
Meteorus pulchricornis Wesm	190
Meteorus japonicus Ashm	190
Limnerium disparis Vier	191
Limnerium (Anilastus) tricoloripes Vier	192
Apanteles fulvipes Hal	193
Secondary parasites attacking *Apanteles fulvipes*	198
Tachinid parasites of the gipsy moth	202
The rearing and colonization of tachinid flies; large cages versus small cages	204
Hyperparasites attacking the Tachinidæ	207
Perilampus cuprinus Först	208
Melittobia acasta Walk	209
Chalcis fiskei Crawf	212
Monodontomerus æreus Walk	212
Miscellaneous parasites	213
Blepharipa scutellata Desv	213
Compsilura concinnata Meig	218
Tachina larvarum L	225
Tachina japonica Towns	227

CONTENTS.

	Page.
Tachinid parasites of the gipsy moth—Continued.	
Tricholyga grandis Zett	228
Parasetigena segregata Rond	229
Carcelia gnava Meig	231
Zygobothria nidicola Towns	232
Crossocosmia sericariæ Corn	232
Crossocosmia flavoscutellata Schiner (?)	234
Unimportant tachinid parasites of the gipsy moth	235
Parasites of the gipsy-moth pupæ	236
The genus Theronia	236
The genus Pimpla	237
Ichneumon disparis Poda	239
The genus Chalcis	240
Monodontomerus æreus Walk	245
The sarcophagids	250
The predaceous beetles	251
The egg parasites of the brown-tail moth	256
The genus Trichogramma	256
Telenomus phalænarum Nees	260
Parasites which hibernate within the webs of the brown-tail moth	261
Pediculoides ventricosus Newp	267
Pteromalus egregius Först	268
Apanteles lacteicolor Vier	278
Apanteles conspersæ Fiske	285
Meteorus versicolor Wesm	286
Zygobothria nidicola Towns	289
Parasites attacking the larger caterpillars of the brown-tail moth	295
Hymenopterous parasites	295
Tachinid parasites	296
Dexodes nigripes Fall	296
Parexorista cheloniæ Rond	297
Pales pavida Meig	300
Zenillia libatrix Panz	302
Masicera sylvatica Fall	303
Eudoromyia magnicornis Zett	303
Cyclotophrys anser Towns	304
Blepharidea vulgaris Fall	304
Parasites of the pupæ of the brown-tail moth	304
Summary and conclusions	305
The present status of the introduced parasites	307
The developments of the year 1910	311

ILLUSTRATIONS.

PLATES.

		Page.
Plate	I. The Calosoma beetles................................. Frontispiece.	
	II. Fig. 1.—View of parasite laboratory at North Saugus, Mass. Fig. 2.—View of parasite laboratory at Melrose Highlands, Mass..	56
	III. Fig. 1.—Roadside oak in Brittany, with leaves ragged by gipsy-moth caterpillars. Fig. 2.—M. René Oberthür, Dr. Paul Marchal; with roadside oaks ragged by gipsy-moth caterpillars...	76
	IV. Fig. 1.—Caterpillar hunters in the south of France, under M. Dillon, 1909. Fig. 2.—Packing parasitized caterpillars at Hyères, France, for shipment to the United States, 1909.....	76
	V. Fig. 1.—View of interior of one of the laboratory structures, showing rearing cages for brown-tail moth parasites. Fig. 2.—Box used in shipping immature caterpillars of the gipsy moth from Japan...	152
	VI. The gipsy moth (*Porthetria dispar*)...........................	156
	VII. The brown-tail moth (*Euproctis chrysorrhœa*)..................	160
	VIII. Fig. 1.—Boxes used in 1910 for importation of brown-tail moth caterpillars, with tubes attached directly to boxes. Fig. 2.—Interior of boxes in which brown-tail moth caterpillars were imported, showing condition on receipt. Fig. 3.—Boxes used in shipping caterpillars of the gipsy and brown-tail moths by mail...	164
	IX. Fig. 1.—Headgear devised by Mr. E. S. G. Titus as a protection against brown-tail rash. Fig. 2.—Show case used when opening boxes of brown-tail moth caterpillars received from abroad...	164
	X. Fig. 1.—Large tube cage first used for rearing parasites from imported brown-tail moth nests and latterly for various purposes. Fig. 2.—Method of packing Calosoma beetles for shipment..	164
	XI. Fig. 1.—Egg of gipsy moth containing developing caterpillar of the gipsy moth. Fig. 2.—Egg of gipsy moth, containing larva of the parasite *Anastatus bifasciatus*. Fig. 3.—Egg of gipsy moth, containing hibernating larva of *Anastatus bifasciatus* which in turn is parasitized by three second-stage larvæ of *Schedius kuvanæ*................................	172
	XII. Fig. 1.—View of cage used for colonization of *Anastatus bifasciatus* in 1910. Fig. 2.—Views of cage prepared for use in colonization of *Anastatus bifasciatus* in 1911................	172
	XIII. Outdoor parasite cage covered with wire gauze................	204
	XIV. Outdoor parasite cages covered with cloth.....................	204
	XV. View of large cage used in 1908 for tachinid rearing work.......	204

		Page.
PLATE	XVI. View of out-of-door insectary used for rearing predaceous beetles in 1910	204
	XVII. Fig. 1.—Wire-screen cages used in tachinid reproduction work in 1909. Fig. 2.—Cylindrical wire-screen cages used in tachinid reproduction work in 1910	204
	XVIII. Fig. 1.—*Blepharipa scutellata:* Full-grown larva from gipsy-moth pupa. Fig. 2.—*Blepharipa scutellata:* Puparia	216
	XIX. Fig. 1.—Importation of gipsy-moth caterpillars from France in 1909; en route to laboratory at Melrose Highlands, Mass. Fig. 2.—Importation of gipsy-moth caterpillars from France in 1909; receipt at laboratory, Melrose Highlands, Mass	216
	XX. Fig. 1.—*Compsilura concinnata:* Puparia. Fig. 2.—*Tachina larvarum:* Puparia. Fig. 3.—*Sarcophaga* sp.: Puparia. Fig. 4.—*Parexorista cheloniæ:* Puparia	220
	XXI. Fig. 1.—View of laboratory interior, showing cages in use for rearing parasites from hibernating webs of the brown-tail moth in 1910–11. Fig. 2.—Sifting gipsy-moth egg masses for examination as to percentage of parasitism	244
	XXII. Map showing sections of its range in New England from which *Monodontomerus æreus* has been collected in hibernating webs of the brown-tail moth, and subsequently reared	248
	XXIII. Map showing distribution of *Monodontomerus æreus* in New England	248
	XXIV. Map showing dispersion of *Calosoma sycophanta* in Massachusetts from liberated colonies	256
	XXV. Map showing distribution of *Pteromalus egregius* in New England	276
	XXVI. Fig. 1.—Riley rearing cages as used at the gipsy-moth parasite laboratory. Fig. 2.—Interior of one of the laboratory structures, showing trays used in rearing *Apanteles lacteicolor* in the spring of 1909	280
	XXVII. View of laboratory interior, showing cages in use for rearing parasites from hibernating webs of the brown-tail moth in the spring of 1908	280
	XXVIII. Fig. 1.—Cocoons of *Apanteles lacteicolor* in molting webs of the brown-tail moth. Fig. 2.—View of laboratory yard, showing various temporary structures, rearing cages, etc	284

TEXT FIGURES.

FIG.	1. *Polygnotus hiemalis*, a parasite of the Hessian fly	21
	2. *Polygnotus hiemalis:* Adults which have developed within the "flaxseed" of the Hessian fly and are ready to emerge	21
	3. *Lysiphlebus tritici* attacking a grain aphis	22
	4. The Australian ladybird (*Novius cardinalis*), an imported enemy of the fluted scale: Larvæ, pupa, adult, work against scales	25
	5. *Rhizobius ventralis*, an imported enemy of the black scale: Adult, larva	31
	6. *Scutellista cyanea*, an imported parasite of the black scale	32
	7. *Pediculoides ventricosus*	34
	8. *Erastria scitula*, an imported enemy of the black scale: Adult, larvæ, pupa	34
	9. The Asiatic ladybird (*Chilocorus similis*), an imported enemy of the San Jose scale: Later larval stages, pupa, adults	37
	10. Rearing cage for tachinid parasites of the brown-tail moth	151

ILLUSTRATIONS. 11

Page.
FIG. 11. Map showing various localities in Europe from which parasite material has been received... 169
12. *Anastatus bifasciatus:* Adult female................................. 170
13. *Anastatus bifasciatus:* Uterine egg.................................. 171
14. *Anastatus bifasciatus:* Hibernating larva............................ 171
15. *Anastatus bifasciatus:* Pupa from gipsy-moth egg.................... 171
16. Diagram showing two years' dispersion of *Anastatus bifasciatus* from colony center.. 173
17. *Schedius kuvanæ:* Adult female...................................... 176
18. *Schedius kuvanæ:* Egg... 179
19. *Schedius kuvanæ:* Third-stage larva still retaining attachment to egg stalk, and anal shield... 180
20. *Schedius kuvanæ:* Pupa.. 180
21. *Schedius kuvanæ:* Egg stalk and anal shield of larva as found in host eggs of gipsy moth from which the adult Schedius has emerged, or in which the Schedius larva has been attacked by a secondary parasite. 181
22. *Schedius kuvanæ:* Larval mandibles.................................. 181
23. *Tyndarichus navæ:* Larval mandibles................................. 181
24. *Pachyneuron gifuensis:* Egg... 182
25. *Pachyneuron gifuensis:* Larval mandibles............................ 182
26. *Anastatus bifasciatus:* Larval mandibles............................ 182
27. Gipsy-moth egg mass showing exit holes of *Schedius kuvanæ*.......... 186
28. *Apantales solitarius:* Adult female and cocoon...................... 189
29. *Limnerium disparis:* Cocoon... 191
30. *Limnerium disparis:* Adult male..................................... 191
31. *Apanteles fulvipes:* Adult... 193
32. *Apanteles fulvipes:* Larvæ leaving gipsy-moth caterpillar............ 194
33. *Apanteles fulvipes:* Cocoons surrounding dead gipsy-moth caterpillar. 195
34. *Apanteles fulvipes:* Cocoons from which Apanteles and its secondaries have issued... 199
35. *Blepharipa scutellata:* Adult female................................ 213
36. *Blepharipa scutellata:* Eggs *in situ* on fragment of leaf............ 214
37. Eggs of *Blepharipa scutellata* and *Pales pavida*................... 214
38. *Blepharipa scutellata:* First-stage larvæ............................ 215
39. *Blepharipa scutellata:* Second-stage larva *in situ*................. 215
40. *Blepharipa scutellata:* Basal portion of tracheal "funnel"............ 216
41. *Compsilura concinnata:* Adult female and details.................... 219
42. Map showing distribution of *Compsilura concinnata* in Massachusetts. 222
43. *Tachina larvarum:* Adult female and head in profile.................. 225
44. *Chalcis flavipes:* Adult... 241
45. *Chalcis flavipes*, female: Hind femur and tibia, showing markings... 242
46. *Chalcis obscurata*, female: Hind femur and tibia, showing markings.. 242
47. *Chalcis flavipes:* Full-grown larva from gipsy-moth pupa............. 243
48. *Chalcis flavipes:* Pupa, side view................................... 243
49. *Chalcis flavipes:* Pupa, ventral view................................ 243
50. Gipsy-moth pupæ, showing exit holes of *Chalcis flavipes*............. 243
51. *Monodontomerus æreus:* Adult female................................. 244
52. *Monodontomerus æreus:* Egg.. 249
53. *Monodontomerus æreus:* Larva.. 249
54. *Monodontomerus æreus:* Pupa, side view.............................. 249
55. *Monodontomerus æreus:* Pupa, ventral view........................... 249
56. Gipsy-moth pupa showing exit hole left by *Monodontomerus æreus*..... 250
57. *Trichogramma* sp. in act of oviposition in an egg of the brown-tail moth. 256

	Page.
FIG. 58. Eggs of the brown-tail moth, a portion of which has been parasitized by *Trichogramma* sp	257
59. Larvæ of *Pteromalus egregius*, feeding on hibernating caterpillars of the brown-tail moth	262
60. Portion of brown-tail moth nests, torn open, showing caterpillars attacked by larvæ of *Pteromalus egregius*	263
61. *Apanteles lacteicolor:* Immature larva from hibernating caterpillar of the brown-tail moth	263
62. *Meteorus versicolor:* Immature larva from hibernating caterpillar of the brown-tail moth	264
63. *Zygobothria nidicola:* First-stage larvæ *in situ* in walls of crop of hibernating brown-tail moth caterpillar	264
64. *Compsilura concinnata:* First-stage larva	265
65. *Pteromalus egregius:* Adult female	269
66. *Pteromalus egregius:* Female in the act of oviposition through the silken envelope containing hibernating caterpillars of the brown-tail moth	274
67. *Apanteles lacteicolor:* Adult female and cocoon	279
68. *Meteorus versicolor:* Adult female and cocoons	287
69. *Zygobothria nidicola:* Adult female and details	290
70. *Pales pavida:* Adult female and details	301
71. *Pales pavida:* Second-stage larva *in situ* in basal portion of integumental "funnel"	302
72. *Pales pavida:* Integumental "funnel," showing orifice in skin of host caterpillar	302
73. *Eudoromyia magnicornis:* Adult female and details	303
74. *Eudoromyia magnicornis:* First-stage maggot and mouth hook	303

THE IMPORTATION INTO THE UNITED STATES OF THE PARASITES OF THE GIPSY MOTH AND THE BROWN-TAIL MOTH:

A REPORT OF PROGRESS,

WITH SOME CONSIDERATION OF PREVIOUS AND CONCURRENT EFFORTS OF THIS KIND.

INTRODUCTION.

By L. O. HOWARD,
Chief, Bureau of Entomology.

As will appear from the opening portion of this bulletin, which gives an account of previous work in the practical handling of natural enemies, carried on in various parts of the world, nothing comparable to the work which is to be described has ever before been undertaken. As will appear also, most of the successful work in this direction has been done with the fixed scale insects. The exceptions to this general statement among the measurably successful efforts have been the introduction of parasites of the sugar-cane leafhopper into Hawaii, some reported work in the introduction of South American natural enemies of fruit flies into Western Australia, and the introduction of one of the many European enemies of the codling moth from Spain into California; but it does not appear that practical results of any very great value have been achieved by the last two introductions, although information from Western Australia is scanty. At the time when the work began nothing practical had been accomplished with the natural enemies of any lepidopterous insects, and in the whole history of the practical handling of parasites no work of this character has ever been attempted upon anything like the large scale with which the present work has been carried on. Some studies had already been made both by the writer and by Mr. Fiske on the subject of the intensive parasitism of two native species of American moths, and for years the bureau had been keeping records of the rearings of parasites of lepidopterous insects as well as of others; moreover, the writer had made a careful study of the records of the rearings of hymenopterous parasites from host insects all over the world and had accumulated an enormous catalogue of such records. Nevertheless the initial work on such a scale was experimental in its character. It seemed to the writer that by attempting to reproduce in New

England as nearly as possible the entire natural environment of the gipsy moth and the brown-tail moth in their native homes, similar conditions of comparative scarcity could surely be reached, and this view he still holds with enthusiasm. Naturally, in the course of the work as it progressed year after year his ideas have been changed as to methods, and very great improvements have been made upon the earlier methods, largely through the intelligence and ingenuity of the junior author of this bulletin. Moreover, the careful, intensive studies which have been made at the gipsy-moth parasite laboratory by the junior author and a corps of trained assistants, aided by abundant material, funds, and supplies, have resulted not only in the ascertainment of very many facts new to science, but in the accumulation of such facts to such a degree as to enable generalizations of a novel character and of a sounder basis than could have been had under other conditions. Many points are brought out in this bulletin which will doubtless be entirely new to the trained scientific reader. Mistakes have been made and wrong conclusions have been drawn from time to time, but these have been corrected, and we are now in a fair way to see a favorable result from the long and expensive work.

The initial idea was that since a large percentage of gipsy-moth caterpillars or brown-tail moth caterpillars in Europe contains parasites each year, therefore if these caterpillars were brought to America in large numbers from every possible place we could not fail to rear from them an abundance of adult foreign parasites. This idea was sound, and in following it out we have constantly improved the methods—methods of collection, of packing, of shipment, and of subsequent rearing. Very large numbers of parasites have been reared.

It was first thought that when parasites had been reared in sufficient numbers they should be widely distributed in small colonies, on the theory that each colony would remain in substantially the same general locality and would increase and spread from that point. This idea was a natural one and was fully justified by previous work which had been done with parasites of other groups of insects, but in this case it proved to be erroneous, and valuable time and valuable specimens were lost. Eventually it was shown to be of prime importance, first to establish a given species of parasite in this country, and not until this has been accomplished to pay any attention to the matter of dispersion. It seems to be the first instinct of many species that have been imported to spread widely. Therefore, if the colony put out be a small one the individuals composing it spread rapidly beyond all means of meeting and of mating, and thus the colonies in many instances were lost. By rearing in the laboratory, however, until colonies of at least a thousand are to be had, such colonies while dispersing are much more likely to remain in touch, mate, and multiply.

By methods based upon the first idea, and by the subsequent modification of the second idea, some of the most important natural enemies of both species have been established in the United States to a certainty. It has been found with several species that they could not be recovered until after three years had elapsed from the time of the original colonization; hence it follows with a reasonable certainty that other species which have not been recovered will ultimately be recovered as a result of colonization one, two, and three, and even perhaps four years ago. It is deemed, however, at this time that nearly as much has been accomplished as can be accomplished by the earlier methods, and subsequent efforts will be devoted to a more specific attempt to import the species still lacking, several of which are known in their original homes to be of very great importance. As will be pointed out elsewhere, attempts will also be made to import the species which, while of lesser importance at home, may here fill in gaps and may possibly multiply to an unprecedented extent in the face of new conditions and a superabundance of host material.

The work has been going on since 1905. Nothing has been published concerning its progress except the short accounts in the annual reports of the writer submitted each year to the Secretary of Agriculture, and except a bulletin on the general subject prepared by the junior author and published by the State forester of Massachusetts. It is hoped that the present account will be deemed a satisfactory reply to all expressed desire for information as to progress.

The joint authorship of the bulletin is deemed desirable by both authors, but the writer takes it upon himself to sign this introduction for the explicit purpose of stating in his own way the conditions under which it has been prepared. The work from the beginning has been under the direct supervision of the writer, and he is therefore to be held responsible for any failures in the speedy accomplishment of results, but the greatest credit in bringing about the results which have been accomplished, he wishes frankly to state, belongs to Mr. Fiske. Following the breakdown in health of Mr. E. S. G. Titus in the spring of 1907, as is shown in the bulletin, Mr. Fiske was stationed at the parasite laboratory and has since been given every freedom in the conduct of its affairs. Nearly every suggestion which he has made, while it has been fully discussed by the two of us, has been adopted. The ingenuity which he has displayed in matters of method and the broad grasp which he has shown of the whole phenomena of parasitism in insects, together with his competent and practical grouping of his ideas, deserve every praise. Such portions of the bulletin as were dictated by the writer have received the editorial criticism of the junior author, and the portions prepared by the latter have received a most careful consideration and editorial pruning of the writer. Mr. Fiske, by virtue of his practical residence at the field laboratory and of his intimate charge of all the field notes and laboratory notes, has prepared all of the matter in this bulletin relating to the laboratory and field end, subject, of course, to the writer's revision. The rest has been prepared by the senior author.

Acknowledgements of assistance should be made by the score. The State authorities of Massachusetts, the admirable corps of laboratory and field assistants, and above all the very numerous foreign officials, voluntary assistants, and paid observers have united to make the undertaking possible. Their individual names are all mentioned in the following pages in connection with the parts they played, but the Governments of Austria, France, Germany, Hungary, Italy, Japan, Portugal, Russia, and Spain should especially be thanked in an official publication like this for the assistance given by the officials of these Governments.

PREVIOUS WORK IN THE PRACTICAL HANDLING OF NATURAL ENEMIES OF INJURIOUS INSECTS.

Two very thorough and careful general papers on the subject of the practical handling of natural enemies of insects, treating the subject from the different points of view, including the historical side, have been published in the last few years. The first of these, entitled "The Utilization of Auxiliary Entomophagous Insects in the Struggle against Insects Injurious to Agriculture," by Prof. Paul Marchal, of the National Agronomical Institute of Paris, was published in 1907,[1] and was partly republished in English in the Popular Science Monthly in 1908.[2] The other, by Prof. F. Silvestri, of the Royal Agricultural School at Portici, Italy, entitled "Consideration of the Existing Condition of Agricultural Entomology in the United States of North America, and Suggestions which can be Gained from it for the Benefit of Italian Agriculture," was published in 1909.[3] This paper was in part translated into English and published in the Hawaiian Forester and Agriculturist for August, 1909. Both of these papers should be consulted by persons wishing to inform themselves thoroughly on this question. For the present purpose, treatment of the subject must be brief.

The study of parasitic and predatory insects is old. Silvestri has pointed out that Aldrovandi (1602) was the first to observe the exit of the larvæ of *Apanteles glomeratus* L. (which he supposed to be eggs) from the common cabbage caterpillar, and that Redi (1668) published the same observation and another on insects of different species born from the same pupa. A later writer, Vallisnieri (1661–

---

[1] Annals of the National Agronomical Institute (Superior School of Agriculture), second series vol. 6, no. 2, pp. 281–354, Paris, 1907.
[2] Popular Science Monthly, vol. 72, pp. 353–370, 407–419, April and May, 1908.
[3] Bulletin of the Society of Italian Agriculturists, vol. 14, no. 8, pp. 305–367, Apr. 30, 1909.

1730) was apparently the first to discover the real nature of this phenomenon and to realize the existence of true parasitic insects. Réaumur (1683–1757) and De Geer (1720–1778) each studied the life histories of living insects with great care and among these worked out the biology of a number of parasites. Very many descriptive works on parasites were published in the closing years of the eighteenth century and the beginning of the nineteenth century, especially by Dalman (1778–1828), Nees ab Esenbeck (1776–1858), Gravenhorst (1777–1857), Walker (publishing from 1833 to 1861), Westwood (publishing from 1827 on through nearly the whole of the century), Förster (publishing from 1841 on), and Spinola (1780–1857).

Many later writers have contributed to the systematic study of these insects, among them Holmgren and Thomson, of Sweden; Mayr, of Austria; Motschulsky, of Russia; Ratzeburg, Hartig, and Schmiedeknecht, of Germany; Wesmael, of Belgium; Haliday, Marshall, and Cameron, of England; Rondani, of Italy; Brullé, Giraud, Decaux, and others in France; Provancher, of Canada; and, in America, Cresson, Riley, Howard, Ashmead, Crawford, Viereck, Brues, Girault, and others.

The best contribution appearing in Europe and devoted to the biology of hymenopterous parasites, and especially consideration of their relations to their hosts, was that by Ratzeburg, whose great work entitled "Die Ichneumonen der Forstinsekten," was a standard for many years. Ratzeburg understood the rôle played by parasites in the control of forest insects, but did not believe that this control could in any way be facilitated by man.

EARLY PRACTICAL WORK.

Froggatt has pointed out that probably the earliest suggestion made regarding the artificial handling of beneficial insects was printed in Kirby and Spence's entomology (1816), where the authors called attention to the value of the common English ladybird as destroying the hop aphis in the south of England. "If we could but discover a mode of increasing these insects at will, we might not only clear our hothouses of aphides by their means, but render our crops of hops much more certain than they are now." As a matter of fact, gardeners and florists in England for very many years have recognized the value of the ladybirds and have transferred them from one plat to another.

Prof. A. Trotter, of the Royal School of Viticulture at Avellino, Italy, has recently pointed out in an interesting paper entitled "Two Precursors in the Application of Carnivorous Insects," published in Redia, in 1908.[1] that probably the first person to make a practical application of the natural enemies of injurious species was Prof.

---
[1] Redia, vol. 5, pp. 126–132, Florence, 1908.

Boisgiraud, of Poitiers, France, in 1840. Prof. Trotter found this reference in a little-known paper by N. Joly, published in 1842, and entitled "Notice of the Ravages which *Liparis dispar* L. has made around Toulouse, followed by some Reflexions upon a Method of Destroying Certain Insects." It seems that Boisgiraud, about 1840, freed the poplars along a road near Poitiers of the gipsy moth by placing upon them the carabid beetle *Calosoma sycophanta* L., and destroyed earwigs in his own garden by placing with them a rove beetle (*Staphylinus olens* Müll). He also experimented against the same insect with the ground beetle *Carabus auratus* L. His experiment must have become rather well known at the time, since Prof. Trotter points out that in 1843 the technical commission of the Society for the Encouragement of Arts and Crafts of Milan offered a gold medal to be given in 1845 to the person who in the meantime should have undertaken with some success new experiments tending to promote the artificial development of some species of carnivorous insects which could be used efficaciously to destroy another species of insect recognized as injurious to agriculture. This offer drew forth a memoir from Antonio Villa, a well-known writer on entomology, who had previously confined himself to the Coleoptera, entitled "The Carnivorous Insects used to Destroy the Species Injurious to Agriculture." This memoir was presented December 26, 1844, and he advocated the employment of climbing carabid beetles for tree-inhabiting forms, rove beetles to destroy the insects found in flowers, and ground beetles for cutworms and other earth-inhabiting forms. The paper of Villa was praised in certain reviews and criticized in others. It seems to have been entirely lost sight of in later years.

A later Italian writer, Rondani, who devoted himself for the most part to systematic work, appreciated the practical importance of parasite work and published tables giving the host relations of different species. His work influenced many arguments in the dispute which sprang up in Italy about 1868 as to the usefulness of insectivorous birds to agriculture, and Silvestri calls attention to the fact that Dr. T. Bellenghi was referring to Rondani when, in 1872, he spoke what Silvestri calls "the prophetic words:" "Entomological parasitism has a future, and in it more than in anything else Italian agriculture must put its faith."

PERMITTING THE PARASITES TO ESCAPE.

The earliest published suggestion as to the practical use of parasites of injurious insects, by permitting the parasites to escape while the host insect is killed, appears to have been made by C. V. Riley when State entomologist of Missouri. Writing of the rascal leaf-crumpler (*Mineola indiginella* Zell.) in his Fourth Report on the

Insects of Missouri,[1] he advocated the collecting of the winter cases of the destructive insect and placing the cases in small vessels in the center of a meadow or field, away from any fruit trees, with the idea that the worms would be able to wander only a few yards and would perish from exhaustion or starvation, while their parasites would escape and fly back to the fruit trees. It is stated that this method was put in practice later by D. B. Wier with success.

A French writer, F. Decaux, the following year made practically the same suggestion with regard to apple buds attacked by Anthonomus. He advised that instead of burning these buds, as was generally done, they be preserved in boxes covered with gauze, raising the latter from time to time during the period of issuing of parasites so as to permit them to escape. In 1880 he put this method in practice, and collected in Picardy buds reddened by the Anthonomus from 800 apple trees, and thus accomplished the destruction of more than 1,000,000 individuals of the Anthonomus, setting at liberty about 250,000 parasites which aided the following year in the destruction of the weevils. The following year the same process was repeated, and, the orchards being isolated in the middle of cultivated fields, all serious damage from the Anthonomus was stated to have been stopped for 10 years.[2]

Practically the same suggestion was made later, in 1877, by J. H. Comstock, in regard to the imported cabbage worm (*Pontia rapæ* L.). Comstock deprecated the indiscriminate crushing of the chrysalids collected under trap boards, on account of the large percentage which contained parasites. He recommended instead the collecting of the chrysalids and placing them in a box covered with a wire screen which should permit the parasites to escape and at the same time confine the butterflies so that they could be easily destroyed. The same author, in his report upon cotton insects,[3] recommended a similar course with the pupæ of the cotton caterpillar (*Alabama argillacea* Hübn.).

Riley later recommended the same plan for the bagworm (*Thyridopteryx ephemeræformis* Haw.); Berlese in Italy recommended it for the grapevine Cochylis, and Silvestri for the olive fly (*Dacus oleæ* Rossi), for *Prays oleellus* Fab., and for *Asphondylia lupini* Silv.

Writing on the Hessian fly, Marchal has pointed out that the destruction of the stubble remaining in the field after harvest may have unfortunate consequences, for if this is done a little late there is a risk that all of the destructive flies will have emerged and abandoned the stubble, exposing to destruction only the parasites whose part would have been to stop the invasion the following year. Marchal also points out that Kieffer has shown that one of the measures

---

[1] Riley, C. V. Fourth Report on the Insects of Missouri, p. 40, 1871.

[2] An excellent article covering these general questions was published by Decaux in the Journal of the National Horticultural Society of France, vol. 22, pp. 158-184, 1899.

[3] Cotton Insects, pp. 230-231, Washington, 1879.

advised for the destruction of the wheat midge (*Contarinia tritici* Kirby), namely, burning the débris after thrashing, has only an injurious effect, for, while it is true that the pupæ of the midge are to be found in this débris, it should be remembered that the healthy nonparasitized larvæ of the midge transform in the ground, while those which remain in the heads are, on the contrary, parasitized.

Still another method of encouraging parasites is pointed out by Marchal and Silvestri. It is to cultivate in the olive groves various plants upon which allied insects live which are parasitized by the same species of parasites as the olive fly. This idea, independently developed in the United States, has been practically used by Hunter in the fight against the cotton-boll weevil. Allied insects feeding in certain weeds along the borders of the cotton fields have parasites capable of attacking the boll weevil. Careful study of the biology of these allied weevils and of their parasites resulted in the gaining of the information that if the weeds are cut at a certain time the parasites are forced to attack the cotton-boll weevil in order to maintain their existence; actual experimentation has resulted in the very considerable increasing of the percentage of parasitism of the cotton-boll weevil in this way.

THE TRANSPORTATION OF PARASITES FROM ONE PART OF A GIVEN COUNTRY TO ANOTHER PART.

In 1872 attempts were made by Dr. William Le Baron, at that time State entomologist of Illinois, to transport *Aphelinus mali* Le Baron, a parasite of the oyster-shell scale of the apple (*Lepidosaphes ulmi* L.) from one part of the State of Illinois to another portion of the same State where the parasite seemed to be lacking. Some slight success was reported, and at the end of the year it was stated that the parasite had become domiciled in the new locality, but, as this parasite subsequently proved to be one of general American distribution, the experiment can not be said to have been worth while except in a very small way.

In France, F. Decaux, above quoted, in 1872, made some experiments in the transportation of parasites from one locality to another.

Riley, in his third report as State entomologist of Missouri (1870), in considering two parasites of the plum curculio, stated that he intended the following year, if possible, to rear enough specimens of *Sigalphus curculionis* Fitch to send at least a dozen to every county seat in the State and have them liberated in someone's peach orchard. There seems, however, to be no record that this was ever done.

In 1880, in his report on the parasites of the Coccidæ in the collection of the Department of Agriculture,[1] the senior author called attention to the fact that with the parasites of scales the matter of trans-

---

[1] Annual Report U. S. Department of Agriculture for 1880, p. 351.

portation from one part of the country to another becomes easy, since all that has to be done is simply to collect twigs bearing the scales, preferably during the winter months, and carry them to nonprotected regions, the parasites being dormant and protected each by the scale of the coccid which it had destroyed; and it was specifically recommended that the important parasite of the black scale (*Saissetia oleæ* Bern.), described in the article as *Tomocera californica*, could be readily carried from California and utilized to destroy Lecanium scales in the Southeast.

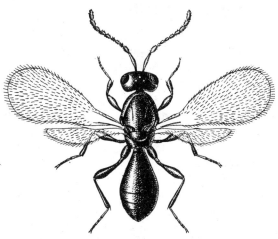

FIG. 1.—*Polygnotus hiemalis*, a parasite of the Hessian fly: Adult. Greatly enlarged. (From Webster.)

FIG. 2.—*Polygnotus hiemalis:* Adults which have developed within the "flaxseed" of the Hessian fly and are ready to emerge. Much enlarged. (From Webster.)

Excellent work in this direction has been done of late years by the Bureau of Entomology. In the study of the Hessian fly (*Mayetiola destructor* Say), under Prof. F. M. Webster, early-sown plats of wheat at Lansing, Mich., and Marion, Pa., in 1906, were very seriously attacked by the Hessian fly, but when examined carefully at a later date fully 90 per cent of the flaxseeds (pupæ) were found to have been stung by a hymenopterous parasite, *Polygnotus hiemalis* Forbes (figs. 1, 2), and to contain its developing larvæ. A field of wheat near Sharpsburg, Md., was found to be infested by the fly, and examination indicated the absence of the parasite. On April 8, 1907, a large number of the parasitized flaxseeds from Marion, Pa., were brought to Sharpsburg and placed in the field. On July 8 an examination of the Sharpsburg field showed that the parasites had taken hold to such an extent that of the large number of flaxseeds taken and brought to the laboratory for investigation not one was found which had not been parasitized. Additional material secured from Sharpsburg in the spring of 1908 in the same locality showed all of the Hessian flies to be parasitized.

In the same way excellent results have been obtained in the investigation of the cotton-boll weevil, under Mr. W. D. Hunter. In the summer of 1906 a number of parasites were taken from Waco, Tex., and liberated in a cotton field near Dallas, Tex., and apparently by this means the mortality rate due to parasites was raised in a few weeks about 9 per cent. Later, parasites were introduced from Texas into Louisiana and increased the mortality of the weevil. Work of this character is still being carried on by Mr. Hunter, and elaborate, although as yet unsuccessful, experiments have been made by Webster in the transfer of the hymenopterous parasite *Lysiphlebus tritici* Ashm. (fig. 3) from southern points into Kansas wheat fields for the destruction of the spring grain aphis or so-called "green bug" (*Toxoptera graminum* Rond.), definite results being prevented by the occurrence of the parasite throughout the range of the destructive insect, parasitic, as it is, upon other species of plant lice.

Prof. S. J. Hunter, of the University of Kansas, however, in the Bulletin of the University (vol. 9, p. 2) states that he was able, in 1908, to hasten the destruction of the Toxoptera in Kansas by the importation of Lysiphlebus from some other point.

FIG. 3.—*Lysiphlebus tritici* attacking a grain aphis. Enlarged. (From Webster.)

In the last two years some very interesting work has been carried on by the State Horticultural Commission of California in the way of collecting Coccinellidæ on a large scale in their hibernating quarters, boxing them, and sending them to different parts of the State for use against plant lice upon truck crops. The biennial report of the commissioner of horticulture for 1907–8, published in Sacramento in 1909, for example, indicates that 50,000 specimens of the ladybird beetles *Hippodamia convergens* Guér. and *Coccinella californica* Mann. had been so collected. This, however, was very small compared to the scale upon which these insects were collected during the winter of 1909–10. Mr. E. K. Carnes, of the commission, writing to the Bureau of Entomology under date of March 14, 1910, makes the following statement:

> We have quite a sight at the insectary now—over a ton of *Hippodamia convergens*, boxed in 60,000 lots each, screened cases, and in our own cold storage. We handle them in large cages, run them into a chute, and handle like grain. They are for the melon growers of the Imperial Valley.

This species collects in large numbers late in summer and early in the autumn at the bases of plants in the mountain valleys and can easily be collected by the sackful. The actual good accomplished by the distribution of these ladybirds among the melon growers has not yet been reported upon, but theoretically speaking the experiment should have excellent results.

THE TRANSFER OF BENEFICIAL INSECTS FROM ONE COUNTRY TO ANOTHER.

Early Attempts.

Dr. Asa Fitch, for many years State entomologist of New York, was probably the first entomologist in America, or elsewhere for that matter, to take into serious consideration the question of the transfer of beneficial insects from one country to another. In 1854, following a disastrous attack upon the wheat crop of the eastern United States by the wheat midge (*Contarinia tritici* Kirby), a species that had been accidentally introduced from Europe during the early part of that century, Dr. Fitch, who had made a careful study of the insect both in this country and from the European records, was struck with the fact that in Europe the insect in ordinary seasons did no damage, and that when occasionally it became so multiplied as to attract notice it was but a transitory evil which subsided soon and was not heard of again for a number of years. He was aware that in Europe certain parasites of this insect were found, and, comparing the insects taken from wheat in flower in France with those taken from wheat in flower in New York, he found that in France the wheat midge constituted but 7 per cent of the insects thus taken, while its parasites constituted 85 per cent; whereas in New York the wheat midge formed 59 per cent of the insects thus captured, and there were no certain parasites. He speculated as to the cause for this extraordinary difference and wrote:

> There must be a cause for this remarkable difference. What can that cause be? I can impute it to only one thing; we here are destitute of nature's appointed means for repressing and subduing this insect. Those other insects which have been created for the purpose of quelling this species and keeping it restrained within its appropriate sphere have never yet reached our shores. We have received the evil without the remedy. And thus the midge is able to multiply and flourish, to revel and riot, year after year, without let or hindrance. This certainly would seem to be the principal if not the sole cause why the career of this insect here is so very different from what it is in the Old World.

Quite naturally after this train of reasoning had entered his brain, Dr. Fitch made an effort to introduce the European parasites of the wheat midge, and in May, 1855, addressed a letter to John Curtis, the famous English economic entomologist, and at that time president of the Entomological Society of London, informing him of the immense amount of damage done by the midge in America and suggesting the manner in which parasitized larvæ could be secured in England and transmitted alive to this country. Mr. Curtis was ill and on the point of starting for the Continent, but laid the letter before the Entomological Society of London, which resulted in the adoption of a resolution to the effect that if any member of the society should be able to find parasitized midges he should send them to Dr. Fitch.

Nothing ever came of this effort, but it is of interest on account of its apparent priority over other experimentation of this kind.

The next international attempt seems to have been made in 1873, when Planchon and Riley introduced into France an American predatory mite (*Tyroglyphus phylloxeræ* Riley) which feeds on the grapevine Phylloxera in the United States. The mite became established, but accomplished no appreciable results in the way of checking the famous grapevine pest.

In 1874 efforts were made to send certain parasites of plant lice from England to New Zealand, but without results of value, although *Coccinella undecimpunctata* L. is said to have become established.

In 1883 Riley imported the braconid *Apanteles glomeratus* into the United States from Europe, where it is an abundant enemy of the imported cabbage worm (*Pontia rapæ* L.). This species has since established itself in the United States and has proved a valuable addition to the North American fauna.

THE AUSTRALIAN LADYBIRD (NOVIUS CARDINALIS MULS (IN THE UNITED STATES).

But all previous experiments of this nature were completely overshadowed by the remarkable success of the importation of (*Vedalia*) *Novius cardinalis* Muls. (fig. 4), a coccinellid beetle, or ladybird, from Australia into California in 1889. The orange and lemon groves of California had for some years been threatened with extinction by the injurious work of the fluted or cottony cushion scale (*Icerya purchasi* Mask.) a large scale insect which the careful investigations of Prof. Riley and his force of entomologists at the United States Department of Agriculture had shown to have been originally imported, by accident, from Australia or from New Zealand, where it had originally been described by the New Zealand coccidologist, the late W. M. Maskell. The Division of Entomology had been for several years engaged in an active campaign against this insect, and had discovered washes which could be applied at a comparatively slight expense and which would destroy the scale insect. It had also in the course of its investigations discovered the applicability of hydrocyanic-acid gas under tents as a method of fumigating orchards and destroying the scale. The growers, however, had become so thoroughly disheartened by the ravages of the insect that they were no longer in a frame of mind to use even the cheap insecticide washes, and many of them were destroying their groves. In the meantime, through some correspondence in the search for the original home of the scale insect, Prof. Riley had discovered that while the species occurred in parts of Australia it was not injurious in those regions. In New Zealand it also occurred, but was abundant and injurious. He therefore argued that the insect was probably introduced from Australia into New Zealand, and that its abundance in the latter country and its relative scarcity in Australia were due to the fact that in its native home it was held in subjection by some parasite or natural enemy, and that in the introduction into New Zealand the scale insect had been brought in alone. The same thing, he argued, had occurred in the case of the introduction into the United States. He therefore, in his annual report for 1886, recommended that an effort be made to study the natural enemies of the scale in Australia and to introduce them into California; and the same year the leading fruit growers of California in convention assembled petitioned Congress to make appropriations for the Department of Agriculture to undertake this work. In February, 1887, the Department of Agriculture received specimens of an Australian parasite of Icerya from the late Frazier S. Crawford, of Adelaide, South Australia. It was a dipterous insect known as *Lestophonus iceryæ* Will., and for some time it was considered, both by Prof. Riley and his correspondents and agents, that the importation of this particular parasite offered the best chances for good results.

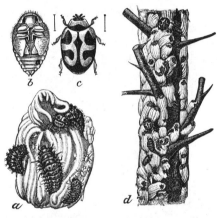

FIG. 4.—The Australian ladybird (*Novius cardinalis*), an imported enemy of the fluted scale: *a*, Ladybird larvæ feeding on adult female and egg sac; *b*, pupa; *c*, adult ladybird; *d*, orange twig, showing scales and ladybirds. *a–c*, Enlarged; *d*, natural size. (From Marlatt.)

Neither the recommendations of Prof. Riley nor of the then commissioner of agriculture, Hon. Norman J. Colman, nor the petitions of the California horticulturists gained the needed congressional appropriations, and, since there appeared at that time annually in the bills appropriating to the entomological service of the Department of Agriculture a clause preventing travel in foreign parts, it became necessary to gain the funds for the expense of the trip to Australia from some other source. A movement was started in California to raise these funds by private subscription, but it was never carried through. In an address given by Prof. Riley before the California State Board of Horticulture at Riverside, Cal., in 1887, he repeated his recommendations. During the summer of 1887 he was absent in Europe, and the senior author, who was at that time the first assistant entomologist of the department, by correspondence secured from Mr. Crawford numerous specimens of Icerya infested by the Lestophonus above mentioned. During the winter of 1887–88 preparations were being made for an exhibit of the United States at the Melbourne Exposition, to be held during 1888, and Prof. Riley, after interviewing the Secretary of State, who had charge of the funds appropriated for the exposition, was enabled to send an assistant, Mr. Albert Koebele, to Australia at the expense of this fund. This result was hastened, and Mr. Koebele's subsequent labors were aided by the fact that the commissioner general of the United States to the exposition was a California man, Mr. Frank McCoppin, and his recommendation, joined to that of Prof. Riley, decided the Secretary of State in favor of the movement. In order to partially compensate the exposition authorities for this expenditure, another assistant in the Division of Entomology, Prof. F. M. Webster, was sent out to make a special report to the commission on the agricultural features of the exposition. Mr. Koebele, who sailed from San Francisco August 25, 1888, was thoroughly familiar with all the phases of the investigation of the cottony cushion scale, and had for some time been stationed in California working for the Department of Agriculture. His salary was continued by the department and his expenses only were paid by the Melbourne Exposition fund. He made several sendings of the Lestophonus parasite to the station of the Division of Entomology of the Department of Agriculture at Los Angeles, where, under the charge of Mr. D. W. Coquillett, a tent had been erected over a tree abundantly infested with the scale insect; but it was soon found that the Lestophonus was not an effective parasite.

On October 15 Mr. Koebele found the famous ladybird (*Vedalia*) *Novius cardinalis* in North Adelaide, and at once came to the conclusion that this insect would prove effective if introduced into the United States. His first shipments were small, but others continued from that date until January, 1889, when he sailed for New Zealand and made further investigations. Carrying with him large supplies of *Vedalia cardinalis*, the effective ladybird enemy, he arrived in San Francisco on March 18, and on March 20 they were liberated under the tent at Los Angeles, where previous specimens which had survived the voyage by mail had also been placed.

The ladybird larvæ attacked the first scale insect they met upon being liberated from the packing cages. Twenty-eight specimens had been received on November 30 by Mr. Coquillett, 44 on December 29, 57 on January 24, and on April 12 the sending out of colonies was begun, so rapid had been the breeding of the specimens received alive from Australia. By June 12 nearly 11,000 specimens had been sent out to 208 different orchardists, and in nearly every case the colonizing of the insect proved successful. In the original orchard practically all of the scale insects were killed before August, 1889, and, in his annual report for that year, submitted December 31, Prof. Riley reported that the cottony cushion scale was practically no longer a factor to be considered in the cultivation of oranges and lemons in California. The following season this statement was fully justified, and since that time the cottony cushion scale, or white scale, or fluted scale, as it is called, has no longer been a factor in California horticulture. Rarely it begins to increase in numbers at some given point, but the Australian ladybirds are always kept breeding at the headquarters of the State Board of Horticulture at Sacramento, and such outbreaks are speedily reduced. In fact, it has been difficult for the State horticultural authorities to keep a sufficient supply of scale insect food alive for the continued breeding of the ladybirds.

The same insect was introduced direct from California into New Zealand at a later date, and the same good results were brought about. The Icerya is no longer a feature in horticulture in New Zealand.

Novius in Portugal.

Still a third striking instance of the value of the Australian ladybird was seen later in the case of Portugal. *Icerya purchasi* was probably introduced into that country in the late eighties or early nineties from her colonies in the Azores, to which point it was probably introduced many years previously from Australia. The insect spread rapidly and threatened the complete destruction of the orange and lemon groves along the banks of the River Tagus. In September, 1896, persons in Portugal applied to the senior author for advice as to the most efficacious means of fighting the scale insect, and a reply was made urging them to make an effort to introduce (*Vedalia*) *Novius cardinalis* and sending information as to the success of the insect in California. In October, 1897, the chief of the bureau was able to secure, through the kindness of the State Board of Agriculture of California, about 60 specimens of the ladybird, which were sent by direct mail from Washington packed in moss. But five reached Portugal alive, but these were so successfully cared for that there was a numerous progeny. Another sending was made on the 22d of November following. These were received on the 19th of December and proved successful. Early in September, 1898, the statement was published in Lisbon newspapers that already colonies or stocks of the Vedalia had been established on 487 estates, whence naturally many others were formed by radiation; gardens and orchards that were completely infested and nearly ruined were already entirely clean or well on the way toward becoming so. Since that time the pest has almost entirely disappeared. The bureau would not have been able to assist the Portuguese Government to this admirable result had it not been for the enlightened policy of the State Board of Horticulture of California in continuing the breeding in confinement of these predaceous beetles long after the apparent great necessity for such work had disappeared in California, and had it not been for the courtesy of the board in promptly placing material at its disposal.

ICERYA IN FLORIDA.

The general effect of the California success on the horticultural world at large was striking, but not wholly beneficial. Many enthusiasts concluded that it was no longer worth while to use insecticidal mixtures, and that all that was necessary in order to eradicate any insect pest to horticulture or to agriculture was to send to Australia for its natural enemy. The fact that the Vedalia preys only upon Icerya and perhaps some very closely allied forms was disregarded, and it was supposed by many fruit growers that it would destroy any scale insect. Therefore the people in Florida whose orange groves were suffering from the long scale (*Lepidosaphes gloveri* Pack.) and the purple scale (*Lepidosaphes beckii* Newm.) sent to California for specimens of the Vedalia to rid their trees of these other scale pests. Their correspondents in California sent them specimens of the beetle in a box with a supply of Iceryas for food. When they arrived in Florida the entire contents of the box were placed in an orange grove. The result was that the beneficial insects died, and the Icerya gained a foothold in Florida, a State in which it had never before been seen. It bred rapidly and spread to a considerable extent for some years, and did an appreciable amount of damage before it was finally subdued.

NOVIUS IN CAPE COLONY.

Prior to the introduction of Novius into Portugal, *Icerya puschasi* having been established at the Cape of Good Hope, the beneficial ladybird was, after an unsuccessful attempt, carried from California to Cape Town by Mr. Thomas Low, member of the Legislative Assembly of Cape Colony, and on the 29th of January, 1892, living specimens were placed in perfect condition in the hands of the department of agriculture of Cape Colony. These specimens multiplied and were reenforced late in 1892 by a new sending from Australia made by Koebele. At the present time the Novius is perfectly naturalized at the Cape.

NOVIUS IN EGYPT AND THE HAWAIIAN ISLANDS.

At the same time, through the United States Department of Agriculture and the courtesy of the State Board of Horticulture of California, the Novius was sent to Egypt to prey upon an allied scale insect, *Icerya ægyptiaca* Dougl., which was doing great damage to citrus trees and to fig trees in the gardens of Alexandria, Egypt. Six adult insects and several larvæ arrived in living condition at

Alexandria. These multiplied so rapidly as to cause an almost complete disappearance of the scales. Later the latter began to increase, but the Novius had not died out and also increased. The Icerya is still held in check in a very perfect way.

In 1890 the Novius had been introduced into the Hawaiian Islands for work against *Icerya purchasi* with the same success.

ICERYA IN ITALY.

In 1900 *Icerya purchasi* was found also in Italy, in a small garden at Portici, upon orange trees. By the autumn of 1900 it had multiplied so abundantly that the owner of the garden tried to stop the trouble by cutting down the trees most badly infested, without bothering himself with the others, so that the infestation continued. When Prof. Berlese's attention was called to it an attempt was first made to destroy it by insecticides without success, and then *Novius cardinalis* was imported from Portugal and from America. The following June the ladybird in both sexes was distributed in the garden, prospered wonderfully, and multiplied rapidly. In July the results were already evident; one could hardly find patches of Icerya which did not show the work of Novius, and at the end of the month it was difficult to find adult Iceryas with which to continue the rearing in the laboratory for food for the reserve supply of Novius. At the present time the multiplication of the scale insect has been reduced to the point of practically no damage, but the original infestation still persists and the area of distribution of the scale insect is slowly enlarging. It is found not only at Portici but in all the little towns around Vesuvius and in the gardens in Naples; but the presence of the ladybird allows the culture of oranges and lemons to go on without interruption.

ICERYA IN SYRIA.

The latest utilization of the beneficial Novius is recorded by Silvestri. It seems that about the year 1905 Icerya made its appearance in Syria, and in July, 1907, Selim Ali Slam wrote to Prof. Silvestri that it had spread so greatly about Beirut that it had almost destroyed the trees. Silvestri sent a shipment of Novius in July, 1907, and another one in August. The result was the same in Syria as it had been in other countries; the Novius multiplied greatly and produced the desired effect.[1]

THE REASONS FOR THE SUCCESS OF NOVIUS.

It thus appears that in the Novius we have an almost perfect remedy against Icerya. There have been no failures in its intro-

---

[1] Since the above was written (in the autumn of 1909) still another success with Novius has been by its carriage from California to Formosa by Dr. T. Shiraki, the entomologist of the Formosan Government, who writes, under date of Jan. 28, 1910: "To-day it has relieved the region from Icerya and has reduced their number to a practically negligible quantity."

duction to any one of the different countries to which it has been carried. Its success has been more perfect than that of any other beneficial insect that has so far been tried in this international work. There are good reasons for this—reasons that do not hold in the relations of many other beneficial insects to their hosts. In the first place, the Icerya is fixed to the plant; it does not fly, and crawls very slowly when first hatched, and later not at all. The Novius, however, is active, crawls rapidly about in the larval state, and flies readily in the adult. In the second place, the Novius is a rapid breeder, and has at least two generations during the time in which a single generation of the host is being developed. In the third place, the Novius feeds upon the eggs of the Icerya. And in the fourth place, it seems to have no enemies of its own. This is a very strange fact, since other ladybirds are destroyed by several species of parasites. For example, as will be shown later, native American ladybird parasites brought about a great mortality in the larvæ of the Chinese ladybird imported from China into America at a later date by Marlatt. The hymenopterous parasites of the widespread genus Homalotylus feed exclusively in ladybird larvæ, which are frequently also fairly packed with the minute hymenopterous parasites of the genus Syntomosphyrum, while the adults are often destroyed by Perilitus, Microctonus, and Euphorus.

The astonishing results of the practical handling of Novius drew attention more forcibly than ever before to the possibilities of this kind of warfare against injurious insects, and although its perfect success as an individual species has never been duplicated, very many efforts in this direction have been made, some of which have met with measurable success and some with very positive results of value.

INTRODUCTION OF ENTEDON EPIGONUS WALK. INTO THE UNITED STATES.

In 1891, with the assistance of Mr. Fred Enock, of London, Riley introduced puparia of the Hessian fly (*Mayetiola destructor* Say) infested with the chalcidid parasite *Entedon epigonus* Walk. into America. These were distributed among several entomologists during the spring of 1891. One American generation was carefully followed by Forbes in Illinois, and four years later (in May, 1895) the species was recovered by Ashmead at Cecilton, Md., where a colony had been placed in 1891. Thus the introduction was apparently successful, but if the species still exists in the United States it must be rare, since extensive rearings of Hessian-fly parasites have been made by agents of the Bureau of Entomology in many different parts of the country during the past few years and not a single specimen of the Entedon has been recognized. The Maryland locality, however, it should be stated, has not been visited by an entomologist since Ashmead's trip in May, 1895.

OTHER INTRODUCTIONS BY KOEBELE INTO CALIFORNIA.

Mr. Koebele took a second trip to Australia, New Zealand, and the Fiji Islands while still an agent of the Department of Agriculture, but at the expense of the California State Board of Horticulture, and in 1893 he resigned from the United States Department of Agriculture and was employed by the State Board of Horticulture of California for still another trip to Australia and other Pacific islands. He sent home a large number of beneficial insects, nearly all of them, however, coccinellids. Several of these species were established in California, and are still living in different parts of the State. The overwhelming success of the importation of *Novius cardinalis* was not repeated, but one of the insects brought over at that time, namely, the ladybird beetle *Rhizobius ventralis* Er. (fig. 5), an enemy of the so-called black scale (*Saissetia oleæ* Bern.), was colonized in various parts of California, and in districts where the climatic conditions proved favorable its work was very satisfactory, notably in the olive plantations of Mr. Ellwood Cooper, near Santa Barbara. Hundreds of thousands of the beetles were distributed in California and in some localities kept the black scale in check. Away from the moist coast regions, however, they proved to be less effective.

INTERNATIONAL WORK WITH ENEMIES OF THE BLACK SCALE.

It will here be convenient to drop the chronological sequence with which the subject in hand has been treated and to refer to the introduction of a very successful parasite of the black scale, whose work against this destructive enemy to olive and citrus culture in California for a time seemed second only to the success of the Novius against the Icerya. In 1859 Motchulsky described, under the name *Scutellista cyanea* (fig. 6), a very curious little hymenopterous parasite reared by Nietner from the coffee scale in Ceylon. Subsequently this parasite became accidentally introduced into Italy and was sent to the senior author for identification by Dr. Antonio Berlese as a parasite of the wax scale, *Ceroplastes rusci* L. As there are wax scales (*Ceroplastes floridensis* Comst. and *C. cirripediformis* Comst.) which are more or less injurious in Florida and the Gulf States, an attempt was made, with Berlese's assistance, to introduce this parasite at a convenient location at Baton Rouge, La., with the further

FIG. 5.—*Rhizobius ventralis*, an imported enemy of the black scale: *a*, Adult ladybird; *b*, larva. Much enlarged. (From Marlatt.)

assistance of Prof. H. A. Morgan at that place. Berlese's sending arrived in good condition, and the parasites issued at Baton Rouge and immediately began to attack the native species. The importation was successful for a time, but the introduced species was finally reduced to an insignificant number, presumably through the attacks of hyperparasites.

In the meantime Prof. C. P. Lounsbury, an American occupying the position of entomologist of the department of agriculture at the Cape of Good Hope, on his arrival at the Cape in 1895 and searching for the usual cosmopolitan scale insects on fruit trees, failed to find the black scale. He commented on this fact in one of his first-published papers, and alluded to the severity of the scale as a pest in California. Shortly afterwards he found the species, and sent the senior author specimens for identification in 1895, together with parasites which he had reared from it. Subsequent correspondence showed other species, and eventually *Scutellista cyanea* was forwarded. Writing to Mr. Lounsbury September 14, 1896, the chief of the bureau made the following suggestion: "I think parasitized black scales could be sent to California to advantage. Mr. Alexander Craw would be the proper person to to whom to send them."

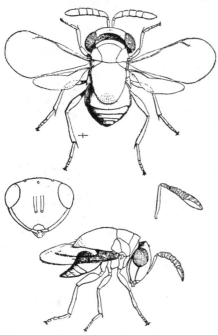

FIG. 6.—*Scutellista cyanea*, an imported parasite of the black scale: Dorsal and lateral views of adult, with enlarged details. Greatly enlarged. (From Howard.)

Mr. Lounsbury made further studies, and commented in his 1898 report on the existence of parasites. When this report met the eye of Mr. E. M. Ehrhorn, of the State horticultural commission, Mr. Ehrhorn wrote Mr. Lounsbury, under date of December 22, 1899, asking him to send a colony of the parasite. Mr. Lounsbury had in the meantime, in a letter to the senior author, suggested that in order to gain authority to spend time over the matter and incur necessary expense it would be desirable for the Secretary of Agriculture of the United States to make a formal request for these parasites to the secretary of agriculture of Cape Colony. This was done, and in May, 1900, Lounsbury secured leave of absence and started for America, carrying with him a box of parasitized scales, and landed at New York on June 2. His box of parasites was at once forwarded to Washington, and the Bureau of Entomology notified Mr. Ehrhorn by telegram, repacked the box, and sent it to California. Mr. Ehrhorn succeeded in temporarily establishing the Scutellista indoors and out around his home at Mountain View, Cal. September 19, 1900, Mr. C. W. Mally, Lounsbury's assistant, sent two more boxes by post direct to California, addressing them to S. F. Leib, of San Jose, notifying the senior author to wire Mr. Ehrhorn to be on the lookout for them. A third lot was sent October 31 of the same year. These later sendings were small, and both failed to yield living parasites. More were requested, and on Lounsbury's return to South Africa a box was shipped in cool chamber to England and thence direct to California by express, Lounsbury's letter of February 28, 1901, to the bureau stating: "To avoid extra delay in transmission the box goes direct to California, but will you kindly have a message sent to Craw to advise him of its coming?" Unfortunately the box was detained by a customs officer at New York, but the bureau secured its release by the Government dispatch agent, Mr. I. P. Roosa. A few parasites emerged after arrival, but failed to propagate. October 1, 1901, Lounsbury started another sending by letter post to insure quick transit and noninterference by customs. These boxes were delivered to Mr. Craw on October 31. Only four females of the Scutellista were reared by Mr. Craw, and probably to these four females are due all of the Scutellistas subsequently occurring in California. This is the full story of the introduction of the species, taken from the letter files of the Bureau of Entomology and the letter files of Mr. Lounsbury in Cape Town.

Mr. Craw was remarkably successful in his rearings, and during the following three years constantly distributed colonies in different portions of California. By July, 1902, he had distributed 25 colonies. It was in the southern part of the State that the parasite did its best work, and there for a time it surpassed the most sanguine expectations of everyone. It was established in every county south of Point Conception and had become very plentiful in Los Angeles, Orange, and San Diego Counties. In the colonization districts by midsummer, 1903, it was estimated that over 90 per cent of the black scale had been destroyed. A year or so later there was great mortality among these parasites caused by a sudden increase in numbers of a predatory mite, *Pediculoides ventricosus* Newp. (fig. 7), which destroyed the larvæ in vast numbers. The Scutellista gradually recovered from this attack, and is at present to be found in very many localities in California, keeping the black scale partly in check.

Another enemy of the black scale was imported in 1901. It is a small moth, *Erastria scitula* Ramb. (fig. 8), the larva of which feeds in the bodies of mature scales, each larva destroying a number of scales. An effort had been made by Riley to import this insect from France in 1892, but without success. In 1901 Berlese sent the senior author living pupæ, which were at once forwarded to Craw and Ehrhorn in California. It was reported in 1902 that the insects had been reared and liberated in Santa Clara, Los Angeles, and Niles, Cal., but if the species was established in the State it has not flourished and has not recently been found.

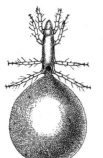

A similar lepidopterous insect, *Thalpochares cociphaga* Meyrick, was brought over from Australia in the summer of 1892 by Koebele and left by him at Haywards, Cal., but the species evidently died out.

THE HAWAIIAN WORK.

FIG. 7.—*Pediculoides ventricosus*. Greatly enlarged. (From Marlatt.)

In 1893 Koebele resigned from the service of the State of California and entered the employment of the then newly established Hawaiian Republic for the purpose of traveling in different countries and collecting beneficial insects to be introduced into Hawaii for the purpose of destroying injurious insects. Before leaving California he had introduced a very capable ladybird, *Cryptolæmus montrouzieri* Muls., which feeds upon mealy bugs of the genus

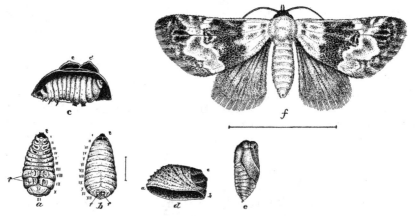

FIG. 8.—*Erastria scitula*, an imported enemy of the black scale: *a*, Larva from below; *b*, same, from above; *c*, same, in case; *d*, case of full-grown larva; *e*, pupa; *f*, moth. Enlarged. (After Rouzaud.)

Pseudococcus. This insect flourished, especially in southern California, and on arrival in Hawaii he found that coffee plants and certain other trees were on the point of being totally destroyed by the allied scale insect known as *Pulvinaria psidii* Mask. He at once introduced this same Cryptolæmus, which is an Australian insect, with the result that the Pulvinaria was speedily reduced to a condition of harmlessness.

It may be incidentally stated that within the past year efforts have been made by the Bureau of Entomology to send the Cryptolæmus to Malaga, Spain, for the purpose of feeding upon a Dactylopius. The first attempt was unsuccessful, and the results of the last attempt have not yet been learned.

Another importation of Koebele's into Hawaii was the ladybird *Coccinella repanda* Thunb. from Ceylon, Australia, and China, which was successful in destroying plant lice upon sugar cane and other crops. Writing in 1896, Mr. R. C. L. Perkins stated that Koebele had already introduced eight other species which had become naturalized and were reported as doing good work against certain scale insects. Among other things he introduced *Chalcis obscurata* Walk. from China and Japan, which multiplied enormously at the expense of an injurious lepidopterous larva (*Omiodes blackburni* Butl.) which had severely attacked banana and palm trees.

Koebele's travels from 1894 to 1896 were through Australia, China, Ceylon, and Japan. In 1899 he left for Australia and the Fiji Islands, and sent many ladybirds and parasites to Hawaii, especially to attack the scale *Ceroplastes rubens* Mask. The Hawaiian Sugar Planters' Association, an organization which was responsible for Koebele's appointment, subsequently employed Mr. R. C. L. Perkins, Mr. G. W. Kirkaldy, Mr. F. W. Terry, Mr. O. H. Swezey, and Mr. F. Muir. By the close of 1902 sugar planters were especially anxious concerning the damage of an injurious leafhopper on the sugar cane, *Perkinsiella saccharicida* Kirk. This insect had been accidentally introduced from Australia about 1897, had increased rapidly, and by 1902 had become a serious pest. Koebele had made an effort to introduce parasites of leafhoppers from the United States into Hawaii, with unsatisfactory results, and consequently in the spring of 1904 Koebele and Perkins visited Australia and collected all possible parasites of different leafhoppers. Altogether they succeeded in finding more than 100 species. Of these the following hymenopterous parasites are said to have become acclimated in Hawaii: *Anagrus* (two species), *Paranagrus optabilis* Perk. and *P. perforator* Perk. and *Ootetrastichus beatus* Perk. These species are all parasitic upon the eggs of the leafhopper. By the end of 1906 observations upon a certain plantation indicated the destruction of 86.3 per cent of the eggs by these parasites. In addition to these egg parasites certain proctotrypid parasites of hatched leafhoppers have apparently become established, namely, *Haplogonatopus vitiensis* Perk., *Pseudogonatopus* (two species), and *Ecthrodelphax fairchildii* Perk. Three predatory beetles, namely, *Verania frenata* Erichs., *V. lineola* Fab., and *Callineda testudinaria* Muls., were also distributed in large numbers.

The practical results of these importations seem to have been excellent. There seems to be no doubt that the parasites have been the controlling factor in the reduction of the leafhoppers.

The good work in Hawaii is still continuing. Koebele is now on a visit to Europe to import the possible parasites of the horn fly (*Hæmatobia serrata* Rob.–Desv.), Muir is trying to find an enemy to a sugar-cane borer (*Rhabdocnemis obscurus* Boisd.), and other similar work is under way.

An Importation of Clerus from Germany.

An early attempt to import beneficial species into the United States was made in 1892 by Dr. A. D. Hopkins, then entomologist to the West Virginia Agricultural Experiment Station and now of the Bureau of Entomology. A destructive barkbeetle, *Dendroctonus frontalis* Zimm., was extremely injurious in that State in the years 1889 to 1892, and Hopkins made the effort to import from Europe another beetle, (*Clerus*) *Thanasimus formicarius* L., from Germany. In Germany he collected more than a thousand specimens of the Clerus, which he took with him to West Virginia and distributed in various localities infested by the barkbeetle. The following year, however, the barkbeetle disappeared almost completely from other causes, and the Clerus has not since been found.

Marlatt's Journey for Enemies of the San Jose Scale.

Another and later expedition was that undertaken by Mr. C. L. Marlatt, of the Bureau of Entomology, in search of the natural enemies of the San Jose scale. The question of the original home of the San Jose scale (*Aspidiotus perniciosus* Comst.) had been a mooted point. As is well known, it started in this country in the vicinity of San Jose, Cal., in the orchard of Mr. James Lick, who had imported trees and shrubs from many foreign countries. Mr. Lick died before the investigation started, and no records of his importations were to be found. The scale was not of European origin, since it does not occur on the continent. In the course of investigation it was found that it occurred in the Hawaiian Islands, in Japan, and in Australia, but in the case of Australia and the Hawaiian Islands it was shown that it had been carried on nursery stock from California. In 1897 plants entering the port of San Francisco from Japan were discovered by Mr. Craw to carry the San Jose scale. Correspondence, however, seemed to point to the conclusion that it had also been introduced into Japan from the United States. In 1901–2 Mr. Marlatt made a trip of exploration in Japan, China, and other eastern countries, lasting more than a year. Six months were spent in Japan, and after a thorough exploration the conclusion was reached that the scale is not a native of that country and that wherever it occurs there it has spread from a center of imported American fruit trees. Finally, as a result of this extended trip, the native home of the San Jose scale was found to be in northern China in a region between the Tientsin-Peking Road and the Great Wall, and its original host plant was found to be a little haw apple which grows wild over the hills. Into that region no foreign introductions of fruit or fruit trees had ever been made, and the fruits in the markets were all of the native sorts. Here in China was found everywhere present a little ladybird, *Chilocorus similis* Rossi (fig. 9), feeding in all stages upon the San Jose scale. One hundred and fifty

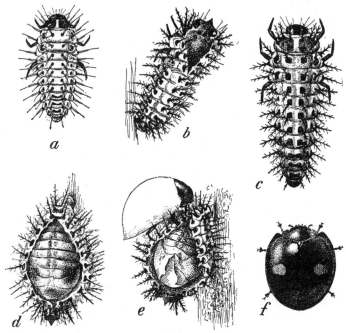

FIG. 9.—The Asiatic ladybird (*Chilocorus similis*), an imported enemy of the San Jose scale: *a*, Second larval stage; *b*, cast skin of same; *c*, full-grown larva; *d*, method of pupation, the pupa being retained in the split larval skin; *e*, newly emerged adult, not yet colored; *f*, fully colored and perfect adult. All enlarged to the same scale. (From Marlatt.)

or two hundred specimens of the beetle were shipped by Mr. Marlatt to Washington alive, but all but two perished during the winter. One at least of the two survivors was an impregnated female, and began laying eggs early in April. From this individual at least 200 eggs were obtained, the work being done in breeding jars. After some hundred larvæ had been hatched from these eggs the beetles were placed on a large plum tree in the experimental orchard and protected by a wire-screen cage covering the tree. The stock increased very rapidly, and during August shipments to various eastern experiment stations were begun, about 1,000 specimens being sent out. At the end of the first summer there remained of the stock at Washington about 1,000 beetles. Among the colonies sent out the best success was obtained in Georgia. An orchard at Marshallville in that State, containing some 17,000 peach trees and covering about 85 acres, adjoined a larger orchard containing about 150,000 trees, all scatteringly infested with the scale. The ladybirds were liberated in August, 1902, in the smaller orchard, and an examination made 11 months later indicated that they were rapidly spreading and would soon cover the orchard. The number of beetles in all stages present was estimated at nearly 40,000. Colonies established in the Northern States perished. In the South the almost universal adoption of the cheap and satisfactory lime-sulphur washes destroyed the possibility of rapid multiplication and destroyed the majority of the beneficial insects. This species has not been found recently, but probably exists in Georgia. The introduction and establishment of the species was successful, but it was practically killed out by the cheap and satisfactory washes in general use. Without the washes the probabilities are that the ladybird would be found at the present time occurring in great numbers in southern orchards.

The Parasites of Diaspis Pentagona Targ.

For a number of years the mulberry plantations of Italy had suffered severely from the attack of the insect known as the West Indian peach scale (*Diaspis pentagona* Targ.). This insect occurs in the United States and is widely distributed in other parts of the world. In the United States, however, it is not especially injurious. In 1905, at the request of Berlese, the writer sent parasitized Diaspis from Washington to Florence, Italy. One of the parasites which issued, *Prospaltella berlesei* How., was artificially reared in Florence by Berlese and his assistants, and at the time of present writing has been so thoroughly established in several localities that the ultimate reduction of the Diaspis to harmless numbers is confidently anticipated by Berlese. Similarly, Silvestri at Portici has introduced the same species from America, and also certain ladybirds, and is making the effort to import the parasites of this species from its entire range.

The Work of Mr. George Compere.

Mr. George Compere, employed jointly by Western Australia and California as a searcher for beneficial insects, for several years has been traveling in different parts of the world in search of beneficial insects which he has either sent or brought to California and Western Australia. One of the most interesting of his achievements was sending living specimens of *Calliephialtes messor* Grav., an ichneumon fly, from Spain to California. This species is a parasite of the codling moth.

In California this ichneumon fly has been reared with great success and has been sent out in large numbers from the headquarters of the State board of horticulture. In the field, however, it is apparently not succeeding, and there is no evidence that the numbers of the codling moth have been at all reduced by it. Nor is it, according to Froggatt, effective in Spain.

Mr. Compere has collected many beneficial species attacking many different injurious insects. He is an indefatigable worker, and his untiring qualities and his refusal to accept failure are well shown in his search for the natural enemies of the fruit fly of Western Australia, *Ceratitis capitata* Wied. He visited the Philippine Islands, China, Japan, California, Spain, returning to Australia, afterwards visiting Ceylon and India, and subsequently Brazil. In Brazil he succeeded in finding an ichneumon fly and a staphylinid beetle feeding upon fruit-fly larvæ. He collected some numbers and carried them to Australia in living condition, prematurely reporting success. The fruit fly is a pest in South Africa, and following the announcement of Compere's importations Claude Fuller and C. P. Lounsbury proceeded from Africa to Brazil to get the same parasites. The result of this journey was discouraging. They did not find the predatory staphylinid, but obtained a braconid parasite, *Opiellus trimaculatus* Spin.; they also concluded from information gained that the fruit fly had been introduced into South America more recently than into South Africa. The material carried home died. Compere left Australia again about the close of 1904; went to Spain for more codling-moth parasites, and then went on to Brazil, collecting more fruit-fly parasites and carrying them to Australia. The Brazilian natural enemies, however, did not succeed, and in 1906 he proceeded to India to collect parasites of a related fly of the genus Dacus, finding several and taking them to Western Australia. He arrived, however, in the middle of winter, and the insects perished. In May, 1907, once more this indefatigable man returned to India, and in a few months collected 70,000 to 100,000 parasitized pupæ, and brought them to Perth, Western Australia, in good condition on the 7th of December. It is reported that the parasites issued from this material in great numbers and in three distinct species. In April, 1908, it was reported that 120,000 parasites had been obtained and distributed, 20,000 of them having been sent to South Africa. The writer has not seen any definite reports of success in the control of the fruit fly by these parasites, but surely Compere deserves great credit for his efforts.

Work with the Egg Parasite of the Elm Leaf-beetle.

In 1905 Dr. Paul Marchal, of Paris, published in the Bulletin of the Entomological Society of France for February 22 a paper entitled "Biological observations on a parasite of the elm leaf-beetle," to which he gave the name *Tetrastichus xanthomelænæ*. In this very interesting article Dr. Marchal called attention to the fact that the elm leaf-beetle had multiplied for several years in a disastrous way about Paris, skeletonizing the leaves in the parks and along the avenues. In 1904 the ravages apparently stopped, and Marchal's observations indicated that this was largely due to the work of this egg parasite. He studied the life history of the parasite carefully during that year at Fontenay-aux-Roses and published his full account the following February.

Visiting Dr. Marchal in June, 1905, after the publication of this interesting article, the senior author asked him whether he had been able to make the further observations promised in the article, and he replied that the elm leaf-beetle had so entirely disappeared in the vicinity of Paris that he had not been able to do so. The visitor urged him to make an effort through his correspondents to secure parasitized eggs of the beetle for sending to the United States in an effort to introduce and establish this important parasite on this side of the Atlantic. It was considered hopeless to attempt the introduction that summer, as the time was so late and it was not then known in what part of France the elm leaf-beetle could be found abundantly. During 1906 practically the same conditions existed. A locality was found, but the parasites did not seem to be present. In 1907, reaching Paris about the 1st of May, the visitor again reminded Dr. Marchal of his desire to import the parasite into the United States, and meeting M. Charles Debreuil, of Melun, the subject was again brought up and M. Debreuil later in the season forwarded eggs of the beetle to the United States, which were promptly sent to the parasite laboratory at North Saugus, Mass., but the time was too late, and the parasites had emerged and died.

In April, 1908, the Entomological Society of France published in its bulletin (No. 7, p. 86) a request from the senior author that eggs of the elm leaf-beetle should be sent to the United States for the purpose of rearing parasites. This notice brought a speedy and effective response. About the 20th of May Prof. Valery Mayet, of Montpellier, France, a personal friend, secured a number of leaves of the European elm carrying egg masses of the beetle, placed them in a tight tin box, and mailed them to Washington. They were received May 28, and at once forwarded to the junior author at the parasite laboratory at Melrose Highlands. On opening the box the junior author found a considerable number of active adults of the parasite. Most of them were placed in a large jar containing leaves of elm upon which were newly deposited masses of the elm leaf-beetle eggs. Probable oviposition was noticed within an hour after the receipt of the sending. There were probably somewhat more than 100 adults received in the shipment and very few emerged from the imported egg masses after the first day. The adults lived certainly for 35 days. Reproduction occurred in the experimental jars, and the adults secured by this laboratory reproduction were liberated in two localities near Boston and parasitized eggs were sent to Prof. J. B. Smith at New Brunswick, N. J., Prof. M. V. Slingerland at Ithaca, N. Y., and others to Washington. The first of the Massachusetts colonies consisted of about 600 parasites inclosed in an open tube tied to a tree in the Harvard yard, Cambridge, Mass., on June 22. Mr. Fiske thinks that more than 100 found their freedom on the same day, and almost certainly all of the rest within a week. A little more than a month later Mr. Fiske found parasitized eggs one-fourth of a mile away from this colony. At Melrose Highlands more than 1,200 were liberated on the 21st of June and the 8th of July; and on the 27th of July fresh native eggs in the neighborhood produced parasites, indicating the development of a generation on American soil. In the summer of 1909 none of the parasites was found, but this by no means indicates that the species has not become established. Both the eggs and the parasites are very small, and the writer expects that even from this first experiment good results will follow. Arrangements had been made for a repetition of the sending in May, 1909, from Montpellier, this southern locality allowing such an early sending as to insure the arrival of the parasitized eggs in the United States at the proper time of the year. Relying upon Prof. Mayet's promises and his great experience as an entomologist, no other arrangements were made. Most unfortunately, however, just before the time arrived Prof. Mayet died, and the introduction was not made. It should be stated that in the death of this admirable man France lost one of its most enlightened and able economic zoologists. It is hoped to repeat the introduction, through the kindness of Dr. Marchal in France and Prof. Silvestri in Italy. Silvestri has promised also to send other natural enemies of the elm leaf-beetle from Italy.

WORK WITH PARASITES OF TICKS.

In 1907 the senior author described the first species of a hymenopterous parasite ever recorded as having been reared from a tick. The name given to it was *Ixodiphagus texanus*, and it had been reared from the nymphs of *Hæmaphysalis leporis-palustris* Pack. collected on a cotton-tail rabbit in Jackson County, Tex., by Mr. J. D. Mitchell. In 1908 he described another, *Hunterellus hookeri*, reared by Mr. W. A. Hooker at Dallas, Tex., from *Rhipicephalus texanus* Banks taken from a Mexican dog at Corpus Christi, Tex., by Mr. H. P. Wood. Inasmuch as a closely allied if not identical tick, *Rhipicephalus sanguineus* Latr., is supposed to be a transmitter of a trypanosome disease in South Africa, sendings of the Hunterellus were made in the autumn of 1908 to Prof. Lounsbury at Cape of

Good Hope and to Mr. C. W. Howard, entomologist to the government of Lourenço Marques, Portuguese East Africa. In June, 1909, Mr. C. W. Howard reared parasites from engorged nymphs of *Rhipicephalus sanguineus* taken from dogs, with which transmission experiments with trypanosomiasis were being made. Examination showed them to be *Hunterellus hookeri*. Mr. C. W. Howard is of the opinion that these 1909 reared specimens could not have been the offspring of those sent over in the autumn of 1908, since, as he writes under date of September 3, 1909, the latter arrived while he was absent in the Zambesi country, and, as he was gone nearly three months, they remained on his desk unopened. When he returned they were all dead. He kept the ticks some time, however, in a sealed jar to see if any more parasites might emerge, but none did so. In his opinion there is absolutely no possibility that the 1909 specimens are the descendants of those sent from Texas. Of course Mr. C. W. Howard is probably correct in his surmise, but a most interesting question arises as to the original home of the parasite. Could it have been carried accidentally from Texas to Africa at an earlier date? As a matter of fact, during the Boer War thousands of horses and mules were shipped from southern Texas to Cape Town, and much of this stock came from the very region in which the Texas Rhipicephalus occurs. Banks, in his revision of the ticks,[1] records this species from horses as well as from dogs, the horse record coming from New Mexico. The suggestion regarding the importation of horses and mules from Texas to Cape Town during the Boer War was made to the writer by Mr. W. D. Hunter, who also suggests that as *Rhipicephalus sanguineus* occurs throughout Africa and Mediterranean Europe, and that as in 1853 several shipments of camels were brought to Texas from Tunis, being turned loose at Indianola and roaming wild throughout the territory around Corpus Christi for some years, it is possible that the Rhipicephalus was brought to Texas on these camels, and the parasite as well. This seems unlikely, however, since the parasite had never been found in Africa or Europe until the specimens referred to were reared by Mr. C. W. Howard in 1909.

Mr. Froggatt's Journey to Various Parts of the World in 1907–8.

As a result of a conference of Australian Government entomologists, held in Sydney, July 9, 1906, and of a conference of State premiers, held in Brisbane, June, 1907, it was agreed that Mr. W. W. Froggatt, entomologist to the Department of Agriculture of the State of New South Wales, should be dispatched to America, Europe, and India, to inquire into the best methods of dealing with fruit-flies and other pests, the expenses of the journey to be shared by Queensland, South

---

[1] U. S. Department of Agriculture, Bureau of Entomology, Technical Series No. 15, p. 35, 1908.

Australia, New South Wales, and Victoria. As a result of the trip following this authorization, Mr. Froggatt has published a report on parasitic and injurious insects, issued in 1909, in which he considers, (1) the commercial value of introduced parasites to deal with insects that are pests; (2) the range and spread of fruit-flies, and the methods adopted in other countries to check them; (3) the value of parasites in exterminating fruit-flies; (4) the habits of cosmopolitan insect pests. On his journey, which began the end of June, 1907, Mr. Froggatt visited Hawaii, the United States, Mexico, Cuba, Jamaica, Barbados, England, France, Spain, Italy, Austria, Hungary, Turkey, Cyprus (spending a day in Smyrna and two days at Beirut on the way), Egypt, India, Ceylon, and thence to Australia, stopping in Western Australia before his return to Sydney. In the course of this trip Mr. Froggatt not only studied the question of parasites and of economic entomology in general, but looked into a large number of matters of agricultural interest, and has given a report which can not fail to be interesting to every one occupied with any branch of agriculture.

With regard to the practical handling of parasites, and especially international work, he is inclined to be rigidly critical. His motive obviously was to look everywhere for accomplished results and where he could not find these to distinctly state the fact. He deprecates all claims that are not or have not been justified by practical results of value. Thus, while admitting the good work of the introduced parasites of the sugar-cane leafhopper in Hawaii, he states that the advocates of the parasite system do not take into account the alteration of methods of cultivation which occurred about the same time, namely, the burning of the refuse (probably containing many eggs and larvæ) instead of burying it as formerly, and the introduction of new varieties of cane more resistant to the leafhoppers. In California, he admits the value of the introduction of the Australian ladybird, but states that his observations show that no good has followed the introduction of the codling-moth parasite from Spain, although it had been claimed previously that this parasite would prove a perfect remedy for the apple pest, and pointing out that when he visited Spain he found that a very large percentage of the apple crop is always infested by the codling moth. He states that the promises of the advocates of the parasite method in California have not been fulfilled; that Western Australian claims that staphylinid beetles destroy the majority of the fruit-fly maggots in Brazil, and that nature's forces in that country control the destructive fruit flies are to be contrasted with the statement of the South African entomologists that only a few months after the visit of the West Australian entomologist to Brazil they found that "all along the Brazilian coast it was difficult to obtain a fruit that had not been punctured by a fly." The statement that nature controls the destructive fruit flies in India he opposes, as a result of his own observations in India. He does not contend that this work has not a great practical value, but insists that it should be done by trained entomologists, and that full information of the habits and life histories of both the pests and their parasites should be understood before liberation is attempted. As already stated, he especially deprecates premature claims, and points out that in New South Wales the passage of the very necessary vegetation diseases bill was delayed for some years by the outcry "Why should we be made to clean up our orchards and spend money, when the department can send out to other countries and get us parasites that will do all that is needed?" In conclusion he states:

> Let the whole question be judged on its results. Allow that one or two experiments have shown perfect results; yet because mealy bugs or scale insects in a restricted locality have once or twice been destroyed by parasites, that can be no reason why the parasite cure alone should be forced upon anyone. Its admirers should be perfectly honest; and if a friendly introduced insect from which, rightly or wrongly, great things had been expected turns out upon further trial to be a failure, they should say so; and they should never proclaim results for a parasite till those results have actually been proved in its adopted country, for the wisest can never be sure of the results of any experiment. Economic entomology is a great commercial science, and those at work for its far-reaching interests could do it no greater harm than by misleading or unproved statements.

OTHER WORK OF THIS KIND.

Reference has already been made to the importations of *Prospaltella berlesei* into Italy to attack the destructive mulberry scale, *Diaspis pentagona*, through cooperative arrangements between the senior author and Prof. Berlese, of Florence. Prof. Berlese has been successful in establishing the species, and believes that it is best to rely upon this species only, and not to attempt to introduce the predatory enemies of the scale, his idea being that coccinellids will feed indiscriminately upon parasitized and unparasitized scales and that thus the Prospaltella will not have a chance to multiply to its limit. The contrary view is taken by Prof. Silvestri, at Portici, in the south of Italy, and he has been making every effort to introduce from all parts of the world all of the enemies, whether parasitic or predatory, of the mulberry scale. He has brought over and has had breeding in his laboratory at Portici, as well as in an experimental olive orchard southeast of Naples, a number of species of Coccinellidæ brought from different parts of the world. At his request, in May, 1910, the senior author carried from Washington a box containing possibly 200 living specimens of *Microweisia misella* Lec. and a few specimens of *Chilocorus bivulnerus* Muls. These were carefully packed with plenty of food in a small paper-covered wooden box, approximating a 10-inch cube. He sailed from New York direct to Naples and, through the kindness of the officers of the Royal Italian Line steamship *Duca di Genova*, was enabled to suspend the box by a cord from a crossbeam in the ordinary cold room of the steamer. After an eleven days' passage, the box was opened in Prof. Silvestri's laboratory in Portici, and practically every coccinellid was found to be alive and in apparently good condition.

Efforts have been made by the Bureau of Entomology, in cooperation with the Pasteur Institute in Paris, to introduce a large bembecid wasp (*Monedula carolina* Fab.) from New Orleans into Algeria to prey upon the tabanid flies concerned in the carriage of a trypanosome disease of dromedaries. The wasps were sent in their cocoons in refrigerating baskets from New Orleans by direct steamer to Havre and from New York by direct steamer to Havre. There they were met by agents of the Pasteur Institute, carried to Marseilles by rail and thence by boat to Algeria, and were planted under conditions as closely as possible resembling those under which they were found in Louisiana, care being taken to simulate not only the character of the soil but the exposure to light, the prevailing wind directions, and the moisture conditions. Adults issued, but the species has not since been recovered, although it is quite possibly established.

In the same way an attempt was made to introduce the common bumblebee *Bombus pennsylvanicus* De Geer of the United States into the Philippine Islands for the purpose of fertilizing red clover. These were sent in refrigerating baskets, carried by hand by Filipino students returning from the United States to the Philippines, and for the most part in the pupal stage. These were properly planted upon arrival and reared, and a few specimens have been recovered.

In the summer of 1910 Dr. L. P. De Bussy, biologist of the Tobacco Planters' Association of Deli, Sumatra, visited the United States for the purpose of investigating damage to the tobacco crop by insects and disease and to make an effort to import into Sumatra the parasites of the destructive tobacco worm known as *Heliothis obsoleta* Fab. Already shipments of an egg parasite, *Trichogramma pretiosa* Riley, have been made to Sumatra via Amsterdam, but information as to the results of these preliminary shipments has not yet reached this country.

Prof. C. H. T. Townsend, an assistant in the Bureau of Entomology, receiving a temporary appointment as entomologist to the Department of Agriculture of Peru, especially to study the injurious work done by the scale insect *Hemichionaspis minor* Mask. on cotton, has during the past year, with the assistance of the bureau, imported a number of shipments of *Prospaltella berlesei* from Washington into Peru. It is too early to announce results.

In July, 1910, Mr. R. S. Woglum, an agent of the Bureau of Entomology, was sent abroad to find the original home of the white fly of the orange, *Aleyrodes citri* R. & H., and to attempt to find parasites or satisfactory predatory enemies. In November, 1910, he found the white fly at Saharampur, India, and discovered that it was killed by a fungous disease (lately determined as a species already occurring in the United States—*Ægerita webberi*—by Prof. H. S. Fawcett, of the Florida Agricultural Experiment Station). He also found that it was attacked by two species of Coccinellidæ (*Verania cardoni* Weise and *Cryptognatha flavescens* Motsch.). A preliminary shipment of the ladybirds by mail was apparently unsuccessful. Later shipments by direct steamer from Calcutta to Boston were also unsuccessful.

At Lahore, India, Mr. Woglum found his first evidence of parasitism by hymenopterous parasites. A certain proportion of *Aleyrodes citri* was found to contain the exit holes of a true parasite. The specimens on leaves sent in by Mr. Woglum were examined with great care. None of the full-grown larvæ or nymphs contained parasites, but five specimens of a very minute aphelinine of the genus Prospaltella were found dead and attached to the orange leaves in close vicinity to the perforated aleyrodids. The size of the specimens was such as to justify the conclusion that they had issued from the aleyrodids, and their juxtaposition and the known habits of the genus confirm this conclusion. The species was described by the senior author as *Prospaltella lahorensis* in the Journal of Economic Entomology for February, 1911, pages 130–132. Efforts will be made to import this parasite into Florida.

The occurrence of a European weevil, *Phytonomus murinus* Fab., in the alfalfa fields of Utah in alarming numbers and the difficulty of fighting the pest by mechanical or cultural means has started an investigation as to its parasites in its original home. Mr. W. F. Fiske, of the Bureau of Entomology, sent from Naples, Italy, on March 17, 1911, a large lot of stems of alfalfa containing eggs of an allied weevil parasitized by a minute mymarid, which at the time of this writing are on their way to Utah.

In the meantime the State board of horticulture of California has been continuing its efforts to import beneficial insects of different kinds. Mr. George Compere returned from a lengthy trip during the summer of 1910, bringing with him a number of interesting species, among them a new coccinellid enemy of mealy bugs in which he has great faith, and which promises to be a valuable addition to the insect fauna of the United States.

Entomologists and horticulturists all over the world have become greatly interested in this aspect of economic entomology and for the immediate future a great deal of experimental work has been planned by the officials of different countries.

EARLY IDEAS ON INTRODUCING THE NATURAL ENEMIES OF THE GIPSY MOTH.

Promptly with the discovery that the gipsy moth had become acclimatized in Massachusetts, in 1889 there was published by Prof. C. H. Fernald a special bulletin of the Massachusetts Agricultural College Hatch Experiment Station, in which he gave popular descriptions of the different stages of the insect and recommended spraying with Paris green. He stated that the insect is generally held in check by its natural enemies in Europe, but occasionally becomes very destructive, and stated that 11 species of hymenopterous parasites and several of dipterous parasites had been noticed in Europe. This bulletin was published in November, 1889. In January, 1890, an illustrated article on the gipsy moth, by Riley and Howard, was published in Insect Life,[1] and a list of 24 European hymenopterous parasites compiled by Howard was published.

Immediately following this publication, there was received at the Department of Agriculture, from Rev. H. Loomis, of Yokohama, Japan, a letter in which he stated that he had seen reports of the ravages of the gipsy moth in Massachusetts and had taken considerable interest in the matter. He also stated that he had seen the gipsy moth caterpillar on a wistaria vine near his house in Yokohama, and that it had been attacked and killed by a parasite. Several of the parasites were sent in an accompanying box, and proved to be Apanteles. Subsequent attempts were made by Mr. Loomis to send this parasite in living condition both to the Department of Agriculture and to the State of Massachusetts, but all arrived dead, for the most part having been killed by secondary parasites.

In March, 1891, a conference was held in the rooms of the committee on agriculture at Boston, at which were present Prof. N. S. Shaler, Gen. F. H. Appleton, and Mr. William R. Sessions, of the State board of agriculture; Prof. C. V. Riley, entomologist of the United States Department of Agriculture; Prof. C. H. Fernald, entomologist of the State Experiment Station; Mr. S. H. Scudder, a well-known entomologist; the mayors of Medford, Melrose, Arlington, and Malden, and others. In the course of the conference, which was held for the purpose of discussing the best measures to be taken against the gipsy moth by the State, Prof. Riley advocated an attempt at extermination by spraying. Mr. Scudder advocated the destruction of the eggs, and in the course of the discussion Prof. Riley made the following remark:

> I would make one other suggestion, and that is, that as an auxiliary method it would be well to spend $500 or $600 in sending one or two persons abroad next summer with no other object than to go to some section of northern Europe to collect and transmit to authorized persons here a certain number of the primary parasites of this species,

---

[1] Insect Life, Division of Entomology, U. S. Department of Agriculture, vol. 2, pp. 208–211, 1890.

which are known to check its ravages over there. The insect was undoubtedly brought over by Trouvelot without any of its natural checks. In my judgment it would be well worth trying to import its parasites from abroad. The advantage would be this: If you failed to exterminate it by spraying, its parasites, seeking for this particular host, would be more apt to find the overlooked or escaped specimens than man would.

No action was taken upon this suggestion, and the State authorities, believing that such an attempt would be useless owing to the fact that their effort for some years was consistently devoted to the aim of absolute extermination of the gipsy moth, perhaps wisely saved the expense of a mission abroad for this purpose. Then, also, there was some hope that the native parasites, particularly the ichneumon flies and the native species of Apanteles, as well as tachina flies and some of the carabid beetles, might gradually accommodate themselves to the imported pest and prove prominent factors in the fight against it.

This last faint hope, however, was not justified. In the course of the careful work done by the State during the next seven or eight years, the better part of which is summarized in the admirable Report on the Gipsy Moth, by Forbush and Fernald, published in 1896, several native parasites and predatory insects were observed to attack the gipsy moth in its different stages, but at no time was the percentage of parasitism sufficiently great to have any value as a factor in the suppression of the pest. At no time was there a greater percentage of parasitism by native parasites than 10, whereas the condition in Europe is such that the percentage reaches frequently well above 80. It may be worth mentioning that parasitism by native species has never exceeded 5 per cent in any collections made since the present laboratory was established. It is nearer 2 per cent on the average.

In discussions among the Washington entomologists it was repeatedly pointed out by E. A. Schwarz and by B. E. Fernow (at that time Chief of the Division of Forestry of the United States Department of Agriculture) that one of the most important of European enemies of the gipsy moth, and the nun moth as well, is one of the tree-climbing ground beetles known as *Calosoma sycophanta* L. There exist a number of species of this same genus Calosoma in the United States, but none of them has the tree-climbing habit developed to the same extent as have *Calosoma sycophanta* of Europe and its relative *Calosoma inquisitor* L. Prof. Fernald, writing to the famous German authority on forest insects, Dr. Bernard Altum, early in 1895, asked his opinion as to the advisability of importing these tree-inhabiting ground beetles, but received the reply that such an importation would not give good results. Prof. Altum considered the services of the hymenopterous parasites of the old genus Microgaster as of much more importance.

In the report just cited Fernald disposed of the question of importing parasites in the following words:

No attempt has been made to import parasites thus far for the reason that the law requires the work to be conducted with direct reference to the extermination of the gipsy moth, and, therefore, the general destruction of the insect would also destroy the parasites. There is no reason why our native hymenopterous parasites may not prove to be quite as effective as those of any other country, since there is no parasite known which confines itself exclusively to the gipsy moth, and, as has been shown, we have several species which attack it as readily as any in its native country.

This position with regard to the nonimportation so long as extermination of the gipsy moth was the end, held until the State of Massachusetts ceased its appropriations, in the year 1900.

CIRCUMSTANCES WHICH BROUGHT ABOUT THE ACTUAL BEGINNING OF THE WORK.

During the five years that elapsed before the State again began to appropriate money for the suppression of the gipsy moth and the brown-tail moth, as is well known, the gipsy moth spread from a restricted territory of 359 square miles throughout an extended range of 2,224 square miles and even more. As soon as the effort to exterminate it was abandoned, owing to the lapse of the appropriations for the year 1900, the project of importing parasites was taken up by the Chief of the Bureau of Entomology, who began correspondence with a number of European entomologists with this end in view. Especial efforts were made to import the Calosomas, but failed, partly owing to a lack of interest in the matter on the part of the Europeans. In 1902 Mr. W. B. Alwood, entomologist of the Virginia Agricultural Experiment Station, went abroad for a series of months and was requested by the chief of the entomological service of the United States Department of Agriculture to endeavor to find, in some well-placed situation in Europe, one or more competent collectors of insects who would undertake systematically to send gipsy-moth parasites to America. This effort also failed, and Mr. Alwood was unable to find the proper persons. Finally, in December, 1904, Congress was asked to make a small appropriation for the distinct purpose of attempting the importation of these parasites, and the sum of $2,500 was appropriated for this purpose in the session of the winter of 1904–5. During the corresponding session of the Massachusetts State Legislature, State appropriations began once more. In 1904 it was apparent to everyone that the old areas had become reinfested and that the insect had spread widely. Private estates and woodlands in June and July of that year were almost completely defoliated. Kirkland wrote:

From Belmont to Saugus and Lynn a continuous chain of woodland colonies presented a sight at once disgusting and pitiful. The hungry caterpillars of both species of moths swarmed everywhere; they dropped on persons, carriages, cars, and automobiles, and were thus widely scattered. They invaded houses, swarmed into living and sleeping rooms, and even made homes uninhabitable * * *. Real estate in the worst-infested districts underwent a notable depreciation in value. Worst of all, pines and other conifers—altogether too scarce in eastern Massachusetts—were killed outright by the gipsy-moth caterpillars, while shade trees and orchards were swept bare of foliage.

There was a general demand upon the State legislature and an excellent bill was prepared and passed with the appropriation of $300,000, $75,000 to be expended during 1905, $150,000 and any unexpended balance during 1906, and $75,000 and any unexpended balance during 1907, up to May 1, 1907, inclusive. And to this appropriation there was added the clause "for the purpose of experimenting with natural enemies for destroying the moths, $10,000 is additionally appropriated for each of the years 1905, 1906, and 1907." There was then available in the spring of 1905 the appropriation of $2,500 by the General Government and that of $10,000 by the State of Massachusetts for work with the natural enemies. Mr. A. H. Kirkland was appointed superintendent for suppressing the gipsy and brown-tail moths, by Gov. Douglas, and immediately following his appointment, and with the approval of his excellency the governor, went to Washington, and by arrangement with the honorable the Secretary of Agriculture, Mr. James Wilson, arranged a cooperation between the State and the Department of Agriculture whereby the Chief of the Bureau of Entomology of the department was practically placed in charge of the details of the attempt to import parasites from abroad, in consultation with Mr. Kirkland.

The reasons which influenced Mr. Kirkland in entering into this cooperation between the State and the United States Department of Agriculture were expressed in his first annual report (p. 117).

At this time for more than 25 years the chief of the bureau had been devoting his especial efforts to the study of the parasitic Hymenoptera, and had especially interested himself in the subject of their biology and host relations. He had accumulated a card catalogue of more than 20,000 entries of records of the specific relations of parasites to specific insects, the great majority of these being European records and covering all of the published information regarding the parasites of the gipsy moth and the brown-tail moth. He also had the advantage of the personal acquaintance of most of the European entomologists interested in this kind of work. These facts were known to Mr. Kirkland and caused his action.

AN INVESTIGATION OF THE INTRODUCTION WORK.

From the beginning of the work, and even before, certain citizens of Boston, impressed by the claims of the State Board of Horticulture of California as to the results said to have been achieved by the agents of the board in the introduction of beneficial insects, urged the employment of these agents in the work of introducing the parasites of the gipsy moth and brown-tail moth. The arguments in favor of this proposal were duly considered by the superintendent of the Massachusetts work, who decided for many reasons to conduct the introduction experiments along the lines just described and not to call in the assistance of the California people. In his third report, submitted January 1, 1908, Mr. Kirkland expressed the situation as follows:

> In spite of all the thought, energy, and skill that have been brought to bear on this most important problem of introducing the natural enemies of the moths—a problem entirely novel in the field of entomology—it was apparent during the winter of 1906–7 that several of our influential citizens had expected immediate results from the importation of the parasites, and were beginning to get restive because such results had not been obtained. Several expressed a doubt if everything possible was being done to secure the successful introduction of the parasites. Others became enthusiastic over the specious proposition put forward by a certain western horticulturist (not an entomologist), who offered to suppress the gipsy moth in Massachusetts by means of parasites for the sum of $25,000, "no cure, no pay." This state of affairs was no doubt a natural outcome of the desire to avoid a repetition of the great damage to property caused by the moth in past years. Again, men without any technical knowledge of entomology or of the life histories of the parasites, not realizing the difficulties in securing, shipping, breeding, and disseminating these beneficial insects, and equally ignorant of how long it takes an imported insect to become established even under the most favorable conditions, might well be pardoned for expecting almost immediate results from the introduction of the relatively small number of parasites—small indeed in comparison with the tremendous numbers of the moths.

Coming before the legislature during the session of 1906–7, this group of Boston citizens stated that it was their opinion that the work with parasites was not progressing with sufficient rapidity, and asked the legislature to appropriate funds and to instruct the superintendent to secure additional counsel and advice in the matter to determine whether the work was going on in the right way. The legislature agreed and appropriated the additional sum of $15,000 to enable the superintendent to secure such advice.

It was first suggested that he consult only with certain California men who had had experience in importing parasites of scale insects. He, however, considered that consultation with men whose experience had been confined to a single group of insects, not to the same group as the gipsy moth and the brown-tail moth, while possibly helpful, would not be broad enough to throw any great light on the Massachusetts problem. To use his own words—

> It seemed much wiser and certainly more thoroughgoing, since this entire work might be called in question at any time, and in view of the large amount of money Massachusetts was expending in securing parasites, to consult not with the trained entomologists of a single State, but with as many entomologists of national or even world-wide reputation as possible. In other words, that a large number of entomologists of the highest possible scientific standing, and particularly those having practical experience in dealing with parasitic insects, should be invited to visit Massachusetts,

learn of our difficult problems on the spot, examine into the methods of importing, rearing, and distributing parasites, and then give us the benefit of their criticism and counsel, based on a full knowledge of the facts at hand. He also suggested that, since by some this movement might be taken as a criticism on his management and on his judgment in placing the direction of the work in the hands of Dr. Howard, it would be well to have some outside board or commission take charge of the matter, so that it should be entirely an ex parte affair. free from any suggestion of influence by the present administration of the work. The suggestion to authorize the superintendent to invite the entomologists was heartily indorsed by the legislative committee which had the matter under consideration, while the arrangement of the entire affair was left in his hands.

In his selection of experts, Mr. Kirkland was aided by Prof. C. H. Fernald, of the Massachusetts Agricultural Experiment Station, one of the oldest and best posted entomologists in the country; and Mr. Kirkland himself, it must be remembered, had been engaged in active entomological work for 15 years and had held official positions in the Association of Economic Entomologists, thus having a very broad personal acquaintance with the best workers. The list selected, as quoted from Mr. Kirkland's report, was as follows:

Prof. Edward M. Ehrhorn, deputy commissioner of horticulture, State of California, a man of large practical experience in importing, breeding, and disseminating insect parasites, particularly those of scale insects, and also a man well trained in applied entomology.

Prof. Herbert Osborn, Ohio State University, one of the country's best known teachers of entomology, and of large experience in investigation and laboratory work.

Dr. John B. Smith, entomologist, New Jersey Agricultural Experiment Station, an investigator of the highest order, a successful teacher, and the author of numerous standard works on insects.

Prof. S. A. Forbes, State entomologist, Illinois, a most successful teacher and investigator, and one of the most prominent entomologists of the Middle West.

Prof. E. P. Felt, State entomologist of New York, a well-known writer on and investigator of insect pests, and particularly ingenious in devising laboratory methods.

Prof. H. A. Morgan, director of the Tennessee Agricultural Experiment Station, of large experience, and one of the best-known entomologists of the Southern States.

Prof. M. V. Slingerland, Cornell University, New York, an investigator with hardly an equal, and one who has had great success in studying life histories of beneficial and injurious insects.

In addition to these, the following well-known foreign entomologists, visiting Boston, were asked to investigate the situation carefully, to study the laboratory and field methods, and to report:

Prof. Charles P. Lounsbury, entomologist, Cape Town, South Africa, one who has had great experience as well as great success in importing beneficial insects.

Prof. Walter W. Froggatt, government entomologist, New South Wales, and also investigator for Victoria and Queensland. Prof. Froggatt's work has been practically along the same line as that of Prof. Lounsbury, and has met with a large measure of success.

Dr. James Fletcher, dominion entomologist, Canada, well known for his success in working out difficult points in the life histories of insects, and more particularly in dealing with a wide range of injurious species.

Prof. R. Blanchard, University of Paris, and member of the Academy of Medicine.

Dr. G. Horvath, director of zoological section, National Hungarian Museum, member of the Academy of Science of Hungary and formerly director of the entomological station of Hungary. The last two gentlemen are entirely familiar with the two moths and their parasites.

Dr. Richard Heymons, extraordinary honorary professor and custodian at the Zoological Museum of the Royal Institute of Berlin. Dr. Heymons has made a large study of the injurious insects of central Europe, and particularly of their natural enemies.

Prof. A. Severin, conservator at the Royal Museum of Natural History of Belgium, and member of the Superior Council of Forests. Prof. Severin's position is naturally that of one of the best posted entomologists, particularly with reference to dangerous forest insects.

In addition to these foreign entomologists, Prof. Filippo Silvestri, of the Royal Agricultural School of Portici, Italy, visiting America on an official mission in the summer of 1908, visited Boston, and was asked to give his professional opinion of the work, his report being printed in the fourth annual report of the superintendent, issued January, 1909, by L. H. Worthley, acting superintendent.

It is worthy of note that Prof. Silvestri had been commissioned by the R. Accademia dei Lincei and by the royal minister of agriculture of Italy to investigate the work in economic entomology being done in the United States, and had visited all portions of the country, including California and Hawaii, studying with especial care all the work being done with parasites. It should be pointed out also that the California claims were perfectly well understood by all of the American experts, Mr. Ehrhorn himself being the second ranking officer in the California service, and the others having either visited California partly for the purpose of investigating this work, or being perfectly familiar with the situation by study of the publications and by correspondence. Moreover, of the foreign experts, Mr. Froggatt had just come from California on an investigating trip for the government of the Federated Colonies of Australia which subsequently carried him around the world, Mr. Lounsbury had visited California for the purpose of studying this work, and Dr. Fletcher had repeatedly visited that State.

The reports of all of these experts, with the exception of that of Prof. Silvestri, are published in the third annual report of the Massachusetts superintendent, Boston, 1908, Prof. Silvestri's report being published, as above stated, in the fourth annual report of the superintendent.

It will be entirely unnecessary to quote from these reports, since they may be found in full in the State documents mentioned. It will suffice to state that the work was commended, it is safe to say, with enthusiasm by every individual. Specific consideration was given to the California suggestion by Mr. Lounsbury, by Mr. Froggatt, and by Prof. Slingerland. Suggestions were made by several of them that the study of the fungous, bacterial, and protozoan diseases of the larvæ should be taken up. Dr. Felt and Dr. Smith recommended the importance of the introduction of the Japanese parasites, and Dr. Felt suggested the importance of careful biological studies of the parasites, not only in America but in Europe. All of these suggestions coincided with plans already made which were about to be entered upon, as indicated in following pages.

The subject of the study of the diseases of the caterpillars does not come under the range of the present bulletin, but since it has been mentioned, it should be stated that the State superintendent has for the past two years been having this subject investigated and that it is now going on under the expert supervision of Dr. Roland Thaxter, of Harvard University, and Dr. Theobald Smith, of the Harvard Medical School.

Mr. Kirkland's summary seems fully justified. It is as follows:

> It will be seen from the foregoing that the work of importing parasites of the gipsy and brown-tail moths in Massachusetts has been thoroughly examined by practically a congress of the world's leading entomological experts. And it is believed that their consensus of opinion, which is, in the main, that everything possible to secure the successful importation of these insects is being done, will be taken as authoritative and final. It would seem that the last word has been said on this matter, and that there should be no further occasion for that kind of adverse criticism, whose sole effect is to harass those who are giving their best thought and most sincere effort to the accomplishment of the desired result. Destructive criticism of scientific work, by the amateur or dilettante, is absolutely valueless. Constructive criticism, such as these reports make on certain minor details of this important work, is helpful and a public good.

NARRATIVE OF THE PROGRESS OF THE WORK.

Down to the time when this work was begun, all attempts at the international handling of beneficial insects had been done either by correspondence or by the sending of an individual collector to search for such insects and to forward them by mail or express or to bring them back himself in comparatively small numbers, the beneficial species being either at once liberated in the field or reared for a time in confinement and then liberated. In planning the present work the normal geographic ranges of both the gipsy moth and the brown-tail moth were well known and most of their parasites had been listed, so that the problem seemed to be a comparatively simple one. Owing to the fact that the most abundant of the Japanese gipsy moths (four of them are listed) presents rather marked differences from the European and New England form—so much so, in fact, as almost to justify the opinion that it is a distinct species—and as the ancestors of the New England gipsy moth came from Europe, it was decided to concentrate the effort, for a time at least and in the main, upon European parasites and natural enemies. From the outset the idea was to secure as many parasites belonging to as many different species as possible from all parts of Europe, in the hope of establishing in New England approximately the natural environment of the gipsy moth and the brown-tail moth in so far as their natural checks are concerned. It was the aim to establish, not one or half a dozen of its natural enemies, but all of them, aiming at the same time to avoid the introduction of hyperparasites—that is, those species that prey upon the true parasites of the injurious forms—thus, if possible, bringing about an even more favorable situation for the primary parasites in New England than exists in Europe.

On account of the enormous numbers in which both gipsy and brown-tail moths existed in Massachusetts, it was considered that the simplest way to secure the true European parasites was to collect caterpillars and chrysalids wherever they could be found in Europe, box them, and ship them directly to Boston; this always with the certainty that a certain percentage, high or low, would contain living parasites which would probably issue in the adult condition on the journey or after arrival in America, in which event they could be cared for, reared until sufficiently multiplied, and then liberated.

A temporary laboratory for the receipt and care of specimens was immediately established by Mr. Kirkland at Malden, Mass., and a careful search was begun for a suitable location for a permanent laboratory for the care of parasites. It was considered desirable that this laboratory should be placed in a region in which both the gipsy moth and the brown-tail moth occurred in abundance, so that there might be plenty of material for food for the parasites at all times; and it was also considered of importance that a considerable area of land should be secured which could be controlled for outdoor experiments. Mr. Kirkland finally found a small farm with buildings in North Saugus, the location easily accessible by electric cars and sufficiently isolated. (See Pl. II, fig. 1.) The house was large enough to give ample room for laboratory use, and at the same time furnished dwelling rooms for the state official in charge. In the immediate vicinity there was a chain of large woodland colonies of the gipsy moth and numerous orchards infested by the brown-tail moth, as well as a large area of scrub-oak land where the brown-tail moth occurred very abundantly. A portion of the building occupied as a laboratory was fitted up by the State with shelves, tables, rearing cages, and all necessary apparatus and supplies, and the State employed Mr. F. H. Mosher, with Mr. E. A. Back and Mr. O. L. Clark as assistants, to help care for the parasites.

While, as just stated and for the reasons given, the main effort was made with Europe, correspondence was begun with the Imperial Agricultural Experiment Station at Nishigahara, Tokyo, Japan, and the Imperial Agricultural College at Sapporo, in order to secure, if possible, the services of expert Japanese entomologists in sending

Japanese parasites, and Prof. S. I. Kuwana immediately prepared an important sending, which, however, was not productive, through accidents in transportation. The method tried by Prof. Kuwana was interesting. A small tree carrying a number of infested gipsy-moth caterpillars was packed in a large wooden case with wire-gauze sides; another case of small elms was shipped with the insects, and they were thus supplied with fresh food from time to time as far as Hawaii. The case, however, shrunk in transit, making openings through which the parasites for the most part escaped.

In May, 1905, the Chief of the Bureau of Entomology visited Boston for conference with Mr. Kirkland, and on June 3 sailed from Boston to Naples. Landing in Naples on June 13, he at once proceeded to the Royal Agricultural School at Portici, some miles away, and held a conference with Prof. F. Silvestri, the entomologist of the college, and his principal assistant, Dr. G. Leonardi. By good fortune, Prof. Silvestri was able to point out a locality in Sardinia where, during 1904, there had been a severe outbreak of the gipsy moth and where, therefore, during 1905 parasites could with almost absolute certainty be predicted to occur in numbers. With true scientific enthusiasm, both Prof. Silvestri and Dr. Leonardi volunteered their assistance, and Dr. Leonardi was at once commissioned by his chief to proceed to Sardinia and to collect such caterpillars as he could find and forward them in tight wooden boxes, with a supply of food, to Boston. His expedition was a success, and there were received from him at Boston, on the 15th of July, 7 boxes, on the 26th of July 24 boxes, and on the 1st of August 7 boxes, all containing valuable material, the most important being a large series of living puparia of certain parasitic tachina flies.

This extremely cordial and profitable reception at Portici by Prof. Silvestri and Dr. Leonardi, both personally known to the chief of the bureau from former visits, was but a foretaste of the encouragement which was to be met at all points, and it may very properly be said in advance that throughout the whole of the work many European and Japanese entomologists, both officials and private individuals, have shown an extreme liberality in their offers of assistance in this great piece of experimental work, and the State of Massachusetts and the United States Government are under great obligations to them for their help and encouragement. For the work done by Dr. Leonardi, just described, and for similar work done in ensuing years, with Prof. Silvestri's permission, no compensation would be accepted, and the State of Massachusetts has paid simply for the expenses, such as packing, postage, small traveling expenses, and items of that general character.

Fig. 1.—View of Parasite Laboratory at North Saugus, Mass. (Original.)

Fig. 2.—View of Parasite Laboratory at Melrose Highlands, Mass. (Original.)

After Portici, Florence was visited, where a conference was held with Prof. A. Berlese, of the Royal Station of Agricultural Entomology, and his assistants, Drs. Del Guercio and Ribaga. It seemed that no occurrences of either the gipsy moth or the brown-tail moth were known that season in Tuscany or adjoining portions of Italy. Prof. Berlese spoke of the destruction of an outbreak of the gipsy moth in southern Italy some years previously by a disease which he considered to be identical with the pébrine of the domestic silkworm. He promised to keep up a watch for occurrences of the pests and wherever possible to assist in the introduction of parasites. A few days were then spent in Lombardy, searching for the larvæ of either of the injurious species, but without success. Then, proceeding to Vienna, the celebrated Natural History Museum was visited and the well-known curator of Lepidoptera, Dr. Hans Rebel, was interviewed. Dr. Rebel stated that both the gipsy moth and the brown-tail moth were to be found rather commonly in parts of Austria, and it was decided to employ a professional collector to assist in the work of shipping larvæ to Boston. Upon Dr. Rebel's recommendation, Mr. Fritz Wagner was employed. Mr. Wagner was and is a resident of Vienna, is well versed in the subject of European butterflies and moths, and perfectly familiar with all the best collecting places for many miles about Vienna. Mr. Wagner accompanied the writer on several expeditions. The first trip was taken to the suburbs of Vienna, and there the first European specimen of the gipsy-moth larva was found. It was resting on the trunk of a locust tree by the side of the street, and further examination showed that there were a hundred or more caterpillars on the trunk and limbs of the same tree. There was some evidence of parasitism, and the white cocoons of a microgaster parasite (*Apanteles fulvipes* Hal.) were found here and there in the crevices of the bark. This particular tree and another one, to be mentioned later, indicate very well the condition of the gipsy moth in Europe. A hundred nearly full-grown larvæ were present, but there was hardly any evidence of defoliation. A trained entomologist walking by the tree would not have noticed that insects had been feeding upon it to any serious extent. On the other hand, a similar tree in any of the small towns about Boston would have carried not 100 larvæ, but probably some thousands, and at that time of the year would hardly have had a whole leaf. These specimens were collected and sent to Boston.

Later a trip was taken into the country to the battlefield of Wagram, and here on two roadside poplars was found another colony of the caterpillars ranging in size from the second stage to full-grown larvæ. There was here more extensive evidence of parasitism by microgaster parasites. Their white cocoons were found abundantly, and here again, although there must have been 250 or more larvæ on the trees, the evidences of defoliation were very slight—so much so that at a rather short distance the trees appeared in full leaf. During the remainder of June and July Mr. Wagner continued the search and sent considerable material to Mr. Kirkland, at Boston.

After Vienna, the city of Budapest was visited. At the Natural History Museum in that city Dr. G. Horvath, the well-known director, and Prof. Alexander Mocsary were consulted, Prof. Mocsary being one of the first authorities in Europe on the subject of parasitic Hymenoptera. Neither of these gentlemen, however, was able to give any new points in connection with the parasites of the gipsy moth and the brown-tail moth. The agricultural experiment station in the suburbs of Pesth was then visited, and Prof. Josef Jablonowski, the entomologist of the station, was consulted. By this time it was the 4th of July, and already the season in Hungary was far advanced, being about two weeks or more earlier there than at Vienna. Prof. Jablonowski stated that gipsy moths had been found in certain localities in Transylvania, but that the adults were already issuing and that the brown-tail moths had been flying for some time. He exhibited, however, a large box full of the previous winter's nests of brown-tail larvæ, and stated that in the early spring he had reared from these nests many hundreds of parasitic insects. This at once seemed to indicate a very easy way of importing such parasites, since these nests could be readily collected in the winter in large numbers and sent to Boston in great packages—a bushel or more in each package—in the late fall or winter season, and Prof. Jablonowski volunteered to make every effort the following winter to send over a large quantity. Taking into consideration the small size of the brown-tail moth caterpillars during hibernation, it seemed very strange that they should be so extensively parasitized as indicated by Jablonowski. The larger caterpillars in the late spring and early summer would seem to be much more likely to be extensively infested. These winter nests, remaining alone on the trees after the leaves have fallen, would seem to be an attractive place for small Hymenoptera of various kinds, in which they might seek shelter for hibernation, and, while of course there was a chance that some of the true parasites of later stages might thus be sheltered, it was with considerable doubts as to the ultimate result that the writer arranged for the importation of these nests in large quantity. Even if unsuccessful, however, it seemed that the experiment must be tried.

From Budapest, Dresden was reached, and, as in Vienna and Budapest, the principal museum (the Zoological Ethnological Museum) was at once visited. Dr. K. M. Heller, at that time acting director of the museum, was asked to recommend a good man who might be employed as a professional collector to undertake work in the same manner as that done by Fritz Wagner in Vienna. Dr. Heller recommended Mr. Edward Schopfer, who was at once engaged. Although at the date of the first visit to him the season was already considerably advanced (July 7), Mr. Schopfer had rearing cages in operation in his rooms, and in these cages were a number of nearly full-grown larvæ of the gipsy moth. He knew the localities about Dresden where these insects were to be found, and at once began sending specimens to Boston. The well-known Forest Academy at Tharandt, near Dresden, was visited, and Prof. Arnold Jacobi and his assistant, Mr. W. Baer, were interested and promised assistance, especially in the matter of sending specimens of *Calosoma sycophanta* (see Pl. I, frontispiece) and *C. inquisitor*. Other trips were made in the vicinity of Dresden, and then the journey was resumed to Zurich, where, through the kindness of Dr. Herbert Haviland Field, director of the Concilium Bibliographicum Zoologicum, the writer met Miss Marie Rühl, editor of the Societas Entomologica, a very well-posted entomologist, especially on matters relating to Lepidoptera, who had and has a large correspondence throughout northern Germany. She was engaged as the official agent of the investigation for that part of Germany and was able, through her own work and that of her correspondents, to send a large amount of material to Boston before the close of the season of 1905, and has since continued the work.

From Zurich the trip was resumed to Paris, where some time was spent in interviewing Dr. Paul Marchal, the entomologist of the agricultural school conducted under the ministry of agriculture, and other entomologists, and in visiting the scientific societies for the purpose of interesting naturalists in the work. Many trips were taken to towns around Paris in search of the pupæ of the gipsy moth and to visit local collectors in search of information, after which the return journey was made to America.

The result of this initial trip was to demonstrate that it is an easy matter and a comparatively inexpensive one to import certain of the parasites of both the gipsy moth and the brown-tail moth in living condition into the United States. The most important part of the European range of the two species was visited, and the entomologists were organized into an active body of assistants.

Mention has already been made of the number of boxes sent in by Dr. Leonardi from Sardinia. Ten boxes were shipped by Fritz Wagner from Vienna, 47 boxes from Schopfer in Dresden, and 36 from Miss Rühl in Zurich, all of these containing parasitized larvæ or pupæ of the gipsy moth or brown-tail moth.

Acting upon Prof. Jablonowski's observations concerning the existence of parasites in the wintering nests of the brown-tail moth, arrangements were made with Miss Rühl, Mr. Schopfer, Prof. A. J. Cook, who was then in Berlin, and several volunteer collectors to send in numbers of the winter nests. During his visit to Paris in July, the chief of the bureau had addressed a meeting of the Entomological Society of France on the subject of his mission and asked the members of the society to assist in the work. The most remarkable response to this request came from Mr. Rene Oberthür, of Rennes, who, although not present at the meeting, read the account in the bulletin of the society, and placed himself and his services entirely at the disposal of the United States authorities. During the autumn of 1905 and the winter of 1905–6 he sent to Boston more than 10,000 winter nests of the brown-tail moth. In all, 117,000 nests were received and cared for during that winter.

In the autumn the laboratory house (Pl. II, fig. 1, p. 56) at North Saugus was taken possession of by Mr. Kirkland, fitted up as previously described, and occupied by Mr. Mosher; the parasite material from Malden was brought over and installed, and arrangements were made for the receipt of the brown-tail winter nests. Very many large boxes were constructed, somewhat on the plan of the California parasite-rearing cage, each one large enough to contain from 500 to 1,000 nests of the brown-tail moth, the front being pierced with auger holes in which were inserted round-bottom glass tubes into which the emerging parasites would come in search of light and through which they might be examined to differentiate between the primaries and the hyperparasites. Much carpenter work was done during the autumn and winter months and on into the spring. Double windows and double doors were provided, and every crack in the laboratory rooms was sealed. Realizing that many different kinds of insects might emerge from this large supply of silken nests, including possibly species injurious to agriculture not previously introduced into the United States, as well as dangerous parasites of beneficial insects, every possible effort was made to prevent the escape of any insect whatever from the laboratory rooms.

On account of the importance of a speedy detection of injurious forms coming from these rearing cages, and on account of the necessity for the most expert supervision of the laboratory end of the experiment, Mr. E. S. G. Titus, an especially well trained expert from the Bureau of Entomology, was assigned in the spring of 1906 to the charge of the laboratory end of the introduction.

In March, 1906, Mr. Titus, with the chief of the bureau and with Mr. Kirkland and Mr. Mosher, visited the parasite laboratory, and for the first time examined the contents of the imported nests. There were in the different cages, well separated as to localities, winter nests from almost the whole of the European range of the brown-tail moth, from Transylvania on the southeast to Brittany on the northwest, and from the Pyrenees on the southwest to the shores of the Baltic on the northeast. In spite of the voluntary assistance of such men as Rene Oberthür and Josef Jablonowski, the expense of getting these nests to Boston had been very considerable, and the moment when this examination was begun was considered to be rather a critical one. No published record of the rearing of parasites from these winter nests was recalled by the senior author or by any of his European correspondents, and the expensive experiment rested solely on the unpublished observation of Jablonowski, and he himself had simply seen parasites emerge from nests in the spring. Would they prove useless? Had the parasitic insects, even if useful, simply crawled into the nests for hibernation? Or were they, some of them, true parasites of the young larvæ? Representative nests were examined from a number of different localities, and the relief and joy were great when parasitic larvæ were found in considerable numbers in each of the nests examined, feeding within the nest pockets externally upon the brown-tail larvæ. This particular experiment was a success, and the expenditure of money and trouble was justified. About April 25 these parasites began to issue from the nests. The nests had been gathered in all from 33 different localities, and from some of them only a small number of parasites was reared. In all, about 70,000 issued, of which about 8 per cent were hyperparasites. In the rearing cages above mentioned it was a comparatively easy matter for Mr. Titus to separate the hyperparasites from the true parasites and to destroy the former. Of the species issuing in that spring—and they continued to issue until about June 15—there were two species which appeared to be important, namely *Pteromalus egregius* Först. and *Habrobracon brevicornis* Wesm. The latter species proved later to have entered the nests for hibernation only.

With the cooperation of Mr. Kirkland, several localities were found in which there was slight danger of forest fire and in which no work against the moths would apparently be undertaken for at least some months to come, and colonies of various sizes—the three principal ones including, respectively, 10,000, 15,000, and 25,000 parasites—were liberated in the open. Outdoor cages had been built over trees, and some smaller colonies of the parasites were placed in these cages. Both the outdoor experiments and the open experiments were seriously hampered, however, by the fact that the season proved to be one of extraordinary humidity, which caused the appearance of a fungous disease which destroyed a large proportion of the brown-tail moth larvæ in the vicinity of Boston.

Coincident with the issuing of these parasites from the nests, as the season grew warm the young larvæ swarmed from the nests and filled the glass tubes in the breeding cages and were constantly being destroyed by the assistants in the laboratory, and when the parasites ceased to issue the remaining nests and larvæ were burned. But later observations showed this destruction to have been a mistake. It was not considered likely that other parasites could be reared from these imported larvæ if they were fed and reared as far as possible, but such proved to be the case, as will be shown later.

During the winter of 1905–6 efforts were made to import in wintering conditions the two large European ground-beetles, *Calosoma sycophanta* (see Pl. I, frontispiece) and *C. inquisitor*. No success in importing living specimens was gained until March, 1906, but from that time on until July small consignments of living adult beetles were received, and in all 690 living specimens of *Calosoma sycophanta* and 172 of *C. inquisitor* arrived at Boston alive, some of them dying soon after arrival. Colonies were started in various localities about Boston. Consideration of the history of these two species will be given in Bulletin 101.

After visiting the parasite laboratory in March and determining the success of the importation of the brown-tail nests, the senior author sailed from New York on the 17th of the month for Europe, returning to America May 17.

Proceeding directly to Paris, Mr. Rene Oberthür was met by appointment, and the whole subject of the summer work was carefully considered. Mr. Oberthür is a man of affairs, proprietor of a large printing business, a learned amateur entomologist, and the possessor of one of the largest insect collections in the world. His advice and assistance throughout the whole work has been most important, and he assures the American representatives that he has highly appreciated the opportunity of being of assistance and of taking part in such an interesting piece of work. At his advice the writer proceeded to the south of France, after interviewing correspondents and agents in Paris, and visited Prof. Valery Mayet at the agricultural school at Montpellier, Dr. P. Siepi, of the Zoological Gardens in Marseilles, and Mr. Harold Powell, of Hyères. Both Prof. Mayet and Dr. Siepi stated that both of the injurious species of insects were rare in their vicinity, but both promised to assist in the importation of the Calosoma beetles. Mr. Powell proved to be a lepidopterist who had been employed professionally by Mr. Oberthür as a collector, and he was engaged to collect parasitized larvæ in Hyères and in the Enghadine district. He sent in much good material, and later, as will be shown in subsequent pages, organized a very efficient service in the summer of 1909. The visit to Prof. Mayet at Montpellier, moreover, was by no means devoid of results, since at a later date he was able to send a few specimens of carabid beetles, and in 1908, as a result of this personal interview, he was able to send to America the first living specimens of the European egg parasite of the imported elm leaf-beetle, *Tetrastichus xanthomelænæ* Marchal, which, as a result of this sending, is now possibly established in New England, although it was not recovered during the summers of 1909 and 1910.

While at Marseilles interviewing Dr. Siepi, April 10, the news was received of the eruption of Vesuvius and the partial destruction by lava flow of Boscatrecase and other villages on the slope of Vesuvius. Having to interview Prof. Silvestri and Dr. Leonardi at Portici, and fearing for their safety, the visitor proceeded at once to Naples, arriving there the day of the great market-house accident in which the roof fell in from the weight of volcanic ash and a number of persons were killed. Everything in Naples was in a state of confusion; the streets were filled with volcanic ash almost knee-deep, and it was with great difficulty that a conveyance could be secured to drive to Portici. Portici is almost on a direct line between Naples and Mount Vesuvius, and the agricultural college was found to be in bad condition; the gardens were utterly destroyed by ashes, and the roof of the old building was deeply covered. The accident happened the week before Easter, and the majority of the faculty and students had, on account of the catastrophe, anticipated their Easter vacations and had departed for their homes, Silvestri and Leonardi among the rest. Letters were forwarded to them, however, giving detailed suggestions as to methods of packing and shipment of parasites.

As in 1905, Florence, Milan, Vienna, Budapest, Dresden, Tharandt, and Zurich were visited. Efforts were made to learn of localities where either the gipsy moth or the brown-tail moth might reasonably be expected to be abundant during the summer of 1906, and a number of such localities were learned and the information given to agents. All of the agents and correspondents were given full instructions regarding the work for the summer of 1906 and the winter of 1907. The experience of 1905 with regard to the best methods of packing and shipment and the best kinds of boxes used was related to all, and these points were fully discussed, with the result that the material received during the summer of 1906 was not only greater in quantity but better in condition than that received during the previous summer.

In Vienna the visitor had the good fortune to find Dr. Gustav Mayr, whom he had missed in the summer of 1905. Dr. Mayr (since deceased) was the European authority on several of the groups of parasites most intimately connected with the work in hand, and the writer had a long consultation with him concerning the systematic position of some of the forms already imported and concerning the practical possibilities of the whole series of Microhymenoptera. Through him was learned the probable importance of certain egg parasites of the brown-tail moth, which he himself had reared in Europe and had described. As a result of this information the agents visited later were instructed to send over egg masses of the brown-tail moth to Massachusetts in midsummer, and later to send over egg masses of the gipsy moth. From the brown-tail moth egg masses parasites were reared by Mr. Titus at North Saugus and were observed to oviposit in native eggs. Mr. Titus reared not only the species referred to by Dr. Mayr, namely, *Telenomus phalænarum* Nees, which came from eggs forwarded by Miss Rühl and collected in Croatia, but he also reared an interesting parasite of the genus Trichogramma from egg masses received from Würtemberg, Dalmatia, and Rhenish Prussia.

At Budapest the visitor was especially glad to be able to announce to Prof. Jablonowski the success of the rearings of parasites from the winter nests of the brown-tail moth, so many of which had been brought over from Europe the previous winter on the basis of Jablonowski's unpublished observations. At the time of this visit Prof. Jablonowski was too busy completing his important work upon the migratory grasshoppers invading Hungary to be able to promise much assistance beyond that of corresponding with foresters and other persons well located in Hungary in order to obtain information as to good places to secure material.

Returning to America about the end of May, the laboratory at North Saugus was again visited, with Mr. Kirkland and Mr. Titus, and the work of preparing indoor cages and field cages was pushed. In the course of the summer a number of outdoor houses were constructed, and in these houses it was hoped to study the breeding habits of the imported insects.

During the summer the number of shipments received from Europe was so large that Mr. Kirkland made no attempt to list them in his Second Annual Report published January 1, 1907. In June, in addition to egg masses previously mentioned, larvæ and pupæ of both the gipsy moth and the brown-tail moth were received in number from many different European localities, and from these a large number of parasites of several different species were reared, the most abundant having been tachina flies. In one lot received from Holland more tachinids were reared than there were gipsy moth caterpillars originally. Nearly 40,000 gipsy-moth larvæ and pupæ were received and more than 35,000 brown-tail moth larvæ and pupæ. The receipt of predatory beetles is recorded in a previous paragraph.

It will be noticed that in the work conducted so far the effort to import parasites was confined to the continent of Europe west of Russia, whereas the well-known occurrence at intervals in large numbers of the gipsy moth in parts of Russia, and especially in southern Russia (a very good account of which will be found in the Third Report on the Gipsy Moth, by Forbush and Fernald), seemed to render it desirable that search should be made in those regions for parasites. The fact, however, that during these two years the writer had been unable to secure answers to letters addressed to correspondents in Russia and the reported unsettled condition of affairs in that country deterred him during the 1905 and 1906 trips from visiting the Russian southern Provinces. In the late summer of 1906, however, advices were received from Prof. J. Porchinsky, of the ministry of agriculture at St. Petersburg, with the information that in the southern part of Russia both the gipsy moth and the brown-tail moth were at that time occurring in sufficiently great numbers to enable the collection of parasites and commending the writer to certain officials, trained entomologists, in Simferopol (Crimea), Kishenef (Bessarabia), and Kief. Prof. Porchinsky wrote that he had apprised these officials of the intended visit, and plans were therefore made to include southern Russia in the itinerary for the spring of 1907.

During the autumn of 1906 egg masses of the gipsy moth continued to be received from parts of Europe, and during the winter hibernating nests of the brown-tail moth were sent in. More than 111,000 nests were received from different portions of the European range of the species. These were placed in the especially constructed cages, and from many of them large numbers of parasites were reared, issuing mainly during the month of May, 1907. As it happened, the month of May in New England, as well as in other parts of the United States, was phenomenally cold and wet. As a result of this unlooked-for condition very many of the parasites refused to leave the nests until they were so weakened as to be unable to survive the close confinement and careful scrutiny to which they were necessarily subjected in order to eliminate the danger of introducing secondary parasites. As a result, a smaller number of *Pteromalus egregius* was colonized in the summer of 1906, but 40,000 specimens were put out in several localities, the principal colonies consisting, respectively, of 13,000, 11,000, and 7,000 individuals. At this time, as well as in the summer of 1906, although this fact has not as yet been stated, a number of important parasites of the genus Monodontomerus issued from the winter nests and were allowed to escape. As will be shown subsequently, this parasite has proved to be more important than the Pteromalus and has made a phenomenal spread.

In this important work with the introduced hibernation nests of the brown-tail moth it was early found most difficult to preserve the health of the laboratory assistants. The irritating and poisonous hairs of the brown-tail moth larvæ, of which the nests are full, soon penetrated the skin of the assistants handling them, entered their eyes and throats, and the atmosphere of the laboratory became almost filled with them. It was necessary that the rooms should be kept thoroughly closed; double windows and screens were used, and the doors of the rooms were doubled, in order that a possible secondary parasite, if accidentally liberated, should have no chance of escape. This made the rooms very warm and increased the irritating effect of the larval hairs. Some of the assistants employed could not stand the work and resigned. One of the best and most experienced helpers was induced to continue the second year only upon the promise that he would be relieved from this especial class of work. Spectacles, gloves, masks, and even headpieces were invented to avoid this difficulty, but these, while greatly increasing the suffering from the heat, were not entirely effective. The most serious result of this trouble was the breaking down in health of Mr. E. S. G. Titus of the bureau, in charge of the laboratory at Saugus, who was obliged to resign in May, 1907, on his physician's advice, in order to save his life. The difficulty in Mr. Titus's case was the intense irritation to his lungs from the entrance of the barbed hairs. Mr. Titus was soon after appointed entomologist of the Utah Agricultural Experiment Station, and the change of work and climate fortunately brought about a speedy recovery. His necessitated departure in the midst of important work, however, threw us into what appeared to be a serious dilemma, but fortunately it so happened that the services of the junior author, then occupying another position in the Bureau of Entomology in Washington, could be spared from the other work upon which he had been engaged, and, since he had made especial studies of the parasitic Hymenoptera and had done a large amount of rearing of parasites in the course of his other work, he was sent on from Washington to replace Mr. Titus in the parasite laboratory and has since had charge of the laboratory.

One of the early points to which the junior author devoted his attention was the invention of new methods of handling the browntail nests in order to avoid the serious effect upon the work of the breaking out of the rash on himself and his assistants. He soon devised an apparatus like the ordinary show cases that are seen in shops, the glass on one side being replaced by cloth with armholes, through which the gloved hands of the worker could be thrust and the brown-tail nests handled in full sight through the top glass. Most of the work with these nests, it has been found, can be done in these cases with a minimum escape of the barbed hairs. There still continued, however, considerable trouble from the rash, since much rearing of brown-tail larvæ must be carried on under conditions in which such cases can not be used, and this difficulty still exists. Miss Rühl, of Zurich, in handling and repacking the large number of nests sent to her by her European correspondents and forwarded by her to Boston, has been a great sufferer from the rash. She has made for herself a complete costume of an especially finely woven cloth, and has made a large light helmet covered with cloth and provided with a cape, the space opposite the eyes being fitted with a sheet of very transparent celluloid. Of course this costume would be very uncomfortable in the summer time on account of the heat, but since she handles her nests for the most part in the autumn and winter, she has been able to reduce the discomfort of the brown-tail rash to a minimum.

Sailing again for Europe on April 20, 1907, the senior author landed at Cherbourg and proceeded directly to Paris, and from Paris to Budapest by the Oriental Express. At Budapest, by prearrangement, he met Mr. Alexander Pichler, whom he had engaged as a guide and courier for the Russian trip. After a conference at Budapest with Dr. Horvath and Prof. Mocsary, of the Natural History Museum, and Prof. Jablonowski, of the agricultural station, he proceeded to Kief, via Lemburg. Prof. Porchinsky, of the ministry of agriculture, had arranged with Prof. Waldemar Pospielow, of the University of Kief, to consult with the Chief of the Bureau of Entomology about future arrangements, and a conference with Prof. Pospielow was held, in the course of which it was agreed that one of Pospielow's assistants, engaged especially for the purpose, at 34 rubles per month, should occupy himself throughout the summer, under Pospielow's directions, in collecting larvæ of the gipsy moth and brown-tail moth, forwarding material to Boston, rearing and studying the parasites, and conducting observations in an orchard in the suburbs of Kief, rented by the writer for the State of Massachusetts for the summer at the rate of 20 rubles per month. This procedure was novel in the work, but was later tried in another locality, as will be shown in subsequent pages.

From Kief, Pichler and the visitor proceeded to Odessa and from Odessa to Kishenef, at which point he had been recommended to Dr. Isaak Krassilstschik by Prof. Porchinsky. Through some misunderstanding as to dates, owing to the difference between the Russian calendar and the one in use in other parts of the world, Prof. Krassilstschik had mistaken the date of arrival announced in the letter sent in advance, and was absent from Kishenef on a brief visit to Germany. Full written instructions, however, were left for him at Kishenef, and the visitor returned to Odessa and thence by boat to Sebastopol, and by train to Simferopol. At Simferopol he was expected by Prof. Sigismond Mokshetsky, the director of the Museum of Natural History at that place and an enthusiastic economic entomologist, through whose efforts American methods in the warfare against insects had been introduced into southern Russia. Prof. Mokshetsky had done some rearing of the Russian parasites of both the gipsy moth and the brown-tail moth, and was able to furnish much valuable information. His hospitality and cordiality were of the most encouraging nature, and after consultation as to the best methods, he promised his hearty support to the work, refusing, however, to accept any compensation from the State of Massachusetts or from the United States Government.

The visitor then proceeded by boat from Sebastopol to Constantinople, but was unable to learn of any person in Turkey having any information on the subject of insect pests, nor was he able in the country about Constantinople to find any indication of the occurrence of either gipsy moth or brown-tail moth.

Leaving Constantinople, the expedition proceeded to Vienna, dropping Mr. Pichler at Budapest. At Vienna the Seventh International Congress of Agriculture was held, beginning May 22, 1907. The visitor met there a number of delegates from the different countries in Europe, with whom he discussed the question of parasite importation, receiving warm assurances of support, especially from Prof. Dr. Max Hollrung, of the Agricultural Department of the University of Halle, Prof. Dr. Karl Eckstein, of the Forest Academy at Eberswalde, and Prof. Dr. J. Ritzema Bos, director of the Phytopathological Station at Wageningen, Holland. While in Vienna arrangements were made with Mr. Fritz Wagner for continuance of the work, and a further consultation on the subject of parasites was held with Dr. Gustav Mayr.

After Vienna, Mr. Schopfer was visited in Dresden, Dr. Hollrung at Halle, Dr. R. Heymons in Berlin, Dr. Eckstein in Eberswalde, Miss Rühl at Zurich, and Prof. G. Severin at Brussels. Prof. Severin is connected with the Royal Natural History Museum at Brussels, is an admirably well-posted entomologist, and is connected with the Forest Conservation Commission of Belgium. He was able to give good advice in the parasite work and promised assistance.

Returning to France, an important conference was held with Mr. Rene Oberthür, and it was arranged to establish during the summer of 1908 a field station at Rennes, to be placed in charge of a special expert, Mr. A. Vuillet, chosen by Prof. Houlbert, of the University of Rennes. Through Mr. Oberthür's courtesy it was arranged to establish field rearing cages at a convenient point near the University of Rennes and to carry on the work in much the same way as it had been arranged for the present summer at Kief. The University of Rennes having a certain connection with the University of Paris, it was considered desirable that the cooperation of the scientific faculty of the University of Paris be gained by direct application. This was readily arranged, through the cordial and sympathetic cooperation of Prof. Alfred Giard, of the faculty of science of the University of Paris (since deceased).

In dealing with the European parasites reared at North Saugus, considerable difficulty was experienced in ascertaining their names. It was very desirable, of course, to have a definite name by which to designate each species, and by which to correlate it with published accounts of observations already made. With the assistance of Dr. O. Schmiedeknecht, of Cassel, Germany, a number of these forms had been named, but with others it seemed practically impossible to bring this about by correspondence. As a result, on the trip in question the writer made an effort, by studying the collections in some of the principal European museums, to determine a few of the unnamed forms reared in America from European material. The difficulty of this search was surprising. The Pteromalus, for example, which had been reared in Boston by scores of thousands and which, therefore, must be a very common European insect, was found to be absolutely unrepresented in the large natural history museums of Vienna, Dresden, Berlin, Brussels, and London; nor did it occur in the type collections of Ratzeburg carefully preserved by Dr. Eckstein at the Forest Academy at Eberswalde, where, on account of Ratzeburg's important work on the parasites of European forest insects, one would naturally expect to find it. At last, in a small special collection in the Museum of Natural History in the Jardin des Plantes at Paris, Mr. H. du Buysson of the museum found in the laboratory a box containing parasites reared many years ago by the French entomologist, Sichel, which had been named for him by the eminent authority on parasitic Hymenoptera, Arnold Förster, of Germany. In this box were specimens of the Pteromalus labeled "*Pt. egregius*" in the handwriting of Förster himself.

Especial efforts were made on the trip to arrange for the importation of large numbers of the egg parasites of both species and to introduce in living condition the important parasites of the genus Apanteles, which, according to the visitor's field observations, are among the most important of the European enemies of the gipsy moth. Previous importations of these parasites had failed, owing to the fact that they emerged and died on the journey. On this trip, however, specific directions were given to agents to send in young larvæ of the second stage, and by this means living specimens in considerable numbers were later reared in the laboratory at North Saugus. These on issuing laid their eggs in the gipsy-moth larvæ of the first stage, and from these caterpillars were secured the cocoons of adults of a second generation which was reared through all of its stages on American soil.

From Kief there were received two species hitherto unknown as parasites of the gipsy moth, and one of these, being a rapid breeder, promised to be of much assistance. This species, belonging to the genus Meteorus, seemed to produce cocoons in about 10 days after egg laying, and will be considered later in this bulletin.

We have previously referred to the destruction in 1906 of the great bulk of brown-tail caterpillars imported from Europe after the early appearance of adult parasites. Mr. Titus, in 1906, tried the experiment of rearing a very few of these imported larvæ, and found that in their later growth they gave out a second lot of parasites entirely different from those reared in May from the very young hibernating larvæ, indicating a delayed development of eggs which must have been laid by adult parasites the previous autumn. Among these were at least two species, one belonging to the genus Apanteles and the other a Meteorus. Before his resignation in 1907 he started an extensive series of rearing experiments with the end in view of securing these parasites in large numbers. Partly on account of his enforced absence from the laboratory during a critical period, and partly through the unsuitable character of the rearing cages which were employed, the project did not meet with entire success. Only about 1,000 of the parasites were reared, of which all but a small percentage were the Apanteles.

The importations of the summer following the trip above described were very large, and reasonably successful, and during June alone 872 boxes were received, many others following during July and into August, shipments of brown-tail eggs and gipsy-moth eggs following, and of brown-tail winter nests in the late autumn and during the winter. As in 1906, tachinids made up the great bulk of the parasites secured through the importation of pupæ and active caterpillars. Notwithstanding the improvement in methods of shipment over previous years, Apanteles invariably hatched en route, and only dead adults or secondary parasites were received.

Before the close of the summer it had become obvious that better quarters for the Massachusetts laboratory were necessary. The heating and lighting arrangements at North Saugus were insufficient; the building was not sufficiently commodious, and the location was not convenient. Therefore, after considerable search, Mr. Kirkland found and leased for a term of years a commodious house at Melrose Highlands (No. 17 East Highland Avenue) (see Pl. II, fig. 2, p. 56.) The building was remodeled so far as necessary to fit it for the work. The grounds back of the house were sufficiently ample to enable the building of several outdoor laboratories, properly screened and ventilated, which were planned and erected under the direction of the junior author. The building is well warmed, lighted with electricity, and, being close to fire protection, possesses many advantages over the old laboratory. Moreover, it is much nearer the central office in Boston, enabling an important saving of time in sending to the laboratory shipments of parasites received from abroad. The rental and the expense of construction were all borne by the State of Massachusetts. The new quarters are also within a stone's throw of a large area of waste land covered with scrub oak.

In planning the work for the season of 1908, several new features were introduced. The parasites constantly sent over by agents belong to three main groups, namely, those of the order Hymenoptera, including the ichneumon flies, the chalcis flies, and others; those of the Diptera, including the tachina flies, and those of the order Coleoptera, including the predaceous ground beetles. The amount of material received had been so great, and the character of the different life histories of the insects involved had been so diverse, that no one expert was able to do the fullest justice to the situation. Therefore, while the junior author was left in general charge of the whole mass of importations and retained his expert supervision of the work on the biology of the parasitic Hymenoptera, Mr. C. H. T. Townsend, of the Bureau of Entomology, was assigned to the work on the biology of the dipterous parasites, and Mr. A. F. Burgess, also of the Bureau of Entomology, was assigned to the expert charge of the ground beetles.

Owing to the fact that the condition of European sendings by mail and express during the summer of 1907 had been by no means uniformly good—those from eastern Europe, subjected to long railway journeys in addition to the sea voyage, frequently arriving in bad condition—the second innovation was made by establishing at Rennes, France, a general laboratory depot in addition to the field cages and rearing station mentioned in a previous paragraph. The expert assistant designated by Prof. Houlbert, of the University of Rennes, was Mr. A. Vuillet, who was placed in specific charge of the general laboratory depot under the general supervision of Mr. Rene Oberthür. Mr. Vuillet placed himself in relations with the steamship company agents at Cherbourg and Havre and was kept informed as to the dates of the sailings of steamers. Nearly all of the European sendings were shipped to Rennes, examined, repacked, and carried personally by Mr. Vuillet to Cherbourg or Havre on the known days of sailing of certain steamers and then placed in the hands of chief stewards of the vessels and carried in the cold rooms to New York, whence they were sent to Boston. Early in the course of the work the honorable the Secretary of the Treasury, upon request of the honorable the Secretary of Agriculture, had issued orders to the collector of the port of New York to admit all such packages without examination and to hasten their departure for Boston through the United States dispatch agent. The steamship officials showed themselves uniformly courteous, and as a result of this new arrangement the average condition of the material received proved to be much better.

With the installation of the new laboratory at Melrose Highlands, and with the added space afforded by the new structures in the garden, the junior author was able to carry out some new ideas with admirable results. The first of these was the carrying on of active winter work with parasites, especially those secured from the imported nests of the brown-tail moth, which began to come in from Europe in December. It was found quite possible to rear these parasites in artificially heated rooms, feeding them upon hibernating native brown-tail larvæ brought in in their nests from out of doors, feeding the latter upon lettuce and other hothouse foliage, and in the early spring securing more normal food for them by sending it up in boxes by mail from Washington and points south. In this way the rearing of the parasites of the genus Pteromalus was carried forward uninterruptedly throughout the winter, and, as during the rearing of successive generations they multiplied exceedingly, it was possible later in the year to liberate a vastly greater number of individuals than had the imported species been allowed to hibernate normally in the nests. In the course of this work the junior author invented a rearing tray which was of the utmost advantage and which has since greatly facilitated parasite rearing work. This tray will be described later.

With the importation of brown-tail moth eggs it often happened that they hatched too soon to be of use in America; or too late, arriving after the American eggs had all hatched. It was ascertained by the junior author during the summer and autumn that native eggs can be kept in cold storage until the arrival of the European egg parasites, which were found to lay their eggs and breed in these cold-storage eggs as freely as in those which they attack in the state of nature. It was found that this process can be carried on for a long time, and that successive generations of these egg parasites may be reared from eggs retarded in their development by cold storage. It was thus shown that it is easy to rear and liberate an almost infinitely greater number of these egg parasites, and under favorable conditions, than would have been possible from a simple importation of European parasitized eggs which would have to arrive in America at a specific time.

In the same way great advance was made in the rearing of the tachinid parasites in Mr. Townsend's charge. This expert devised methods and made observations that greatly added to our knowledge of the biology of these insects and resulted in the accumulation of a store of information of the greatest practical value, not only in the prosecution of the present undertaking but in any problem of parasite introduction or control that may arise later. Extraordinary and almost revolutionary discoveries were made in the life histories of certain of these flies, and without this knowledge the greatest success in handling them practically could not have been reached. Certain of these facts regarding the most important of these parasites are related in a later part of this bulletin, and many of them have been described in some detail in Technical Series No. 12, Part VI, Bureau of Entomology, United States Department of Agriculture (1908), by Mr. Townsend.

Similarly Mr. Burgess, in charge of the Coleoptera, succeeded in a very perfect way in rearing and liberating the important European predatory beetle, *Calosoma sycophanta*, as well as some other insects of the family Carabidæ.

While these extensive importations from Europe were going on, Japan had by no means been lost sight of. While it seemed probable that the European parasites in themselves would succeed in reestablishing the balance of nature in New England, and in spite of the somewhat dangerous nature of Japanese importations on the ground that the Japanese gipsy moth is probably a different species and might prove in New England even more voracious and destructive than the European moth, there was at no time any intention to neglect Japan in the search for effective parasites. Continuous correspondence had been carried on with Japanese entomologists, and some shipments had been made by correspondents which resulted unsuccessfully. For some time the Apanteles previously mentioned was the only gipsy-moth parasite known to occur in Japan. Later information was received from Prof. U. Nawa, of Gifu, Japan, to the effect that there exists in Japan an important egg parasite of the gipsy moth. During the previous annual trips of the Chief of the Bureau of Entomology to Europe the European service of collectors, agents, and advisers had been well organized and instructed, and the work during 1908 was reasonably sure to be well continued without further personal consultation; it was therefore decided to interrupt the European trip for 1908 and to send a skilled agent to Japan. In considering the appointment of such an agent, Prof. Trevor Kincaid, of the University of Washington at Seattle, was at once suggested to the mind of the writer, primarily on account of his extraordinary skill as a collector, as indicated in the remarkable results of his work on the Harriman expedition to Alaska in 1899, and also on account of his comparative proximity to Japan and the fact that he was personally acquainted with many persons in Japan. He was therefore recommended to the State officials of Massachusetts for appointment, and was commissioned by the State to undertake the expedition. At the same time he was formally appointed a collaborator of the Bureau of Entomology of the United States Department of Agriculture, and the Japanese Government was formally notified by the honorable the Secretary of Agriculture, through the Department of State, of the intended visit, the writer having also notified by personal correspondence some of the well-known Japanese entomologists. Prof. Kincaid sailed from Seattle on March 2, and the results of his expedition far more than justified the expense involved. A very large amount of parasite material was received from him in good condition at Boston, and very many parasites from Japan were colonized in the woodlands in New England. Prof. Kincaid was received with the most extreme courtesy and cordiality by the Japanese Government and by official and private entomologists everywhere. His work was facilitated in every possible way; assistants were placed at his disposal and in this way a large number of individuals occupied themselves in the collection of parasitized material. After consultation with the Japanese entomologists, whose great cleverness in manipulation and ingenuity in devising methods are well known, Prof. Kincaid was able to pack his shipments in such a way as to bring about a minimum of mortality on the journey. The steamship companies showed him every courtesy, and much of his material arrived at Melrose Highlands in better condition than corresponding sendings-received from Europe. A single indication of the value of Prof. Kincaid's work may be mentioned: From one shipment of cocoons between 40,000 and 50,000 adults of the Japanese Apanteles were reared and were liberated directly in the open in Massachusetts, and this is the species which, although repeatedly sent by correspondents, had never arrived in New England in such condition that a single living adult could be reared.

The European importations in the meantime continued to arrive in numbers, and at the close of the summer it was found that the actual number of beneficial insects liberated had been far in excess of that for 1906 or 1907, and that the list included several species of apparently great importance and promise that had never before been received at the laboratory in living condition.

The successful European importations all came from western Europe, and unfortunately the few shipments sent from Russia arrived in very bad condition. This is considered to have been most unfortunate, since several of the Russian parasites were very promising, and the subject of improving the Russian service was taken into consideration.

With the great success of the summer's Japanese work, and the question of the great desirability of similar work in Russia in his mind, the senior author, visiting the Pacific coast in the autumn of that year (1908) on a tour of inspection of the field laboratories of the Bureau of Entomology, called on Prof. Kincaid at Seattle and discussed with him at length the plans for 1909. Although Kincaid expressed himself as charmed with Japan and anxious to repeat his visit to that most interesting country, his innate honesty compelled him to state that he considered the expense of the trip unnecessary; that he had found the Japanese entomologists, officials, and others so intelligent and so thoroughly competent, and at the same time so heartily interested in the experiment, that he considered them not only perfectly able, but perfectly willing to carry on the work by themselves. After this authoritative expression of opinion from one who knew the ground so well, the visitor asked Mr. Kincaid whether he would care to spend the early summer months of 1909 in Russia, and, upon his affirmative reply, later recommended his reappointment to the Massachusetts State authorities for that purpose.

During the autumn and winter shipments of eggs of the gipsy moth were received from Japan, principally from Prof. Kuwana. From these eggs were reared numerous specimens of *Anastatus bifasciatus* Fonsc., a previously known European parasite of these eggs, and of another parasite belonging to a genus and species new to science (since named by the senior author *Schedius kuvanæ*) which has turned out to be an important primary parasite and which is considered in later pages. During the winter, also, Prof. Jablonowski, of Budapest, sent over several thousand egg masses of the gipsy moth collected in various localities in Hungary. After they arrived in Massachusetts there were reared from them and liberated under the most favorable conditions more than 75,000 adult individuals of *Anastatus bifasciatus*. This was a surprising thing to the laboratory workers, since less than 1,000 parasites of this species had been received from all localities, the earlier ones having come from southern Russia and from Japan.

The winter of 1908–9 was spent at the laboratory, in additional rearing operations, some of them on a large scale, and in studying the parasites already reared, and planning for the coming summer.

As it happened, during the winter the brown-tail moth was introduced into the United States upon nursery stock from France in large numbers. Shipments of nursery stock bearing winter nests of this insect were sent to many States of the Union. Fortunately this was discovered early in the winter, and through prompt action and the cooperation of the customs officials and the railroads probably every sending was traced to its ultimate destination, and was there inspected and the nests destroyed either by State officials or by persons appointed for this purpose by the United States Department of Agriculture.

In the spring of 1909 it seemed necessary for the chief of the bureau to proceed to Europe for the purpose of making an investigation of the European methods of growing nursery stock, with a view to the prevention of similar introductions in the future either by general legislation by the United States Government or in some other way. On this trip he utilized the opportunity to consult further with European agents in the importation of the parasites and to arrange for the summer's work.

In the meantime Prof. Kincaid, whose appointment had been made by the State of Massachusetts, and who had again been made an official collaborator of the Bureau of Entomology of the United States Department of Agriculture, securing leave of absence from the University of Washington, proceeded to Russia, and stationed himself in Bessarabia for the purpose of collecting and sending parasitized material from that country to the United States. It had been noticed by Mr. Vuillet at Rennes during the preceding summer that all material coming from Russia had been opened on the journey and had deteriorated in consequence. Before Prof. Kincaid's departure from America, Russian officials had been communicated with through correspondence between the chief of the Bureau of Entomology and Prof. Porchinsky, of the ministry of agriculture, and also directly between the United States Department of State and the American ambassador at St. Petersburg through the instigation of the honorable the Secretary of Agriculture. The United States Government was assured that the Russian Government would welcome the expedition and would facilitate the sending of material in every way possible.

The chief of the bureau landed at Cherbourg May 12. He proceeded immediately to Paris, where a conference had been arranged in advance with M. Oberthür, M. Vuillet, and Mr. Henry Brown, the latter an English entomologist resident in Paris. At this conference it was decided to abandon the forwarding laboratory at Rennes and to station Mr. Vuillet, during the forwarding season, at Cherbourg. He was instructed to engage quarters at that seaport and to arrange for cold-storage facilities, with the intention that shipments from France, Switzerland, and Italy should be forwarded to him to be kept in cold storage until the date of sailing of vessels, and then should be transferred to the cold room of the next steamer, thus practically keeping all living specimens dormant from the time of arrival in Cherbourg until the time of arrival in New York, making the exposure to summer temperature practically only 24 hours or less in Europe and 24 hours or less in the United States. In the meantime Mr. Oberthür was authorized to arrange for an extensive service in the south of France, through Mr. H. Powell, of Hyères, one of the agents for the year 1906. The preparation of the requisite boxes was intrusted, as in previous years, to the superintendence of Mr. Oberthür, and Mr. Powell was authorized to engage as many collectors as the material would seem to need, with full instructions as to packing and shipping to Cherbourg.

The visitor then proceeded to Wageningen, Holland, where he arranged for further assistance from Prof. Dr. J. Ritzema Bos. From there he went to Hamburg, where he arranged with the American Express Co. to care for shipments coming from Germany, Russia, and Austria-Hungary, arrangements being made to keep the material on ice until the next steamer should sail, and in case of the breakage

Fig. 1.—Roadside Oak in Brittany, with Leaves Ragged by Gipsy-Moth Caterpillars. (L. O. Howard, June, 1909.) (Original.)

Fig. 2.—M. Rene Oberthür (in Center), Dr. Paul Marchal (at Right), with Roadside Oaks (Behind) Ragged by Gipsy-Moth Caterpillars. (L. O. Howard, June, 1909.) (Original.)

Fig. 1.—Caterpillar Hunters in the South of France, under M. Dillon, 1909. (Original.)

Fig. 2.—Packing Parasitized Caterpillars at Hyères, France, for Shipment to the United States, 1909. (Original.)

or other bad condition of packages arrangements were made with Dr. L. Reh, of the Hamburg Museum, to act as expert adviser of the express company.

From Hamburg he proceeded to Berlin for a short consultation with Dr. R. Heymons, and thence to St. Petersburg. At St. Petersburg he was assured by Mr. Montgomery Schuyler, the secretary of the embassy, that all arrangements had been made with the Russian Government, and the same assurance was given by Prof. Porchinsky. The Russian officials insisted that none of the 1908 packages going out of Russia had been opened by the Russian postal authorities, and stated that in their opinion the opening must have been done at the German frontier by German officials. A strong letter was then written to the Hon. David J. Hill, United States ambassador to Germany, reciting the facts, dwelling upon the importance to America of these importations, and urging him to secure from the German Government orders to postal officials to pass without opening boxes of these parasites addressed to the American Express Co. in Hamburg. Later, in Dresden, a reply was received from Ambassador Hill, stating that the German Government consented to issue the necessary instructions, but still later, in Paris, an additional communication from the ambassador requested detailed information as to the points on the German frontier where these sendings would enter the Empire. By telegraphic communication with Prof. Kincaid, in southern Russia, and the Austrian agents, this information was furnished, but there seems still to have been some opening of the Russian boxes with resulting damage to their contents.

After Russia, Dresden, Tetschen, Vienna, Budapest, Innsbruck, Zurich, and Paris were consecutively visited, and agents were instructed concerning the new arrangements for shipping material. At Innsbruck the visitor met for the first time Prof. K. W. von Dalla Torre, the author of the great catalogue of the Hymenoptera of the world, and got his views on the subject of the parasitic Hymenoptera and their practical handling.

From Paris he took a trip into Normandy and Brittany with Dr. Paul Marchal, of the ministry of agriculture of France, and Mr. René Oberthür, for the pupose of examining into the export nursery industry, and at the same time with a view of observing gipsy-moth and brown-tail moth conditions in that part of France. (See Pl. III, fig. 2.) It transpired that both of the injurious insects were unusually abundant in portions of this territory, and by good fortune a small oak forest covering some hundreds of acres was found not far from Nantes, in which there had been an outbreak of the gipsy-moth more serious than either Dr. Marchal or Mr. Oberthür had ever seen or had ever heard of in France. Practically every tree was defoliated (see Pl. III, fig. 1), and at the time of the visit, the last week in June, the larvæ were about full grown and making ready to spin. The natural enemies of the gipsy-moth were not abundant in this forest, although a few were seen on trees along the highway in this general region. Nevertheless the invariable experience in Europe is that following such an outbreak as this parasites congregate in the region the following year and multiply in enormous numbers. The finding of this area, therefore, seemed fortunate, since during the season of 1910 it seemed probable that parasites would be abundant at that point. This hope was not fulfilled, however, and in 1910 practically no gipsy-moth larvæ were to be found in that general region.

In the meantime the honorable minister of agriculture for Japan had at the request of the honorable the Secretary of Agriculture of the United States designated Prof. S. I. Kuwana, of the Imperial Agricultural Station at Tokyo, to be the official representative of the Japanese Government in the parasite work to be carried on during the spring and summer of 1909, and to conduct his operations in cooperation with and in correspondence with the chief of the Bureau of Entomology of the United States Department. Prof. Kuwana has shown himself in this, as in his previous work, a man of extraordinary intelligence and activity, and has sent in a number of interesting and valuable lots of parasitic material which were received at Melrose Highlands in uniformly good condition. This was due to the great care and intelligence shown by Prof. Kuwana in its collection and in his methods of packing and shipping.

The most nearly perfect European service during the summer of 1909 was secured in France, owing to the arrangement made at the May conference in Paris. In the south of France very many people were employed under Mr. Powell, and several thousand boxes of good material were received at the parasite laboratory from this region. (See Pl. IV, fig. 2.) In quantity it exceeded the total of all the importations of a similar character made since the inception of the work, and from it have been reared a greater number of important tachinid parasites than have been reared from all other importations of similar character taken together. The size of the French shipments is largely due to the intelligent energy of Mr. M. Dillon (see Pl. IV, fig. 1), with whom the bureau was placed in relations by Mr. Powell.

Quantities of miscellaneous material were also received, as formerly, from numerous collectors in Germany, Austria, Italy, Holland, Belgium, and Switzerland.

Prof. Kincaid's account of his Russian observations is as follows:

At the request of Dr. L. O. Howard, Chief of the Bureau of Entomology, United States Department of Agriculture, the writer visited the provinces of Russia bordering upon the Black Sea during the summer of 1909 with a view to the introduction into America of the parasites of the gipsy moth reported to exist in that part of Europe. Proceeding to St. Petersburg via New York and Paris, an interview was had with Prof. Porchinsky, of the Russian Bureau of Entomology, who supplied valuable information and suggestions for the furtherance of the investigation. Leaving the Russian capital on April 28, a journey of 48 hours brought the writer to the city of Kishenef and after making a survey it was decided to establish a base of operations in the forest of Gauchesty, an area of wooded hills adjacent to a village of that name about 30 versts [1] northwest from Kishenef. Since the accommodations in the village of Gauchesty were of an unsatisfactory character, Mr. Artemy Nazaroff, the manager of the estate of Prince Manook Bey, on the lands of which the more important infested areas existed, invited the writer and his interpreter to become his guests during the progress of the investigation. A suite of rooms in the guest house of Gauchesty castle was placed at our disposal, and Mr. Nazaroff did all in his power to forward our interests and to make agreeable our stay in that part of Russia. An outbuilding upon the farm of the estate was transformed into a laboratory in which was erected a set of rearing frames for the rearing of the parasites. During the first week of April systematic exploration of the adjacent wooded areas was begun. The forest cover was found to consist almost exclusively of young oaks, with a few scattering trees of other species. The ground beneath the trees was fairly free from underbrush and was carpeted with a rich profusion of shrubs and flowers. At a distance of 7 versts from Gauchesty was an area covered with trees of considerable age among which the underbrush was comparatively dense.

From the forester in charge of the timbered areas upon the estate it was learned that the gipsy moth had done great damage to the forest during the previous season, large areas having been completely defoliated. This statement was borne out by the immense number of egg masses attached to the trees. At the time we commenced our investigations the caterpillars had emerged from the eggs but were still resting upon the bark. Few signs of previous parasitic activity were observed beyond the discovery of a number of empty cocoons of *Apanteles solitarius* Ratz. attached to the bark of the trees. In the ancient forest mentioned above the egg masses were very numerous, but the number of larvæ upon the bark was remarkably small. From the abnormal appearance of most of these egg masses, and from the fact that several Microhymenoptera were discovered in them, it seemed probable that a considerable number of the eggs had been destroyed through this agency. In other parts of the forest no evidence was secured indicating the presence of egg parasites.

The brown-tail moth seemed to be practically absent from the forested areas, but in the open rolling country between Kishenef and Gauchesty many wild pear growing in cultivated fields were found to be completely defoliated. A large number of the larvæ were placed in rearing frames but yielded no parasites, not even Meteorus making its appearance.

By June 1 the caterpillars of the gipsy moth had passed into the second stage and the oak trees were showing obvious signs of damage, but up to this date there was no indication of the emergence of

---

[1] Verst: Russian measure of distance=3,500 English feet; 6 versts=approximately 4 English miles.

hymenopterous parasites either in the field or from the thousands of larvæ reared in rearing frames. It became apparent that the conditions were unfavorable for the purposes in mind of assembling parasites for export, and it was decided to shift our headquarters to a more promising locality.

On June 5 a new base of operations was established at the town of Bendery on the Dniester River. Quarters were selected in the principal hotel, the Petersburgia, and in a remote corner of the extensive grounds of the hostelry a temporary laboratory was constructed in which several tiers of rearing frames were erected. The forest conditions in this district were much more diversified than at Gauchesty. Tó the northeast of the town at a distance of 7 versts was the forest of Gerbofsky, occupying a dry elevated area of about 5,000 acres and consisting almost exclusively of mature oak trees. To the southward, on the banks of the river, was the forest of Kitzkany, composed largely of black poplar, maple, and willow. In both of these forests the caterpillars of the gipsy moth were found in immense numbers, and evidence of attack by both hymenopterous and dipterous parasites was readily obtained, although nowhere in the abundance hoped for. For two weeks the two forests, as well as the extensive orchards in the vicinity of Bendery and the neighboring town of Tiraspol, were scoured for parasites. A number of Russian boys were pressed into service and trained to assist in making collections, at which they became quite expert. Except for a few clusters of cocoons derived from *Apanteles fulvipes* Hal., the only hymenopterous parasite to appear in considerable abundance was *Apanteles solitarius*. Caterpillars of the gipsy moth attacked by this species crawl down to the trunk or lower branches of the tree and collect in colonies on the lower side of the branches, under bark, in cavities and other sheltered places. Here the larva of the parasite emerges and spins its cocoon beneath the body of its host. The task of collecting these scattered cocoons was a tedious one, since it was necessary to remove each one carefully from the bark without undue pressure and also to disentangle it from the hairy body of its host.

In the forest of Kitzkany, where the conditions were favorable for bacterial infection owing to excessive dampness, the caterpillars of the gipsy moth were swept away in vast numbers by a bacterial disease before any extensive defoliation took place. The search for hymenopterous parasites in this district soon become a vain one, since very few of the caterpillars appeared to have escaped the infection.

The forest of Gerbofsky, owing to its being elevated, open, and well drained, was not favorable for bacterial infection and no trace of disease was observed. This forest was therefore almost completely defoliated by the caterpillars, and multitudes of the insects, failing to find any further nourishment upon the oaks, descended to the ground, where they died in great numbers, apparently from starvation. Hymenopterous parasites seemed to play a relatively small part in the destruction of the caterpillars, since the attacks of *Apanteles solitarius* were of the most scattering character. In the shrubbery growths adjacent to the main forest, where new plantations had been recently established by the forester in charge, a considerable number of Calosoma were found at work destroying the caterpillars, but their operations did not appear to extend into the main forest, where the open grass-covered ground did not offer sufficient concealment for the beetles.

The principal check to the depredations of the caterpillars of the gipsy moth in this forest came with the advent of the tachinids, the latter appearing upon the scene after the trees had been almost or entirely defoliated. Chalcid flies also appeared at this time, but not in considerable numbers. The species of Limnerium, a few specimens of which had been previously received from Russia, and of which it had been hoped to secure a supply for transfer to America, proved to be exceedingly rare, only three specimens being found. The larva of this parasite on emerging from its host spins an elongated silken thread, at the end of which it spins a cocoon and transforms to the pupal state.

Considerable numbers of the cocoons of *Apanteles solitarius* were collected from the forest, from the extensive orchards of the neighborhood, and from clumps of willow bushes conmonly found at the edges of fields. For several weeks shipments were made almost daily to Hamburg, from which port the packages were shipped in cold storage to New York. Many difficulties arose in attempting to make rapid shipments. The postal connections were very unsatisfactory and caused annoying delays, while at the German frontier another cause for loss of time developed through the formalities of the customs authorities of the German Government.

The brown-tail moth seemed to be quite uncommon in the region about Bendery, and no parasites were observed upon the small number of larvæ collected at this point.

Since it seemed desirable to cover as extensive a territory as possible during the season, the writer, leaving an assistant in charge of the laboratory and collecting organization at Bendery, journeyed northward on June 17 and established a new center of exploration at the city of Kief, in the province of the same name. Through the courtesy of Prof. Waldemar Pospielow the writer was furnished with much valuable information in regard to the forests of this portion of Russia and concerning the areas in which the gipsy moth was known to exist. Several immense forested areas were traversed, but as they were for the most part purely coniferous in character the gipsy moth appeared to be quite a rare insect. Through information supplied by Prof. Pospielow it was ascertained that at Mechnigori, a monasterial institution on the banks of the Dnieper, several hours by steamer from Kief, an area of woodland existed which was infested to a moderate extent by caterpillars of the gipsy moth, among which the parasites were reported to be much in evidence. A visit to the locality showed an interesting condition. The monastery was surrounded by beautiful groves of elm and oak trees in which the gipsy moth had made considerable inroads, but the parasites had developed to a sufficient extent to practically clear the foliage of caterpillars. Almost the sole agency in bringing about this condition was *Apanteles rufipes*, which attacks the larvæ of the gipsy moth in a manner closely resembling *Apanteles japonicus*, as observed during the preceding season in Japan, but in the case of the latter the caterpillars usually die upon the leaves of the trees, whereas in the former the caterpillars descend to the trunk and lower branches to form colonies. On emerging from the caterpillars the parasites spin cocoons beneath the host, which are also attached ventrally to the bark of the tree, and as numerous caterpillars die in a restricted area a mass of Apanteles cocoons, often of considerable thickness, is formed. Such masses standing out as white patches against the dark tree trunks on which they rest may be seen for considerable distances. Cocoons of *Apantales solitarius* were also observed in the forest of Mechnigori, but were comparatively rare, so this species evidently did not represent a very important element in the control of the gipsy moth.

In the forested areas about Kief the caterpillars of the brown-tail moth were rarely met with, but in several of the parks on the outskirts of the city they were found in abundance. In the grounds of the military school a large number of magnificent oak trees were almost denuded of foliage, and some of the other deciduous trees and shrubs, such as poplars, rose bushes, and Crataegus, were severely damaged. The usual brown-tail parasites were found at work, the most effective being Meteorus. Almost every branch of the injured trees bore the suspended cocoons of this parasite. Tachinids were also active, so it was obvious that very few of the caterpillars would reach maturity.

On departing from Kief on July 9 the season was practically over, and gipsy moths were in flight.

Returning to Bendery, it was found that the season was over so far as *Apantales solitarius* was concerned, but large numbers of tachinid puparia were in evidence. As many as possible of these were assembled and shipped to America. The chrysalides of the gipsy moth were also forwarded in considerable numbers in the hope of securing pupal parasites.

These lines of work were continued till July 16, by which time the season was so advanced that the moths were beginning to deposit their eggs for the succeeding season. From the abundance of moths in flight it was obvious that unless the natural parasites multiplied sufficiently to control the situation the region would experience another visitation of the same character during the following year.

Leaving Bendery on July 16, the writer returned to Paris via Odessa, Constantinople, and Naples, arriving in New York August 28.

Owing to various unforeseen conditions, and principally owing to the deficient transportation facilities, the material received as the result of Prof. Kincaid's expedition proved to be unsatisfactory on the whole.

In May and June, 1910, the senior author went to Europe once more, visited agents and officials in Italy and France, and, through the courtesy of the Spanish and Portuguese Governments, was able to start new official services in each of these countries for the collection and sending of parasitized gipsy-moth larvæ to the United States. In Italy Prof. Silvestri at Portici and Dr. Berlese at Florence were visited and informed as to the latest ideas of the laboratory regarding methods of shipment. In Spain Prof. Leandro Navarro, of the Phytopathological Station at Madrid, volunteered his services with the approval of the minister of agriculture. In Portugal Senhor Alfredo Carlos Lecocq, director of agriculture, placed the visitor in relation with Prof. A. F. de Seabra, of the Phytopathological Station at Lisbon, and the latter gladly consented to act as the agent of the bureau in this work in Portugal. In France arrangements were made with Mr. Dillon as during the previous year in the south of France, and arrangements were renewed with Miss Rühl in Zurich and Mr. Schopfer in Dresden. The distributing agency in Hamburg was continued, and a new distributing agency was started at Havre, France, on account of its convenient proximity to the American line steamers starting from Southampton. In order to insure the best results, Mr. Dillon accompanied certain large shipments from Hyères to Havre, and personally saw that they were placed upon the channel steamer the night before the sailing of an American line steamer from Southampton.

Sendings from Japan were continued in the same manner as during the previous year. The minister of agriculture for Japan, at the request of the Secretary of Agriculture of the United States, again designated Prof. S. I. Kuwana, of the Imperial Agricultural Experiment Station at Tokyo, to be its official representative in this work, and he continued his extremely valuable sendings.

The amount received during the summer was larger than ever before, but the results obtained, owing partly to the condition of the material on receipt and owing to curious seasonal fluctuations and differences in the countries of origin and in the infested territory in America, the results by no means corresponded with the increased material. The work carried on in the laboratory during the season and the results obtained are mentioned later.

In the autumn the junior author visited France and Russia for the purpose of studying certain important points regarding the question of alternate hosts of the parasites and methods of hibernation. The results of his observations will be given in detail in the later section headed "The extent to which the gipsy moth is controlled through parasitism abroad."

At the close of the season of 1910, and in part owing to the preparation of the present bulletin, a general review of the whole work was undertaken, and a summing up of present conditions seemed to indicate that nearly as much had already been accomplished by present methods as could be expected. The great need at this time seemed to be a careful study in the countries of origin of the species of apparent importance which have been sent over but have not become established, in order to ascertain the reasons for the apparent failure; and, further, to see on the spot what can be done with regard to the importation of parasites of apparently lesser importance, but which, through the fact that they may fill in gaps in the parasitic chain and may at the same time increase beyond their native wont when confronted with American conditions, may be very desirable. Accordingly the junior author was commissioned to visit France, Italy, and Russia in the winter and early spring of 1911, and subsequently to spend the breeding season if found desirable, in Japan. He was given authority to employ the necessary agents in each of these countries. He sailed January 5, 1911.

KNOWN AND RECORDED PARASITES OF THE GIPSY MOTH AND OF THE BROWN-TAIL MOTH.

When the work of introducing the parasites of the gipsy moth and of the brown-tail moth was begun in 1905, the available assets consisted of generous appropriations by the State of Massachusetts and the Federal Government, an abundant faith in the validity of the theory which was to be put to test, and a long bibliographical list of the parasites which were recorded as attacking these insects in Europe and Japan. Of these, the appropriations have withstood most effectively the ordeal of the years which have since passed. Our faith in the validity of the principle at stake has also stood out wonderfully well, when the numerous trials to which it has been subjected are taken into consideration. It is not too much to say that at the present time it is stronger than ever, notwithstanding that a good many facts have come to light in this period which are more or less flatly in contradiction to the theory of parasite control as generally accepted at the beginning. It has more than once been necessary to modify beliefs and ideas as previously held, in order to make them conform to the actual facts. To take a pertinent example, it was necessary to place an entirely different value upon the bibliographical list above mentioned than that which was placed upon it when the work was begun, and when the policies of the laboratory were first determined.

Nearly thirty years ago the present head of the Bureau of Entomology undertook the compilation of a card catalogue of references to the host relations of the parasitic Hymenoptera of the world. For more than twenty years the work was continued until some 30,000 such references were accumulated. From among them those in which the gipsy moth was mentioned as the host were collected and a list of gipsy-moth parasites was published in Insect Life.[1] With the exception of a comparatively few recent additions this list forms the basis of that which follows. That of the parasites which have been recorded as attacking the brown-tail moth is largely from the same source.

---

[1] U. S. Department of Agriculture, Division of Entomology, Insect Life, vol. 2, pp. 210–211, 1890.

KNOWN AND RECORDED PARASITES. 85

HYMENOPTEROUS PARASITES OF THE GIPSY MOTH (*Porthetria dispar* L.).

BRACONIDÆ.

Reared at laboratory.
Apanteles fulvipes (Hal.).
Apanteles solitarius (Ratz.).

Recorded as parasites.
Apanteles fulvipes (Hal.).[1,2]
Apanteles solitarius (Ratz.).[1,2]
Microgaster calceata Hal.[1,2]
Apanteles tenebrosus (Wesm.).[1]
Microgaster tibialis Nees.[1]
(*Microgaster*) *Apanteles fulvipes liparidis* (Bouché).[1,2]
Apanteles glomeratus (L.).[1,2]
Apanteles solitarius var. *melanoscelus* (Ratz.).[1]
Apanteles solitarius? *ocneriæ* Svanov.

Meteorus versicolor (Wesm.).
Meteorus pulchricornis (Wesm.).
Meteorus japonicus Ashm.[3]

Meteorus scutellator (Nees).[1]

ICHNEUMONIDÆ.

PRIMARY.

Pimpla (*Pimpla*) *instigator* (Fab.).
Pimpla (*Pimpla*) *porthetriæ* Vier.[3]
Pimpla (*Pimpla*) *examinator* (Fab.).
Pimpla (*Pimpla*) *pluto* Ashm.[3]
Pimpla (*Apechthis*) *brassicariæ* (Poda).
Pimpla (*Pimpla*) *disparis* Vier.[3]
Theronia atalantæ (Poda).
Limnerium (*Hyposoter*) *disparis* Vier.
Limnerium (*Anilastus*) *tricoloripes* Vier.
Ichneumon disparis (Poda).

Pimpla (*Pimpla*) *instigator* (Fab.).[1,2]

Pimpla examinator (Fab.).[1]

Theronia atalantæ (Poda).[1,2]
Campoplex conicus Ratz.[1]
Casinaria tenuiventris (Grav.).[1]
Ichneumon disparis (Poda).[1,2]
Ichneumon pictus (Gmel.).[1,2]
Amblyteles varipes Rdw.[2]
Trogus flavitorius [sic.] *lutorius* (Fab.)?[1,2]
(*Cryptus*) *Aritranis amœnus* (Grav.).[1]
Cryptus cyanator Grav.[1]

PROBABLY SECONDARY BUT RECORDED AS PRIMARY.

Mesochorus pectoralis Ratz.[1,2]
Mesochorus gracilis Brischke.[1,2]
Mesochorus splendidulus Grav.[1,2]
Mesochorus confusus Holmgr.[1]
Mesochorus semirufus Holmgr.[1]
(*Hemiteles*) *Astomaspis fulvipes* (Grav.).[1,2]
= *A. nanus* (Grav.) according to Pfankuch.
Hemiteles bicolorius Grav.[2]
Pezomachus hortensis Grav.[2]
Pezomachus fasciatus (Fab.)[1] = *Pezomachus melanocephalus* (Schrk.).

---

[1] Recorded by the senior author in a card catalogue of parasites kept in the Bureau of Entomology.
[2] Recorded by Dalla Torre in Catalogus Hymenopterorum.
[3] Japanese species.

CHALCIDIDÆ.

Reared at laboratory.

Eupelmus bifasciatus Fonsc.
Monodontomerus æreus Walk.
Chalcis flavipes Panz.
Chalcis obscurata Walk.[4]
Schedius kuvanæ How.[4]

Recorded as parasites.
Pteromalus halidayanus Ratz.[1]
Pteromalus pini Hartig.[1]
Dibrachys boucheanus Ratz.[1] (Secondary.)
Eurytoma abrotani Panzer [1] [2]=appendigaster Swed. (Secondary.)
Eupelmus bifasciatus Fonsc.[1] [2]

Chalcis callipus Kby.[3]

HYMENOPTEROUS PARASITES OF THE BROWN-TAIL MOTH (*Euproctis chrysorrhœa* L.).

BRACONIDÆ.

Reared at laboratory.
Meteorus versicolor (Wesm.).

Apanteles lacteicolor Vier.

Recorded as parasites.
Meteorus versicolor (Wesm.).[5]
Meteorus ictericus (Nees).[5]
Apanteles inclusus (Ratz.) [2] [5]
Apanteles ultor Reinh.[1] [2] [3]
Apanteles difficilis (Nees).[5]
Apanteles liparidis (Bouché).[5]
Apanteles vitripennis (Hal.).[5]
Apanteles solitarius (Ratz.).[5]
Microgaster consularis (Hal.) [5]= Microgaster connexa Nees.
Microgaster calceata Hal.[1]
Rogas geniculator Nees.[2] [5]
Rogas testaceus (Spin.).[1]
Rogas pulchripes (Wesm.).[1]

ICHNEUMONIDÆ.

PRIMARY.

Pimpla (Pimpla) examinator (Fab.).
Pimpla (Pimpla) instigator (Fab.).
Pimpla (Apechthis) brassicariæ (Poda).
Theronia atalantæ (Poda).

Pimpla (Pimpla) examinator (Fab.).[2] [5]
Pimpla (Pimpla) instigator (Fab.).[1] [2] [3]

Theronia atalantæ (Poda).[1] [2] [3]
Campoplex conicus Ratz.[5]
(Campoplex) Omorgus difformis (Gmel.).[5]
Cryptus moschator (Fab.).[1]
(Cryptus) Idiolispa atripes (Grav.).[1]
Ichneumon disparis (Poda).[5]
Ichneumon scutellator (Grav.).[2]

---

[1] Recorded by the senior author in a card catalogue of parasites kept in the Bureau of Entomology.
[2] Recorded by Dalla Torre in Catalogus Hymenopterorum.
[3] Reared by Dr. S. I. Kuwana.
[4] Japanese species.
[5] Recorded by Emelyanoff.

PROBABLY SECONDARY, BUT RECORDED AS PRIMARY.

Reared at laboratory.	Recorded as parasites.
	Mesochorus pectoralis Ratz.[1,2,3]
	Mesochorus dilutus Ratz.[2,3]
	Hemiteles socialis Ratz.[3]

CHALCIDIDÆ.

Pteromalus sp.	Pteromalus rotundatus Ratz.[3] = Pt. chrysorrhœa D. T.[1,2]
Pteromalus nidulans Thoms. = Pt. egregius Först.	Pteromalus processioneæ Ratz.[1,2]
Diglochis omnivora Walk.	Pteromalus nidulans Thoms.[1,3]
	Pteromalus puparum L.[3]
	Dibrachys boucheanus (Ratz.).[3] (Secondary.)
	Chalcis scirropoda Först.[1]
Monodontomerus æreus Walk.	Torymus anephelus Ratz.[3] = Monodontomerus æreus Walk.[1,2]
	Monodontomerus dentipes Boh.[1,2]
	Anagrus ovivorus Rondani.[2]
Trichogramma sp. I.	
Trichogramma sp. II.	

PROCTOTRYPIDÆ.

Telenomus phalænarum Nees (?).	Telenomus phalænarum Nees.[1,2,3]

DIPTEROUS PARASITES OF THE GIPSY MOTH (*Porthetria dispar* L.).

The following are lists of the dipterous parasites reared and recorded from *Porthetria dispar* L. and *Euproctis chrysorrhœa* L. Each list is supplemented by a list of recorded hosts for each species enumerated.

These lists have been compiled from various sources, the principal being the "Katalog der Paläarktischen Dipteren," Brauer & Bergenstamm's "Die Zweiflügler des Kaiserlichen Museums zu Wien," Fernald and Forbush's "The Gipsy Moth," and the senior author's "List of parasites bred from imported material during the year 1907" (3d annual report of the superintendent for suppressing the gipsy and brown-tail moths).

In the choice of names of the foreign tachinids the Katalog der Paläarktischen Dipteren has been followed with the exception of a few cases in which other names have been in use at the Gipsy Moth Parasite Laboratory; in these few cases, to avoid confusion, no change has been made.

---

[1] Recorded by Emelyanoff.
[2] Recorded by Dalla Torre in Catalogus Hymenopterorum.
[3] Recorded by the senior author in a card catalogue of parasites kept in the Bureau of Entomology.

Foreign Tachinid Parasites on Porthetria dispar.

Reared.
- *Blepharipa scutellata* R. D.
- *Carcelia gnava* Meig.
- *Compsilura concinnata* Meig.
- *Crossocosmia sericariæ* Corn.
- *Dexodes nigripes* Fall.
- *Parasetigena segregata* Rond.
- *Tachina larvarum* L.
- *Tachina japonica* Towns.
- *Tricholyga grandis* Zett.
- *Zygobothria gilva* Hartig.

Recorded.
- *Argyrophylax atropivora* R. D.
- *Carcelia excisa* Fall.
- *Compsilura concinnata* Meig.
- *Echinomyia fera* L.
- *Epicampocera crassiseta* Rond.
- *Ernestia consobrina* Meig.
- *Eudoromyia magnicornis* Zett.
- *Exorista affinis* Fall.
- *Histochæta marmorata* Fab.
- *Lydella pinivoræ* Ratz.
- *Meigenia bisignata* Schin.
- *Parasetigena segregata* Rond.
- *Phryxe erythrostoma* Hartig.
- *Ptilotachina larvincola* Ratz.
- *Ptilotachina monacha* Ratz.
- *Tachina larvarum* L.
- *Tachina noctuarum* Rond.
- *Zenillia libatrix* Panz.
- *Zygobothria gilva* Hartig.
- *Zygobothria bimaculata* Hartig.

N. B.—It is interesting to note that only four species are common to both lists.

Recorded Hosts of Foreign Tachinid Parasites of Porthetria dispar Reared at the Gipsy Moth Parasite Laboratory.

Blepharipa scutellata R. D.:
  Acherontia atropos L.; *Vanessa antiopa* L.

Carcelia gnava Meig.:
  Malacosoma neustria L.; *Orgyia antiqua* L.; *Stilpnotia salicis* L.

Compsilura concinnata Meig.:
  See list of recorded parasites of *P. dispar*

Crossocosmia sericariæ Corn.:
  Antheræa yamamai Guér.; *A. mylitta* Moore; *Sericaria mori* L.

Dexodes nigripes Fall.:
  Ascometia caliginosa Hb.; *Agrotis candelarum* Stgr.; *Bupalus piniarius* L.; *Cucullia asteris* Schiff.; *Deilephila euphorbiæ* L.; *Eurrhypara urticæ* L.; *Heliothis scutosa* Schiff.; *Hybernia* sp.; *Mamestra pisi* L.; *Miana literosa* Hw.; *Ortholitha cervinata* Schiff.; *Phragmatobia fuliginosa* L.; *Plusia gamma* L.; *Porthesia similis* Fussl.; *Tapinostola elymi* Tr.; *Tephroclystia virgauriata* Dbld.; *Thaumetopœa pinivora* Tr.; *Vanessa io* L.; *V. polychlorus* L.; *V. urticæ* L.; *Lophyrus* sp.; *Nematus ribesii* Scop.

Parasetigena segregata Rond.:
  (See list of recorded parasites of *P. dispar*.)

Tachina larvarum L.:
  (See list of recorded parasites of *P. dispar*.)

Tachina japonica Towns.:
  Porthetria dispar L.

Tricholyga grandis Zett.:
  Arctia caja L.; *Mamestra oleracea* L.; *M. pisi* L.; *Saturnia pavonia* L.; *S. pyri* Schiff.; *Sphinx ligustri* L.; *Thaumetopœa pityocampa* Schiff.; *Vanessa io* L.

KNOWN AND RECORDED PARASITES. 89

RECORDED HOSTS OF FOREIGN TACHINID PARASITES RECORDED ON PORTHETRIA DISPAR.
ARGYROPHYLAX ATROPIVORA R. D.:
  P. dispar L.; Acherontia atropos L.; Notodonta trepida Esp.; Vanessa io L.
CARCELIA EXCISA Fall.:
  Abrostola tripartita Hufn.; A. triplasia L.; Arctia caja L.; A. hebe L.; A. villica L.; Bupalus piniarius L.; Callimorpha dominula L.; Cucullia scrophulariæ Cap.; Dasychira pudibunda L.; Endromis versicolora L.; Hyloicus pinastri L.; P. dispar L.; P. monacha L.; Malacosoma castrensis L.; M. neustria L.; Orgyia antiqua L.; Phragmatobia fuliginosa L.; Pterostoma palpina L.; Pygæra curtula L.; Saturnia pyri Schiff.; Sphinx ligustri L.; Stilpnotia salicis L.; Thalpochæres pannonica Frr.; Thaumetopœa processionea L.
COMPSILURA CONCINNATA Meig.:
  Abraxas grossulariata L.; Acronycta aceris L.; A. alni L.; A. cuspis Hb.; A. megacephala F.; A. rumicis L.; A. tridens Schiff.; Araschinia levana L.; A. prorsa L.; Arctia caja L.; Attacus cynthia L.; Catocala promissa Esp.; Craniophora ligustri Fab.; Cucullia lactucæ Esp.; C. verbasci L.; Dasychira pudibunda L.; Dilina tiliæ L.; Dilobia cæruleocephala L.; Dipterygia scabriuscula L.; Drymonia chaonia Hb.; Euproctis chrysorrhœa L.; Hyloicus pinastri L.; Libytha celtis Laich.; Porthetria dispar L.; P. moncha L.; Macrothylacia rubi L.; Mamestra brassicæ L.; M. oleracea L.; M. persicariæ L.; Malacosoma neustria L.; Oeonistis quadra L.; Phalera bucephala L.; Pieris brassicæ L.; P. rapæ L.; Plusia festucæ L.; P. gamma L.; Pœcelocampa populi L.; Porthesia similis Füssl.; Pygæra anachoreta Fab.; Pyrameis atalanta L.; Smerinthus populi L.; Spilosoma lubricipeda L.; S. menthastri Esp.; Stauropus fagi L.; Stilpnotia salicis L.; Tæniocampa stabilis View.; Thaumetopœa processionea L.; T. pityocampa Schiff.; Timandra amata L.; Trachea atriplicis L.; Vanessa antiopa L.; V. io L.; V. urticæ L.; V. xanthomelas Esp.; Yponomeuta padeila L.; Cimbex humeralis Fourcr.; Trichiocampus viminalis Fall.
ECHINOMYIA FERA L.:
  Agrotis glareosa Esp.; Arctia aulica L.; Leucania obsoleta Sb.; Porthetria dispar L.; P. monacha L.; Mamestra pisi L.; Oeonistis quadra L.; Panolis grieovariegata Goeze.
EPICAMPOCERA CRASSISETA Rond.:
  Porthetria dispar L.; Thaumetopœa processionea L.
ERNESTIA CONSOBRINA Meig.:
  Cucullia artemisiæ Hufn.; Porthetria dispar L.
EUDOROMYIA MAGNICORNIS Zett.:
  Agrotis sp. ind.; Hadena adusta Esp.; Porthetria dispar L.
EXORISTA AFFINIS Fall.:
  Acronycta alni L.; Arctia caja L.; Porthetria dispar L.; Pachytelia villosella O.; Saturnia pavonia L.; S. pyri Schiff.
HISTOCHÆTA MARMORATA Fab.:
  Arctia caja L.; A. quenselii Payk.; A. villica L.; Cucullia verbani L.; Malacosoma neustria L.; Porthetria dispar L.; Goniarctena rufipes Payk.
LYDELLA PINIVORÆ Ratz.:
  Porthetria dispar L.:
MEIGENIA BISIGNATA Schin.:
  Porthetria dispar L.:
PARASETIGENA SEGREGATA Rond.:
  Porthetria dispar L.; P. monacha L.; Lophyrus pini L.
PHRYXE ERYTHROSTOMA Hartig:
  Dendrolimus pini L.; Haloicus pinastri L.:

PTILOTACHINA LARVINCOLA Ratz.:
　　Porthetria dispar L.:
PTILOTACHINA MONACHA Ratz.:
　　Porthetria dispar L.:
TACHINA LARVARUM L.:
　　Acronycta rumicis L.; *Agrotis præcox* L.; *Arctia caja* L.; *A. villica* L.; *Catocola fraxini* L.; *Cosmotriche potatoria* L.; *Cucullia prenanthis* B.; *Dasychira fascellina* L.; *Deilephila gallii* Rott.; *D. euphorbiæ* L.; *Dendrolimus pini* L.; *Gastropacha quercifolia* L.; *Lasiocampa quercus* L.; *Porthetria dispar* L.; *P. monacha* L.; *Macroglossa stellatarum* L.; *Macrothylacia rubi* L.; *Malacosoma castrensis* L.; *M. neustria* L.; *Mamestra brassicæ* L.; *Melitæa didyma* O.; *Melopsilus porcellus* L.; *Ocneria detrita* Esp.; *Olethreutes hercyniana* Tr.; *Orgyia ericæ* Germ.; *O. gonostigma* F.; *Orthosia humilis* F.; *Panolis griseovariegata* Goeze; *Papilio machaon* L.; *Plusia iota* L.; *Saturnia pyri* Schiff.; *Stilpnotia salicis* L.; *Vanessa antiopa* L.; *Vanessa io* L.; *V. polychloros* L.; *V. urticæ* L.; *Yponomeuta evonymella* L.; *Lophyrus pini* L.; *Pamphilius stellatus* Christ.
TACHINA NOCTUARUM Rond.:
　　Cosmotriche potatoria L.; *Porthetria dispar* L.:
ZENILLIA LIBATRIX Panz.:
　　Abrostola asclepiadis Schiff.; *Brephos nothum* Hb.; *Dasychira pudibunda* L.; *Larentia autumnalis* Strom.; *Porthetria dispar* L.; *Malacosoma neustria* L.; *Pygæra pigra* Hufn.; *Thaumetopœa processionea* L.; *Yponomeuta evonymella* L.; *Y. padella* L.
ZYGOBOTHRIA GILVA Hartig:
　　Porthetria dispar L.; *Stauropus fagi* L.; *Lophyrus laricis* Jur.; *L. pallidus* Klug.; *L. pini* L.; *L. rufus* Latr.; *L. variegatus* Hartig.
ZYGOBOTHRIA BIMACULATA Hartig:
　　Lymantria monacha L.; *Lophyrus pallidus* Klug.; *L. pini* L.; *L. rufus* Latr.; *L. socius* Klug.; *L. variegatus* Hart.; *L. virens* Klug.

NATIVE DIPTERA REARED FROM PORTHETRIA DISPAR.

Tachinidæ: [1]
　　Exorista blanda O. S.　　*Exorista fernaldi* Will.
　　Exorista pyste Walk.　　*Tachina mella* Walk.

Other than Tachinidæ: [2]
　　Aphiochæta setacea Aldr.　　*Phora incisuralis* Loew
　　Aphiochæta scalaris Loew.　　*Sarcophaga* sp.
　　Gaurax anchora Loew.

---

[1] These have only been reared very occasionally at the Gipsy Moth Parasite Laboratory.

[2] At the Gipsy Moth Parasite Laboratory these have been recorded only as scavengers and not as parasites.

DIPTEROUS PARASITES OF THE BROWN-TAIL MOTH (*Euproctis chrysorrhœa* L.).

Foreign Tachinid Parasites of Euproctis chrysorrhœa.

Reared.
Blepharidea vulgaris Fall.
Compsilura concinnata Meig.
Cyclotophrys anser Towns.
Dexodes nigripes Fall.
Digonichæta setipennis Fall.
Digonichæta spinipennis Meig.
Eudoromyia magnicornis Zett.
Masicera sylvatica Fall.
Nemorilla sp.
Nemorilla notabilis Meig.
Pales pavida Meig.
Parexorista cheloniæ Rond.
Tachina larvarum L.
Tricholyga grandis Zett.
Zenillia libatrix Panz.
Zygobothria nidicola Towns.

Recorded.
Compsilura concinnata Meig.
Echinomyia præceps Meig.
Erycia ferruginea Meig.
Pales pavida Meig.
Tachina latifrons Rond.
Zenillia fauna Meig.
Zenillia libatrix Panz.

N. B.—It is interesting to note that only three species are common to both lists.

Recorded Hosts of Foreign Tachinids Reared from Euproctis chrysorrhœa at the Gipsy Moth Parasite Laboratory.

Blepharidea (Phyrxe) vulgaris Fall.:
    Abraxas grossulariata L.; *Adopæa lineola* O.; *Aporia cratægi* L.; *Araschinia levana* L.; *A. prorsa* L.; *Argynnis lathonia* L.; *Arctia hebe* L.; *Boarmia lariciaria* Dbld.; *Brotolomia meticulosa* L.; *Calymnia trapezina* L.; *Cosmotriche potatoria* L.; *Cucullia anthemidis* Gn.; *C. asteris* Schiff.; *C. verbasci* L.; *Dendrolimus pini* L.; *Ephyra linearia* Hb.; *Epineuronia cespitis* F.; *Euchloë cardamines* L.; *Euplexia lucipara* Hb.; *Hybernia defoliaria* Cl.; *Hyloicus pinastri* L.; *Hylophila prasinana* L.; *Leucania albipuncta* F.; *L. lythargyria* Esp.; *Mamestra advena* F.; *M. persicariæ* L.; *M. reticulata* Vill.; *Melitæa athalia* Rott.; *Metopsilus porcellus* L.; *Nænia typica* L.; *Parasemia plantaginis* L.; *Pieris brassicæ* L.; *P. daplidice* L.; *P. rapæ* L.; *Plusia gamma* L.; *Thamnonona wavaria* L.; *Thaumetopœa pityocampa* Schiff.; *T. processionea* L.; *Toxocampa pastinum* Tr.; *Vanessa antiopa* L.; *V. io* L.; *V. urticæ* L.; *V. xanthomelas* Esp.; *Zygæna achilleæ* Esp., ab. *janthina; Z. filipendulæ* L. (?); *Procrustes coriaceus* L.

Compsilura concinnata Meig.:
    See host list of tachinid parasites of *P. dispar*.

Cyclotophrys anser Towns.:
    No records other than at the Gipsy Moth Parasite Laboratory.

Dexodes (Lydella) nigripes Fall.:
    Ascometia caliginosa Hb.; *Agrotis candelarum* Stgr.; *Bupalus piniarius* L.; *Cucullia asteris* Schiff.; *Deilephila euphorbiæ* L.; *Eurrhypara urticata* L.; *Heliothis scutosa* Schiff.; *Hybernia* sp.; *Mamestra pisi* L.; *Miana literosa* Hw.; *Ortholitha cervinata* Schiff.; *Phragmatobia fuliginosa* L.; *Plusia gamma* L.; *Porthesia similis* Füssl.; *Tapinostola elymi* Tr.; *Tephroclystia virgaureata* Dbld.; *Thaumetopœa pinivora* Schiff.; *Vanessa io* L.; *V. polychloros* L.; *V. urticæ* L.; *Lophyrus* sp.; *Nematus ribesii* Scop.

Digonichæta setipennis Fall.:
    Grapholitha strobilella L.; *Notodonta trepida* Esp.; *Pheosia tremula* Cl. (?); *Forficula auricularia* L.

DIGONICHÆTA SPINIPENNIS Meig.:
  Lasiocampa quercus L.; *Panolis griseovariegata* Goeze.
EUDOROMYIA MAGNICORNIS Zett.:
  Agrotis sp.; *Hadena adusta* Esp.; *Porthetria dispar* L.
MASICERA SYLVATICA Fall.:
  Apopestes spectrum Esp.; *Cucullia verbasci* L.; *Deilephila euphorbiæ* L.; *D. gallii* Rott.; *D. vespertilio* Esp.; *Dilina tiliæ* L.; *Gastropacha quercifolia* L.; *Lasiocampa quercus* L.; *Nonagria typhliæ* Thbg.; *Pieris brassicæ* L.; *Saturnia pavonia* L.; *S. pyri* Schiff.; *S. spini* Schiff.; *Sphinx ligustri* L.
NEMORILLA NOTABILIS Meig.:
  Notocœlia uddmanniana L.; *Plusia festucæ* L.; *Sylepta ruralis* Scop.; *Tachyptylia populella* Cl.
PALES PAVIDA Meig.:
  Acronycta tridens Schiff.; *Agrotis stigmatica* Hb.; *A. xanthographa* F.; *Attacus cynthia* L.; *A. lunula* Fab.; *Eriogaster catex* L.; *Emphytus cingillum* Klug; *Euproctis chrysorrhœa* L.; *Orgyia ericæ* Germ.; *Panolis griseovariegata* Goeze; *Plusia gamma* L.; *Thaumetopœa processionea* L.
PAREXORISTA (EXORISTA) CHELONIÆ Rond.:
  Ammoconia cæcimacula Fab.; *Arctia caja* L.; *A. hebe* L.; *A. villica* L.; *Hadena secalis* L.; *Macrothylacia rubi* L.; *Orthosia pistacina* Fab.; *Phragmatobia fuliginosa* L.; *Rhyparia purpurata* L.; *Spilosoma lubricipeda* L.; *Stilpnotia salicis* L.; *Cimbex femorata* L.; *Pamphilius stellatus* Christ.
TACHINA LARVARUM L.:
  See host list of foreign tachinids recorded from *P. dispar*.
TRICHOLYGA GRANDIS Zett.:
  Arctia caja L.; *Mamestra oleracea* L.; *M. pisi* L.; *Saturnia pavonia* L.; *S. pyri* Schiff.; *Sphinx ligustri* L.; *Thaumetopœa pityocampa* Schiff.; *Vanessa io* L.
ZENILLIA LIBATRIX Panz.:
  See host list of foreign tachinids recorded as parasites of *P. dispar*.
ZYGOBOTHRIA NIDICOLA Tn..
  No record other than at the Gipsy Moth Parasite Laboratory.

RECORDED HOSTS OF FOREIGN TACHINIDS RECORDED AS PARASITIC ON EUPROCTIS CHRYSORRHŒA.

COMPSILURA CONCINNATA Meig.:
  See list of recorded hosts of foreign tachinids recorded as parasitic on *P. dispar*.
ECHINOMYIA PRÆCEPS Meig.:
  Hemaris fuciformis L.; *Euproctis chrysorrhœa* L
ERYCIA FERRUGINEA Meig.:
  Euproctis chrysorrhœa L.; *Melitæa athalia* Rott.; *M. aurinia* Rott.; *Porthesia similis* Füssl.; *Vanessa io* L.
PALES PAVIDA Meig.:
  See list of recorded hosts of foreign tachinids reared from *E. chrysorrhœa* L.
TACHINA LATIFRONS Rond.:
  Euproctis chrysorrhœa L.; *Zygæna filipendulæ* L.
ZENILLIA FAUNA Meig.:
  Acronycta rumicis L.; *Cossus cossus* L.; *Euproctis chrysorrhœa* L.; *Smerinthus ocellatus* L.
ZENILLIA LIBATRIX Panz.:
  See list of recorded hosts of foreign tachinids reared from *E. chrysorrhœa* L.

NATIVE (AMERICAN) TACHINIDS REARED FROM EUPROCTIS CHRYSORRHŒA L., AT THE
GIPSY MOTH PARASITE LABORATORY.

Blepharipeza leucophrys Wied.   ? *Phorocera leucaniæ* Coq.
Euphorocera claripennis Macq.   *Sturmia discalis* Coq.
Exorista griseomicans V. de Wulp. *Tachina mella* Walk.

N. B.—The above species have only been reared very occasionally. The species, however, doubtfully referred to *Phorocera leucaniæ* Coq. has been reared through to the pupal stage in considerable numbers. These pupæ have always been imperfect and "larviform" and at the time of writing none has been reared through to the adult.

The compilation of the catalogue of parasites was originally undertaken in the expectation that it would prove of great service upon exactly such occasions as the present, when the application of the theory of control by parasites should be put to the test. Its value naturally depended upon the accuracy of the original records, and it was only right to suppose that in the majority of instances these could be depended upon. It was equally natural to suppose that the parasitic fauna of such common, conspicuous, and widely distributed insects as the gipsy moth and the brown-tail moth would be well represented in these lists, which were based upon a thorough overhauling of European literature, and it was not expected that any parasites of particular importance would be found which were not thus recorded, unless, indeed, they were confined to Continental Asia or to Japan.

In the fall of 1907, as soon as the turmoil of his first summer's work permitted, the junior author attempted to make use of the numerous bibliographical references for the purpose of learning as much as possible of the insects with which he was to deal. One after another, various species were taken up, until he was in possession of practically all of the published information concerning perhaps half of the Hymenoptera listed. Then he stopped, because the information thus gained was obviously not worth the labor. It was not so much that recorded information was scanty, or lacking in interest, but it was because in a great many instances it was contradictory to the results of the actual rearing work which had been carried on in the laboratory throughout the summer. It was obviously impossible to accept everything at its face value, and apparently next to impossible to choose between the true and the false. But one thing remained to be done, and that was to determine at first hand everything which it was necessary to know concerning the numerous species of parasites which it was desired to introduce into America.

If the list of parasites which have been reared at the laboratory from imported eggs, caterpillars, and pupæ of the gipsy moth and the brown-tail moth be compared with the lists which have already been given, the numerous and obvious differences which are immediately apparent will serve better than words to illustrate the situation which confronted us at the close of the season of 1907.

ESTABLISHMENT AND DISPERSION OF THE NEWLY INTRODUCED PARASITES.

In the beginning we were very far from accrediting to that phase of the project which has to do with the establishment and dispersion of the newly introduced parasites the importance which it deserved. Many widely diverse species of insects were known to have been introduced from the Old World and firmly established in America. Presumably they were accidentally imported, as was the case with the gipsy moth and the brown-tail moth; presumably, also, they had spread and increased from a small beginning, at first very gradually and later more rapidly, until they had become component parts of the American fauna over a wide territory. The circumstances under which the gipsy moth was imported were well known, and a good guess had been made as to those which resulted in the introduction of the brown-tail moth. But these were and are rare exceptions in this respect, and for the most part the preliminary chapters in the story of each of the insect immigrants never have been and probably never will be written.

Because the two very conspicuous instances of the gipsy moth and the brown-tail moth were constantly and automatically recurring whenever the probable future of the intentionally introduced parasites was considered, it was, perhaps, taken a little too much for granted, that they were to be considered as typical and significant of what to expect. In each instance the invasion started from a small beginning, and while the subsequent histories were different, the more rapid spread of the brown-tail moth was directly due to the fact that the females were capable of flight, and the relatively slow advance of the gipsy moth into new territory to the reverse. Even the brown-tail moth was for some years confined to a comparatively limited area, and it was rather expected that the parasites, if they established themselves at all, would remain for a similar period in the immediate vicinity of the localities where they were first given their freedom.

Accordingly, in accepting this theory without submitting it to a test, attempts were made to encompass the rapid dissemination of the parasites coincidently with their introduction. In 1906 and 1907 the parasites which were reared from the imported material were mostly liberated in small and scattered colonies. In a few instances this procedure was the best which could have been adopted; in others the worst. Small colonies of Calosoma, for example, remained for several years in the immediate vicinity of the point where the parent beetles were first liberated before any material dispersion was apparent (see Pl. XXIV), and the small colony was thus justified. The gipsy-moth egg parasite Anastatus, as was later determined, spreads at a rate of but a few hundred feet per year, and if it is to become generally distributed throughout the gipsy-moth-infested area within a reasonable time, natural dispersion must be assisted by artificial.

These, however, are both exceptions. In the case of Monodontomerus, and perhaps of other parasites, gregarious in their habit, it is not only conceivable but probable that a single fertilized female would be sufficient to establish the species in a new country, because the union between the sexes is effected within the body of the host in which they were reared. No matter how far a female may range and no matter how widely separated the victims of her maternal instincts, her progeny will rarely die without each finding its mate. Species having such habits are eminently well fitted to establish themselves wherever they secure foothold, even in the smallest numbers, and the small colony is again justified.

Many of the hymenopterous parasites, and very likely all of them, are capable of parthenogenetic reproduction, and here again is a factor which becomes of considerable importance in this connection. Some few of these are thelyotokous (bearing females only) and as such are eminently well fitted to establishment in a new country under otherwise unsatisfactory conditions. Most are arrhenotokous (bearing males only), and such are probably better fitted to establishment than would be the case if the species were wholly incapable of parthenogenetic reproduction. It has been proved, for example, that a single female of a strictly arrhenotokous species, may, through fertilization by her own parthenogenetically produced offspring, become the progenetrix of a race the vigor of which appears not to be immediately affected by the fact that their continued multiplication must be considered as the closest form of inbreeding.

Whenever opportunity has offered the ability of the various species to reproduce pathenogenetically has been studied, and many interesting and some peculiar facts have been discovered which, it is hoped, will serve as the subject for a technical paper later on. This power appears to be confined to the Hymenoptera, however, and the tachinid parasites, like their hosts, are rarely or perhaps never parthenogenetic.

When continued existence of an insect in a new country is dependent upon the mating of isolated females it is at once evident that it is also dependent upon the rapidity of dispersion and upon the number of individuals which are comprised in the original colony. One of the most constant sources of surprise is in the rapidity with which the parasites disperse. One, Monodontomerus, has undoubtedly extended its range for more than 200 miles in the course of the five years which have elapsed since its liberation, and there is no reason to believe that others among the introduced species will not disperse at an equal rate, once they are sufficiently well established.

But Monodontomerus is eminently well fitted for dispersion, and its case is altogether different from that of a tachinid which is dependent upon sexual reproduction for the continuation of the species. A few hundred individuals, spreading rapidly toward all points of the compass, soon become widely scattered, and it is, and will remain for a long time, a question just how rare an insect may be and each individual still be able to find its mate. That the individuals of the first colonies of many of the tachinid parasites scattered so widely as to make the mating of the next generation purely a matter of chance and of rare occurrence is now accepted as well within the bounds of probability.

The first serious doubts as to the wisdom of the policy of the small colony were felt in 1907, and beginning with June of that year larger colonies were planted in the instance of every species than had been the practice up to that time. In the fall of 1908 the recovery of Monodontomerus over a wide territory lent strength to these half-formed convictions, and when, during 1909 and 1910, one after another of the various parasites were recovered under circumstances which were in most cases essentially similar, all doubts vanished as to the wisdom of the course finally adopted. At the present time there is no more inexorable rule governing the conduct of the laboratory than that establishment of a newly introduced parasite is first to be secured, while dispersion, if later developments prove that it can be artificially aided, comes as a wholly secondary consideration. For the most part, however, dispersion may be left to take care of itself.

An even larger appreciation of the necessity for strong colonies has been reached during the present winter (1910–11), coincidently with the results of the scouting work for Monodontomerus and Pteromalus in the brown-tail moth hibernating nests. (See maps, Pls. XXII, XXV.) The details will not be given in this immediate connection, but they will be found later on in connection with the discussion of these species. It is sufficient at this time to say that the circumstances under which the Pteromalus was recovered after the lapse of two years following its colonization were such as to cast doubts upon the conclusions which had been tentatively reached concerning the inability of certain other species to exist in America, and their possible significance had something to do with the decision to continue the work of parasite importation along wholly different lines in 1911. It may be, after all, that 40,000 individuals of *Apanteles fulvipes* are not enough to make one good colony.

DISEASE AS A FACTOR IN THE NATURAL CONTROL OF THE GIPSY MOTH AND THE BROWN-TAIL MOTH.

In continuing this work consideration must be given to the probable effect which the prevalence of disease would possibly have in the reduction of the gipsy moth and the brown-tail moth to the ranks of ordinary rather than of extraordinary pests. In America, as is generally well known, the brown-tail moth is annually destroyed to an extraordinary extent as the result of an epidemic and specific fungous disease, while the gipsy moth is frequently subjected to very material diminution of numbers through a much less well known affection popularly known as "the wilt," apparently similar to the silkworm disease "flacherie."

In more respects than one the prevalence of these diseases has been inimical to the prosecution of the parasite work. In the beginning, when it was expected that the parasites would remain in the immediate vicinity of the localities where they were first given their freedom, great pains were taken to provide colony sites in situations where the caterpillars were not only common but where there was reason to believe that they would remain healthy for at least one or two years. This was an exceedingly difficult matter, and one which was the cause of more troubles, doubts, and fears during 1907 and 1908 than almost any other phase of the parasite work.

With the final recognition of the great superiority of the large colony, which came about through a better knowledge of the powers of rapid dispersion possessed by the parasites, this seeming obstacle to success wholly disappeared, except in the case of such parasites as Anastatus, which actually did remain in the spot where they were placed, and which could not travel beyond a certain limited radius, no matter how great the necessity.

At the present time the association with the parasite problem of the otherwise wholly separate question of disease as a factor in the control of the gipsy moth and the brown-tail moth is entirely confined to speculations as to the probable future of these pests, provided their control is left to disease alone. If, as is conceivable, effective control is exerted through disease, further importation of parasites is rendered not only needless but wholly undesirable. If, on the contrary, such control is likely to be inefficient, from an economic standpoint, every effort should be exerted to make the parasite work a success. In other words, the decision as to the adoption of a policy for the future conduct of the activities of the laboratory depended very largely upon whether or not disease seemed likely to become effective in the case of the more important of the two pests. The fact that the present plans provide for the continuation of the work along even more energetic lines than in the past indicates sufficiently well the character of the decision finally reached.

This is not the place for, nor are the writers prepared to enter into, a discussion of the caterpillar diseases of the gipsy moth and the brown-tail moth, but it is perhaps not out of place to recount some of the incidents which have been taken into consideration in the present instance. In this, as in many others similar in character, the brown-tail moth has largely been ignored, owing to its being generally considered as the lesser pest of the two. As frequently before, the work upon the brown-tail moth parasites, although pursued quite as actively as that upon the parasites of the gipsy moth, was relegated to a secondary position.

Very little has been published concerning the gipsy-moth caterpillar disease previously to 1907, when the junior author first had opportunity to familiarize himself with the situation at first hand. It was to him a novelty when, early in the summer of that year, wholesale destruction of the half-grown caterpillars was first noticed in numerous localities before they had succeeded in effecting the complete defoliation of the trees and shrubs upon which they were feeding. In all its essential characters the disease was similar to that which had swept over the army of tent caterpillars which were defoliating the apple and cherry trees in southern New Hampshire in 1898, as recounted in the bulletin upon the parasites of that insect, published as No. 5 in the Technical Series of the New Hampshire Agricultural Experiment Station. It was believed of this disease, at the time when these investigations were being conducted, that it was infectious, since the inhabitants of whole nests would all perish simultaneously. At the same time, its infectious or contagious nature was not established.

On the supposition that the disease of the tent caterpillar was infectious, and that that of the gipsy-moth caterpillars was similar in character, it looked for a time as though the parasite work was destined to an untimely end through the destruction of the gipsy-moth caterpillars before the parasites had opportunity to establish themselves and increase to the point of efficacy. Neither was there anything observed during the summer of 1907 to render this supposition untenable, except (and from an economic standpoint the exception was one of grave importance) the fact that, taking the infested area as a whole, there was a tremendous increase in the number of egg masses of the gipsy moth in the fall of 1907 over the number which had been present the previous spring.

There did not seem to be any particular reason why the disease should not increase in effectiveness as time passed on, however, and when in the spring of 1908 myriads of caterpillars in the first stage were found "wilting" in the forests in Melrose, and when just a little later practically every caterpillar was destroyed in one particular locality which had been selected as a good place for the very first colony of *Apanteles fulvipes*, there seemed to be reason to hope for speedy relief through disease. About this time these hopes were rudely shattered by the failure of several attempts to demonstrate the infectious or contagious nature of the disease through experiments carried on at the laboratory. Its noncontagious nature was further indicated by the fact that it did not spread across a narrow roadway near the laboratory, one side of which was swarming with dying caterpillars, while the other was peopled with an alarming but not destructive abundance of healthy ones. It appeared, after all, as though the views often expressed by Mr. A. H. Kirkland (at that time superintendent of the moth work in Massachusetts) to the effect that the disease was nothing more than the natural concomitant of overpopulation, and that an insufficient or unsuitable food supply was the true explanation of its prevalence, were right. That it was not to be depended upon for immediate results was certain when, at the close of 1908, a further alarming and apparently an unaffected increase in the distribution and abundance of the gipsy moth in Massachusetts and New Hampshire was found to have taken place wherever conditions were not such as to render destruction through disease the only thing which saved the gipsy moth from extinction through starvation, or where active hand suppression work had not been undertaken.

In 1909, and again in 1910, observations upon the progress of the disease were made almost daily throughout the caterpillar season. It was no longer looked upon as a serious obstacle to the success of the parasite work, except as it interfered (as it frequently did most seriously) with the work of colonizing Anastatus, and to a lesser extent Calosoma. It was also, as ever, the cause of serious trouble whenever attempts were made to feed caterpillars in the laboratory in confinement.

The disease acquired new interest, however, through the gradual accumulation of evidence tending to support the theory that it was either transmissible from one generation to another through the egg or that a tendency to contract it was thus transmitted.

Recognition of this characteristic through cumulative evidence resulting from more occasional or specific observations than it would be possible to review at this time, was accompanied by the almost equally apparent fact that the disease was becoming slightly more effective at a somewhat earlier stage in the progress of a colony of the gipsy moth following its establishment in a new locality. It was found, for example, in New Hampshire in colonies which had barely reached the stripping stage. A few years before the caterpillars composing such colonies would naturally have migrated from the stripped trees to others in the vicinity, and it is an unmistakable fact that such migrations, which have several times been mentioned in the earlier reports of the State superintendent of moth work, are now decidedly less frequent, even without taking into consideration the greater territory throughout which the moth is now present in destructive abundance. Although the junior author has personally visited large numbers of outlying colonies of the moth in the course of 1908, 1909, and 1910, he has yet to see one in which the disease had not appeared coincidently with the development of the colony to the stripping stage, if not slightly in advance of that time.

It is probably safe to say that such conditions as are described in the first annual report of the superintendent of moth work as prevailing over a large territory in the old infested section during 1904 and 1905, will probably not immediately recur in the history of the gipsy moth in eastern Massachusetts. That something approaching this may result in parts of New Hampshire is well within the bounds of probability, and that the conditions will be very bad in that State during the course of the next few years as well as in some of the towns in Massachusetts may be accepted as most probable. Whatever may be the condition presented by the older infested sections in eastern Massachusetts five or ten years from now, the only hope of preventing an ever-increasing wave of destruction from spreading over western Massachusetts, across New Hampshire and Vermont, and over the border into the State of New York, seems to lie, as always, in an increasing expenditure for hand suppression or in the success of the experiment in parasite introduction. Through the methods now in operation it is probable that the pest will very largely be prevented from making long "jumps," which would otherwise have been of frequent occurrence, but the slower and more steady natural spread, through the agency of wind, and probably, when the headwaters of the Connecticut, Hudson, and Ohio are reached, by water, must be considered in every attempt to discount the future. It was taken into consideration when the future of the parasite work was decided upon.

In the course of the studies of the parasites and parasitism of native insects which have been undertaken in connection with those of the parasites of the gipsy moth and the brown-tail moth, no less than three species have been encountered which are controlled to some extent by a disease which bears a very close superficial resemblance to the "wilt" of the gipsy moth. These are the white-marked tussock moth, the tent caterpillar, and the "pine tussock moth."

The white-marked tussock moth (*Hemerocampa leucostigma* S. & A.) is well known as a defoliating pest in cities, and has been so abundant as at times to become a rival of the gipsy moth in its destructive capacities in certain of the larger cities in southeastern New England.

It is very subject to a "wilt" disease, and no colony has been observed which has reached such proportions as to threaten the defoliation of street trees in which the disease has not appeared. In one instance the disease was so prevalent as to destroy practically all of the caterpillars, and, as in the case of the gipsy moth, the scattering caterpillars which hatched from the eggs deposited by the few survivors were seriously affected the following year, notwithstanding the presence of an abundance of food. Furthermore, caterpillars hatching from eggs collected in this and similar colonies removed to the country where the few native caterpillars to be found have always been remarkably healthy, perished through the "wilt" exactly as though they had hatched in the city.

Nevertheless, the white-marked tussock moth has been for long, is now, and probably will remain the worst defoliating insect enemy of such trees as the horse chestnut, maple, sycamore, etc., in strictly urban communities in the Eastern States generally. It does not appear unreasonable to suppose that the gipsy moth may similarly continue to be a pest in spite of the disease. As a matter of fact, every observation which has been made upon either the fungous disease of the brown-tail moth caterpillars, the wilt disease of the gipsy-moth caterpillars, or diseases of other defoliating caterpillars, such as that of the white-marked tussock moth, the tent caterpillar, and the "pine tussock moth," has tended to confirm the conclusion that such insect epidemics rarely play more than the one rôle in the economy of nature. They do not *prevent* an insect from increasing to an extent which renders it a pest, but they may, and frequently do, render very efficient service in effecting a wholesale reduction in the abundance of such insects when other agencies fail. When the insect in question is ordinarily controlled by parasites, as appears to be the case with the white-marked tussock moth, the "pine tussock moth," etc., it is probable that a long time will elapse before it will again encounter the combination of favorable circumstances which make possible abnormal increase.

When, as with the white-marked tussock moth in cities, the tent caterpillar in southeastern New England, or the brown-tail moth and gipsy moth in America, adequate control by parasites is lacking, reduction in numbers through disease is not likely to result in more than temporary relief. The more complete the destruction wrought by the insect the longer the period which must necessarily elapse before it again reaches the state of destructive abundance and, looking at it from this standpoint, it is not unlikely that the gipsy moth is much more abundant at the present time than it would have been had it not been for the prevalence of disease. There are, each year, an abundance of localities where the destruction of a great majority of the caterpillars by disease has been the only thing which has saved the whole race from complete extinction in that locality through consumption of the entire supply of available food before growth was completed. Under such circumstances the disease has been of positive benefit to the gipsy moth, rather than the reverse.

STUDIES IN THE PARASITISM OF NATIVE INSECTS.

Among a considerable number and variety of native insects studied at the laboratory which resemble the gipsy moth in habit, or which are more or less closely allied to it in their natural affinities, no two have been found in the economy of which parasitism has played an exactly similar rôle. There is this to be said, however, that only one amongst them, and this the tent caterpillar, appears to be ineffectually controlled by parasitism, except under unusual circumstances.

Several very beautiful examples of control by parasites have been encountered in the course of these investigations, and, comparatively speaking, the exceptional instances in which parasites lose control through one reason or another are exceedingly rare. Such instances are usually, if not inevitably, accompanied by a conspicuous outbreak of the insect in question.

The destructiveness of the white-marked tussock moth in cities is apparently due to the fact that it is peculiarly adapted to life under an urban environment. It is an arboreal insect, and one which is prevented through the winglessness of its females from dispersing over the country as the brown-tail moth, for example, would do under similar circumstances. Its parasites, on the other hand, are not always fitted for a peculiarly arboreal existence. Many of them are partially terrestrial, and in addition they are strong upon the wing.

Most of the introduced parasites of the gipsy moth and brown-tail moth which are known to have established themselves in America are known to be dispersing at a rapid rate. Several of them have been reared as parasites of the white-marked tussock moth from caterpillars or pupæ collected under urban surroundings, and since we have positive proof of their wandering habits there is every reason to believe that the native parasites of the tussock moth possess similar characteristics. That is to say, instead of staying within the limited area in which their host abounds, they are likely to scatter throughout the country immediately following the completion of their transformations. They are neither fitted for continued existence in the city to the degree which is characteristic of their host, nor are they compelled, like it, to accept it when they find themselves city-born through chance ancestral wanderings.

Every season's observations (and for four consecutive years the tussock moth has received more than a modicum of attention) has added arguments to support the contention that the white-marked tussock moth is controlled in the country through parasitism and not by birds or other predators. In any event it is controlled to such an extent as to have made a study in parasitism under strictly rural conditions very difficult, except when eggs or caterpillars have been artificially colonized for the purpose.

The outbreak of the Heterocampa in New Hampshire and Maine is another exceptional instance. In many respects the results of the relatively limited study given to this insect were the most remarkable of any, since there was offered what, to the writers, was the unique spectacle of unrestricted increase being checked through starvation without the intervention of disease. Notwithstanding the fact that the abundance of this insect was so great as to bring about complete defoliation of its favored trees over a very wide area, not a sign of disease was observed in the fall of 1909 in forests where millions of caterpillars were literally starving to death. The final, thoroughly effective, and miraculously complete subjugation of the outbreak, which resulted in the insect dropping from the abundance above mentioned to what is perhaps less than its normal numbers in the course of a single year, has already been described in a paper which appeared in a recent number of the Journal of Economic Entomology. There is every reason to believe that it was entirely the result of insect enemies, including both parasites and a predaceous beetle, which latter, through its ability to increase abnormally at the direct expense of that particular insect, played a rôle exactly comparable to that of the true facultative parasites. Such another outbreak of Heterocampa has never been known, and it is probable that it will be very many years before a combination of conditions makes its repetition possible. It is altogether probable that during this period the parasites will remain in full control.

A third exceptional instance is the present outbreak of the "pine tussock moth" in Wisconsin. This interesting and, as it has proved itself, potentially destructive insect is decidedly rare in Massachusetts, but notwithstanding its scarcity a sufficient number was collected in 1908 and 1909 to make possible a study of its parasites. Parasitism to an extent rarely exceeded amongst leaf-feeding Lepidoptera was found to be existent, and it is safe to say that had it not been for its parasites the host would have increased at least fivefold or sixfold in 1909 over the numbers which were present in 1908. Such a rate of increase, if continued, would have placed it among the ranks of destructive insects in a very few years, and it appears that something of this sort actually occurred in northern Wisconsin. There, some years ago, it reached a stage of abundance which resulted in partial or complete defoliation of pine throughout a considerable territory and, as was expected, a relatively small percentage of the caterpillars and pupæ were found to be parasitized. Existing parasitism in 1910 was not sufficiently effective to prevent its increase to a point which would have made complete defoliation of its food plant, and consequently its death through starvation, an accomplished fact had its abundance not been reduced through the prevalence of a disease superficially similar to the "wilt" of the gipsy moth.

The fall webworm is generally a common and abundant insect in New England, but rarely as common or abundant as it frequently becomes in the South. An elaborate study of its parasites and the effect which parasitism apparently played in effecting its control was made in the fall of 1910, with interesting results. It was found that the prevailing percentage of parasitism was sufficient to offset an increase of no less than fourfold annually, and even at that there is reason to believe that our results err on the side of conservatism. The elimination of these parasites for a very short period of years would undoubtedly be followed by an increase of the host comparable to that of the gipsy moth.

The one insect studied at the laboratory which appears habitually and under its normal environment to become so unduly abundant as to invite destruction through disease at regular intervals is the tent caterpillar.

In the report upon its parasites,[1] it was contended that they played a part subservient to that taken by the disease, and this statement drew forth some criticism at the time of its publication. It is a satisfaction to note that the original contention appears to be upheld by the results of studies conducted at the gipsy-moth parasite laboratory. These results seem to justify the further contention that the present status of the tent caterpillar is, in a way, prophetic of that which would result were the gipsy moth to be left to the control of its disease.

At frequent but irregular intervals the tent caterpillar increases to such an extent as to become a pest, and unless artificially checked it defoliates fruit trees in southern New England. That it never reaches the destructiveness characteristic of the gipsy-moth invasion is seemingly due to difference in habit. As is well known, the gipsy-moth caterpillar is almost an omnivorous feeder and the female moth is incapable of flight. Its eggs are deposited indiscriminately in every conceivable place to which a caterpillar or moth can gain access. The adult of the tent caterpillar is in no way restricted to the immediate vicinity of the locality where it chose to pupate as a caterpillar, but, instead, uses what really amounts to an unwise amount of discretion in its selection of a place for oviposition. Cherry first and then apple is selected in preference to all other food plants, and with the exception of a limited number of other rosaceous trees and shrubs, its eggs are almost never found elsewhere. As a result, when it is at all abundant its caterpillars, which have not the

---

[1] Technical Bulletin 5, New Hampshire Agricultural Experiment Station.

wandering characteristics of those of the gipsy moth, but rather the opposite, find themselves crowded in excessive numbers upon a limited variety of shrubs and trees; complete defoliation of these comparatively few host plants quickly follows, and weather conditions being favorable to the development of disease, wholesale destruction is all that intervenes between an unnatural migration or starvation. Such reduction is followed by a period of years during which the parasites check but do not overcome the tendency to increase, and it is only a little while before the process is repeated.

There were, in certain localities in eastern Massachusetts in the summer of 1910, continuous strips of roadside grown up to a variety of trees and shrubs, the most of which were defoliated by tent caterpillars, all of which had hatched from eggs deposited upon the occasional wild-cherry tree which was present. Several such strips were visited at about the time when the caterpillars elsewhere were beginning to pupate, and not a single living caterpillar or pupa could be found amongst the thousands of dead and decomposing remains of the victims of overpopulation. These were but a repetition of conditions as observed a few miles north in New Hampshire 12 years ago. How frequently similar conditions occurred during the intervening period is not known.

In addition to those species mentioned in the preceding pages, quite a number of other leaf-feeding Lepidoptera have been more or less casually studied in a less comprehensive but at the same time a careful manner.

PARASITISM AS A FACTOR IN INSECT CONTROL.

In reviewing the results of these studies, the fact is strikingly evident that parasitism plays a very different part in the economy of different hosts. Some habitually support a parasitic fauna both abundant and varied, while others are subjected to attack by only a limited number of parasites, the most abundant of which is relatively uncommon. No two of the lepidopterous hosts studied, unless they chanced to be congeneric and practically identical in habit and life history, were found to be victimized by exactly the same species of parasites. Neither are the same species apt to occur in connection with the same host in the same relative abundance, one to another, year after year in the same locality, nor in two different localities the same year.

At the same time there are certain features in the parasitism of each species which are common to each of the others, whether these be arctiid, liparid, lasiocampid, tortricid, saturniid, or tineid, one of the most common of which is that each host supports a variety of parasites, oftentimes differing among themselves to a remarkable degree in habit, natural affinities, and methods of attack. Departures from this rule have not been encountered among the defoliating Lepidoptera as yet, and while exceptions will probably be found to exist, they will doubtless remain exceptions in proof of the rule. From this the rather obvious conclusion has been drawn, that to be effective in the case of an insect like the gipsy moth or the brown-tail moth, parasitic control must come about through a variety of parasites, working together harmoniously, rather than through one specific parasite, as is known to be the case with certain less specialized insects, having a less well-defined seasonal history. To speak still more plainly, it is believed that the successful conclusion of the experiment in parasite introduction now under consideration depends upon whether or not we shall be able to import and establish in America each of the component parts of an effective "sequence" of parasites. This belief is further supported by the undoubted fact, that in every locality from which parasite material has been received abroad, both the gipsy moth and the brown-tail moth are subjected to attack by such a group or sequence of parasites, of which the component species differ more or less radically in habit and in their manner of attack.

In the case of the gipsy moth and the brown-tail moth abroad, as well as in that of nearly every species of leaf-feeding Lepidoptera studied in America, there are included among the parasites species which attack the eggs, the caterpillars, large and small, and the prepupæ and pupæ, respectively. Frequently, but not always, there are predatory enemies, which, through their ability to increase at the immediate expense of the insect upon which they prey, whenever this insect becomes sufficiently abundant to invite such increase, are to be considered as ranking with the true facultative parasites when economically considered.

It is, therefore, our aim to secure the firm establishment in America of a sequence of the egg, the caterpillar, and the pupal parasites of the gipsy moth and brown-tail moth as they are found to exist abroad, and until this is either done or proved to be impossible of accomplishment through causes over which we have no control, we can neither give up the fight nor expect to bring it to a successful conclusion.

It was stated a page or two back that some species of insects support a parasitic fauna both numerous and varied, while others are subjected to attack by only a limited number of parasites, none of which can be considered as common. Notwithstanding the fact that somewhat similar differences are discernible between the parasitic fauna of the same insect at different times or under different environment, it is perfectly safe to elaborate the original statement still further and to say that some species are habitually subjected to a much heavier parasitism than others. Unquestionably the average percentage of parasitism of the fall webworm in eastern Massachusetts, taken over a sufficiently long series of years to make a fair average possible, is the same as the average would be over another similar series of years in the same general region. This could be said of the larvæ of any other insect as well as of that of the fall webworm, but the average percentage of parasitism in another would most likely not be the same, but might be very much larger or very much smaller. To put it dogmatically, each species of insect in a country where the conditions are settled is subjected to a certain fixed average percentage of parasitism, which, in the vast majority of instances and in connection with numerous other controlling agencies, results in the maintenance of a perfect balance. The insect neither increases to such abundance as to be affected by disease or checked from further multiplication through lack of food, nor does it become extinct, but throughout maintains a degree of abundance in relation to other species existing in the same vicinity, which, when averaged for a long series of years, is constant.

In order that this balance may exist it is necessary that among the factors which work together in restricting the multiplication of the species there shall be at least one, if not more, which is what is here termed facultative (for want of a better name), and which, by exerting a restraining influence which is relatively more effective when other conditions favor undue increase, serves to prevent it. There are a very large number and a great variety of factors of more or less importance in effecting the control of defoliating caterpillars, and to attempt to catalogue them would be futile, but however closely they may be scrutinized very few will be found to fall into the class with parasitism, which in the majority of instances, though not in all, is truly "facultative."

A very large proportion of the controlling agencies, such as the destruction wrought by storm, low or high temperature, or other climatic conditions, is to be classed as catastrophic, since they are wholly independent in their activities upon whether the insect which incidentally suffers is rare or abundant. The storm which destroys 10 caterpillars out of 50 which chance to be upon a tree would doubtless have destroyed 20 had there been 100 present, or 100 had there been 500 present. The average percentage of destruction remains the same, no matter how abundant or how near to extinction the insect may have become.

Destruction through certain other agencies, notably by birds and other predators, works in a radically different manner. These predators are not directly affected by the abundance or scarcity of any single item in their varied menu. Like all other creatures they are forced to maintain a relatively constant abundance among the other forms of animal and plant life, and since their abundance from year to year is not influenced by the abundance or scarcity of any particular species of insect among the many upon which they prey they can not be ranked as elements in the facultative control of such species. On the contrary, it may be considered that they average to destroy a certain gross number of individuals each year, and since this destruction is either constant, or, if variable, is not correlated in its variations to the fluctuations in abundance of the insect preyed upon, it would most probably represent a heavier percentage when that insect was scarce than when it was common. In other words, they work in a manner which is the opposite of "facultative" as here understood.

In making the above statement the fact is not for a moment lost to sight that birds which feed with equal freedom upon a variety of insects will destroy a greater gross number of that species which chances to be the most abundant, but with the very few apparent exceptions of those birds which kill for the mere sake of killing they will only destroy a certain maximum number all told. A little reflection will make it plain that the percentage destroyed will never become greater, much if any, as the insect becomes more common, and, moreover, that after a certain limit in abundance is passed this percentage will grow rapidly less. A natural balance can only be maintained through the operation of facultative agencies which effect the destruction of a greater proportionate number of individuals as the insect in question increases in abundance.

Of these facultative agencies parasitism appears to be the most subtle in its action. Disease, whether brought about by some specific organism, as with the brown-tail moth, or through insufficient or unsuitable food supply without the intervention of any specific organism, as appears at the present time to be the case with the gipsy moth, does not as a rule become effective until the insect has increased to far beyond its average abundance. There are exceptions to this rule, or appear to be, but comparatively only a very few have come to our immediate attention. Finally, famine and starvation must be considered as the most radical means at nature's disposal, whereby insects, like the defoliating Lepidoptera, are finally brought into renewed subjugation.

With insects like the gipsy moth and the brown-tail moth disease does not appear to become a factor until a degree of abundance has been reached which makes the insect in question, *ipso facto*, a pest. Whether in the future methods will be devised for artificially rendering such diseases more quickly effective, remains to be determined through actual experimental work continued over a considerable number of years.

In effect, the proposition is here submitted as a basis for further discussion that only through parasites and predators, the numerical increase of which is directly affected by the numerical increase of the insect upon which they prey, is that insect to be brought under complete natural control, except in the relatively rare instances in which destruction through disease is not dependent upon superabundance.

The present experiment in parasite introduction was undertaken and has been conducted on the assumption that there existed in America all of the various elements necessary to bring about the complete control of the gipsy moth and the brown-tail moth, except their respective parasites. Believing that this stand was correctly taken, much time has been devoted to a consideration of the extent to which these pests are already controlled through natural agencies already in operation. The fact that both insects have increased steadily and rapidly in every locality in which they have become established and where adequate suppressive measures have not been undertaken, until they have reached a stage of abundance far in excess of that which prevails in most countries abroad, renders superfluous further comment upon the present ineffectiveness of these agencies. The difference between the rate at which they have averaged to increase in localities where they have become established and their potential rate of increase as indicated by the number of eggs deposited by the average female should indicate very accurately the efficiency of such agencies, and the difference between the actual rate of increase and no increase similarly indicates the amount of additional control which must be exerted by the parasites if their numbers are to be kept at an innocuous minimum.

THE RATE OF INCREASE OF THE GIPSY MOTH IN NEW ENGLAND.

The potential rate of increase as determined by the number of eggs deposited by the average female of the gipsy moth varies considerably under different circumstances, and affords an interesting example of a phase of facultative control not touched upon in the last chapter. When the exhaustive studies into its life and habits were conducted under the general supervision of the Massachusetts State Board of Agriculture during the final decade of the last century, it was determined that the number was between 450 and 600.

In the opinion of some, the fecundity of the gipsy moth has distinctly decreased during the 14 years which have elapsed since the publication of the report in which these figures were given, and in order to determine the point a considerable number of egg masses was collected during the winter of 1908–9 and the eggs carefully counted. It was found that in those from the older infested territory or from outlying colonies where the moth was particularly abundant, the number of eggs to a mass averaged considerably less than 300. In egg masses from outlying districts where the infestation was new, and where the moth had never reached its maximum abundance, the average in a few masses counted was slightly in excess of 500. The number is, however, very variable, and the character of food and the meteorological conditions during the feeding period of the caterpillars are doubtless important features. Hot weather during June forces the development of the caterpillars and they do not become large. Small moths deposit fewer eggs rather than smaller eggs. It is possible that there is actually a decrease in the fecundity of moths brought about by our short and ardent summers, but for the present it is not proved, and it is believed that whenever abundance of the insect is sufficiently reduced the original rate of multiplication will prevail. The point is one well worthy of further investigation, but for the present the potential rate of increase, provided no controlling factors whatever are operative, will be considered as 250-fold annually.

The best information available as to the rate of increase of the gipsy moth actually prevailing in Massachusetts is contained in the report entitled "The Gypsy Moth," by Mr. Edward H. Forbush and Dr. C. H. Fernald, which was published under the direction of the board of agriculture. These authorities, in their discussion of the matter, say as follows:

The study of the increase and dissemination of the gipsy moth in Massachusetts is most interesting. Perhaps there never has been a case where the origin and advance of an insect could be more readily traced. As the moth appears to be confined as yet to a comparatively small area, and as the region has been examined more or less thoroughly for five successive years, the opportunities offered for the study of the multiplication and distribution of the insect have been unequaled.

When it is considered that the number of eggs deposited by the female averages from 450 to 600, that 1,000 caterpillars have been seen to hatch from a single egg cluster, and that at least one egg cluster has been found containing over 1,400 eggs, there can be no doubt that the reproductive powers of the moth are enormous. Mr. A. H. Kirkland has made calculations which show that in eight years the unrestricted increase of a single pair of gipsy moths would be sufficient to devour all vegetation in the United States. This, of course, could never occur in nature, and is mentioned here merely to give an idea of the reproductive capacity of the insect.

It seems remarkable at first sight that an insect of such reproductive powers, which had been in existence in the State for 20 years, unrestrained by any organized effort on the part of man, did not spread over a greater territory than 30 townships, or about 220 square miles. Some of the causes which at first checked its increase and limited its diffusion in Medford have already been set forth. Most of the checks which at first served to prevent the excessive multiplication of the gipsy moth in Medford operate effectively to-day wherever the species is isolated. True, it has now become acclimated. But any small isolated moth colony still suffers greatly from the attacks of its natural enemies and from the struggle with other adverse influences which encompass it. The normal rate of increase in such isolated colonies as are found to-day in the outer towns of the infested district seems to be small. The annual increase can be readily ascertained by noting the relative number of egg clusters laid in successive years, the unhatched or latest clusters being easily distinguished from the hatched or "old" clusters, and the age of these latter, whether one, two, three, or more years, being indicated by their state of preservation. The ratio of the average annual increase of 10 such colonies was found to be 6.42; that is, six or seven egg clusters on an average may be found in the second season to one of the first season.

If the number of eggs deposited by the average female moth be set at 500, and if the sexes of her progeny are equally divided, a potential increase of 250-fold for each annual generation is provided for. Under complete control only one pair of moths would average to be produced from each mass of eggs deposited, and since each egg represents an individual embryo, all but 2 of each 500 must fail to reach full maturity. Reduced to percentage this is equivalent to the survival of 0.4 per cent and the destruction of 99.6 per cent of the gipsy moths in one stage or another every year. Since the total number of gipsy moths in any locality can not possibly be computed, the only method by which mortality through any cause may be expressed is on this basis.

It will surprise many who have not given the matter consideration to learn what an extraordinary *apparent* mortality it requires to offset a potential increase of 250-fold. The gipsy-moth caterpillars molt five or six times after they hatch and before they change to pupæ, making the number of caterpillar stages six or seven. If through natural controlling agencies 50 per cent of the young caterpillars were destroyed in the first stage before they had molted, and this was followed by similar destruction of another 50 per cent in the second stage, and so on through the third, fourth, fifth, sixth, and seventh stages, respectively, and in addition 25 per cent of the pupæ and 25 per cent of the adults before depositing their eggs were similarly destroyed, it would still permit of a slight annual increase.

The following table (if the incongruity of fractions as applied to insects may be overlooked) indicates the number of survivors of each stage resulting from the hatching of a mass of 500 eggs:

	Stage.	Number.	Loss.	Number remaining.	Potential increase.
			Per cent.		
Eggs		500	0	500	250 fold.
Caterpillars	First	500	50	250	125 fold.
Do	Second	250	50	125	62 fold.
Do	Third	125	50	62	31 fold.
Do	Fourth	62	50	31	15.5 fold.
Do	Fifth	31	50	15.5	7.75 fold.
Do	Sixth	15.5	50	7.75	3.875 fold.
Do	Seventh	7.75	50	3.875	2.906 fold.
Pupæ		3.875	25	2.906	2.179 fold.
Adults		2.906	25	2.179	1.634 fold.

To give another illustration: The life of the gipsy-moth caterpillar is approximately seven weeks. If beginning on the first day after hatching and on every day thereafter during this period a decrease in numbers of 10 per cent should be brought about through natural causes, there would still be enough survivors to permit of a substantial increase in the abundance of the insect.

Twelve gipsy moths (6 pairs) from each egg mass would be sufficient to provide for a sixfold annual increase. If reference be made to the preceding table, it will be seen that if all destruction ceased after the caterpillars had reached the fourth stage, the survivors would permit of slightly in excess of sixfold increase; that is, the mortality during the first to the fourth stage, inclusive, with a part of that which resulted during the fifth stage, would be sufficient to account for all of the control at present exerted by natural agencies in New England, and this gives, at the same time, an idea as to the amount of additional control which the parasites must accomplish if they are to become effective.

The conditions under which the gipsy moth was studied at the time when the material for the report just quoted was accumulated were, for the most part, abnormal. In only relatively few localities was it allowed to increase undeterred, and there were relatively very few examples of unrestricted increase to the point when defoliation resulted. This, in part, explains what seems to be an element of indecision concerning the character of conditions which favored more rapid increase of the moth, as quoted below, from the same source.

CONDITIONS FAVORING RAPID INCREASE.

When any colony under average normal conditions has grown to a considerable size and then received an added impetus from exceptionally favorable conditions, its power of multiplication and its expansive energy are greatly augmented, and its annual increase arises above all calculations.[1] Under such influences hundreds of egg clusters will appear in the fall where few were to be seen in the spring, and thousands are found where scores only were known before. It is probable that the season of 1889 was particularly favorable for the moth's increase. The season of 1894 and that of 1895 appear also to have furnished conditions especially favorable for an abnormal multiplication of the insect.

The operation of the causes of these sudden outbreaks is not understood. It is evident, however, that the warm, pleasant spring weather of the past two years (1894 and 1895) hastened the development of the caterpillars, thereby shortening their term of life. The length of life of the caterpillars varies from six to twelve weeks. During cold, rainy weather the caterpillars eat little and grow slowly. During warm, dry weather they consume much more food and grow with great rapidity. In the unusually warm spring and early summer of 1895 many of the caterpillars molted a less number of times than usual, and their length of life did not exceed six or seven weeks. Under these conditions they proved more quickly injurious to foliage than in a more normal season, and were more completely destructive within any given area in which their numbers were great. And they were not so long exposed to the attacks of their

---

[1] The increase of these large colonies seems to be limited only by the supply of food. Whenever food becomes scarce many of the moths are less prolific. The larvæ which do not find sufficient food either die or develop early, and the female moths lay fewer eggs than those which transform from well-nourished caterpillars.

enemies. While it may be true that the parasitic enemies of the moth will also develop rapidly under conditions that hasten the growth of their host, birds and other vertebrate enemies will secure fewer of the moths in 6 or 7 weeks than in 10 or 12. It is believed that dry weather is unfavorable for vegetable parasites of insects, but to what extent the caterpillars are affected by them in a humid season it is impossible to say.

The past two years have been "cankerworm years" in the infested region. Many of the birds which habitually feed on the caterpillars of the gipsy moth have been largely occupied during May and the early part of June in catching cankerworms, which they seem to prefer, turning their attention to the gipsy-moth caterpillars in the latter part of June and July, when the cankerworms have disappeared. The birds, therefore, have not been as useful in checking the increase of the gipsy moth as in years when the cankerworms were less numerous.

A few of the restraining influences which have been less active than usual during the past two years have been mentioned, and possibly many others have escaped observation, but those given serve in a measure to explain the unusual increase of the moth. It is during such seasons that its destructiveness is most apparent. It is then that the groves and forests are stripped of their leaves, and whole rows of trees in orchards and along highways appear to have been stripped in a single night.

The conditions as described seem to be comparable to those prevailing at the present time, and at the same time to be inadequately explained. Repeatedly personal observations have been made which indicate beyond the shadow of a doubt that under certain circumstances the gipsy moth has increased at a rate very far in excess of sixfold annually at the present time. Counts of old egg masses as compared with those newly laid, in several localities, in the spring of 1908 and each spring subsequently, have shown positively that an increase of at least twentyfold was not uncommon. In fact, unless an unduly large number of old egg masses was concealed, it could be said with equal certainty that increase sometimes amounted to fiftyfold in the course of a single generation. The arguments presented by Forbush and Fernald, who evidently observed something very similar, and who were inclined to credit it to seasonal or climatic conditions (in part at least), do not stand, in view of the fact that the rate of increase differs extraordinarily in localities nearly adjacent to where the conditions are practically identical, saving only the varying abundance of the moth; this latter, it may be noted, has in each instance corresponded roughly and in direct ratio to the rate of increase. The fact was not considered to be of more than coincidental interest at first, but later, when an attempt was made to classify according to their manner of operation, the various factors which were already responsible for the partial control of the gipsy moth in New England, the correlation between relative abundance and rate of increase recurred and seemed to afford excellent support to the contention which has been made as to the part which birds and most other predators play in bringing this about.

Without attempting to go into the details of rather elaborate calculations, which were made for the purpose of bringing out this point more graphically, attention is merely called to the three divisions into which the elements operative in the natural control of any insect naturally fall as they were outlined in the preceding section. These are, first, the catastrophic (storms, etc.), which result in the destruction of a certain fixed percentage, irrespective of the abundance of the insect; second, that represented by the birds, most other predators, and a part of the parasites which encompass the destruction of a certain gross number, rather than of any given percentage each year or generation; third, the facultative agencies, of which certain parasites are considered to be typical, which increase in efficiency as the insect increases in abundance.

The elements composing this last group are absent in New England, or, rather, those elements which are present (disease and starvation) and which do not properly belong to it, are inoperative until a state of extreme abundance is attained.

Such control as is effected by existing agencies would therefore fall into one or the other of the first two groups mentioned, and since both groups together are obviously inefficient, even when the moth is scarce, that due to the operation of the elements falling in the second group would become relatively less efficient as the time went on and the moth increased. This, it is believed, is actually what has happened and what is happening each year in each of the very numerous outlying colonies of the gipsy moth throughout the more recently infested territory, and thus the larger rate of increase is explained.

As a matter of fact, there is reason to believe that the average rate of increase during the first few years *immediately* following the introduction of the moth in a new locality is actually less than sixfold annually, and that it may even be as low as threefold, or perhaps less. In any case, there is a stage in the progress of the moth in which the average is no greater than that recorded by Forbush and Fernald, and there is no longer room for doubt that the lowest rate of increase is in localities where the moth is relatively a rare or uncommon insect for the time being, while the highest occurs in localities where the moth is rapidly approaching its maximum abundance.

AMOUNT OF ADDITIONAL CONTROL NECESSARY TO CHECK THE INCREASE OF THE GIPSY MOTH IN AMERICA.

It was evident in 1907, as it is now, that the problem of the introduction of parasites was far from being as simple as it might appear to be upon its surface and as it evidently did appear to be to some who were at that time agitating for a radical change in the methods adopted for its solution. It was plain that the expense incident to the actual work of importation was going to be considerably more than had been expected two years before and that practical results could not possibly be achieved until long after the time originally predicted. Additional information upon the biology and habits of each of the several parasites, if not necessary in every instance, was necessary in some and desirable in all, and here again additional expenditures became imperative. Furthermore, the situation was such as to make of very doubtful advisability the indiscriminate importation of very large quantities of parasite material before a better knowledge of the parasites themselves had been secured. The repetition of the very large shipment of brown-tail hibernating nests winter after winter, as will be described in another chapter, is an instance in point. Had we been in possession of a complete knowledge of the parasites hibernating in those webs at the beginning, perhaps one winter's importation would have been sufficient.

There was no certainty that the results of the technical studies as conducted at the American laboratory would be sufficiently full and complete to answer our purposes and make possible the intelligent continuation of the work. Should we fail in this respect, the only alternative to a discontinuation of the introduction work in advance of its logical conclusion was the establishment of a laboratory abroad, at a considerable expenditure.

With these several reflections, it was inevitable that the advisability of continuing the work beyond the time limit originally set should come into question. Accordingly, in anticipation of the necessity for making a decision when the time for it should arrive, the whole proposition was subjected anew to the closest sort of scrutiny from every point of view.

The successful consummation of the work involved, first of all, the establishment in America of a group of parasites or other natural enemies sufficiently powerful to meet and offset the prevailing rate of increase of the gipsy moth. This, as determined by Forbush and Fernald, was at least sixfold annually; as determined by actual observation in the field, it was often far in excess of sixfold. Before the continuation of the work could be recommended it was absolutely necessary to arrive, first, at some conclusion as to the amount of parasitism (gauged on the percentage basis) which would be required in order to offset this increase and maintain the gipsy moth at an innocuous minimum; and, second, whether parasitism to such an extent actually prevailed abroad or whether natural control in those localities where it was obviously effected was due to the increased efficiency of other agencies.

In so far as the first proposition was concerned, it was obvious from the beginning that if enough egg masses could be destroyed each fall so that the number remaining would be no greater than that which had been present in the spring the insect could never, by any possibility, increase beyond the abundance then prevailing. If the increase each year was 6 egg masses for each 1 of the year before, it was merely necessary to destroy 5 out of every 6 in order to maintain the status quo. If the increase was tenfold, the destruction of 9 out of every 10 egg masses would be required, etc. The same would obtain if 5 out of every 6 or 9 out of every 10 eggs in each and every mass were similarly destroyed.

Reduced to percentages, this would be equivalent to the destruction of 83.33 per cent or of 90 per cent, respectively, a rate of parasitism which was physically an impossible accomplishment for the egg parasites alone. Additional parasitism of the caterpillars or pupæ would be a requisite to success, and such parasitism would of necessity be similarly limited in many instances through circumstances as completely beyond our control as the physical inability of Schedius or Anastatus to parasitize more than the uppermost layer of eggs in each mass attacked. Without attempting to go into any of the details of the processes by which conclusions were reached, it was finally determined, beyond any doubts arising through arguments which have been presented up to the present time, that an aggregate parasitism of 83.33 per cent would be absolutely necessary if a sixfold increase was to be met, but that it made no difference whether this was brought about by one species or two or a dozen, or whether they attacked the egg, the caterpillar, or the pupæ. It was also determined that the aggregate percentage necessary could not be secured by simply adding together the figures representing the parasitism resulting through attack by each of two or more species. It was going to be necessary to combine these several aggregates in a different manner. To illustrate: A 50 per cent parasitism of the eggs, if it could possibly be secured, followed by another 50 per cent parasitism of the caterpillars, could not by any possibility be considered as resulting in 100 per cent parasitism or complete extinction, but only in 50 per cent parasitism added to 50 per cent of what remained, which amounted, in effect, to 25 per cent of the whole. In this manner an aggregate of 75 per cent only is secured.

As is illustrated by the table on page III, it requires the combination of an imposing array of figures representing relatively small percentages of parasitism in each instance to acquire a sufficiently large aggregate.

It was further determined that any specific amount of parasitism, as 20 per cent of the eggs, was neither more nor less, but exactly as effective as 20 per cent parasitism of the caterpillars or pupæ, in so far as its value in constructing the final aggregate was concerned.

It can not be denied that when the validity of these conclusions became established and when in addition the possibility that a much greater rate of increase than sixfold would have to be met and offset before the much-to-be-desired consummation could reasonably be expected the prospects looked rather discouraging. Recognition of the correlation which existed between increased abundance and rate of increase served more than anything else to allay the doubts which these reflections created. Field work in 1908, 1909, and 1910 showed pretty conclusively that a rate of increase of not in excess of sixfold and possibly considerably less prevailed whenever the moth was in that state of innocuousness incident to the scarcity which it was hoped to bring about and maintain through the introduction of the parasites. An aggregate parasitism of 85 per cent will almost certainly be sufficient, and it may well be that 80 per cent or even 75 per cent will answer equally well. Much less than 75 per cent will probably not be effective.

THE EXTENT TO WHICH THE GIPSY MOTH IS CONTROLLED THROUGH PARASITISM ABROAD.

While it is true that the work which has been done for the purpose of determining the prevailing rate of increase of the gipsy moth in America leaves considerable to be desired in the way of exactness, in the main the statements made by Forbush and Fernald, as confirmed and modified by later observation, may be accepted as essentially accurate. It is fortunate that the situation is no worse than it appears to be, for if it were necessary to undertake the work of parasite introduction with the idea that the maximum rate of increase exhibited by the gipsy moth must be met by the parasites, such an unreasonable percentage of parasitism would be demanded as to make the proposition of introducing them a decidedly difficult task. As it is, there seems to be good reason to believe that a parasitism amounting to 75 per cent will be sufficient, provided that it can be maintained during the periods when the moth is relatively rare. It may be that less than that will answer equally well, but it would require actual test or else a much more careful study of the actual rate of increase of the moth under favorable conditions to justify such prophecy. In any event, 85 per cent will probably be amply efficient, if it can be established and maintained during all stages in the abundance of the moth. Such a rate would undoubtedly prevent the moth from increasing to destructive abundance in new territory or from regaining ground lost through the activities of disease in older infested regions.

Granted that it is sufficient, the question naturally arises as to whether such a degree of parasitism is to be found abroad in countries where the gipsy moth is present without being considered as a serious pest.

There is very little published information at hand bearing upon this subject, and that which is available is general rather than definite in its tenor. Anyone who for the first time encounters a tree covered with caterpillars of the gipsy moth dead and dying through the effects of the "wilt" disease is very apt to think that at last the gipsy moth has met its Waterloo, and disillusionment has only come in the present work as the result of several years' consecutive observations in the same or similar localities. In like manner the observations of foreign entomologists, or of American entomologists traveling abroad, as to the actual effectiveness of the parasites in accomplishing the control of the moth have to be taken with a grain of conservatism. Parasitism by a species as conspicuous as *Apanteles fulvipes* to the extent of 50 per cent would undoubtedly create a most favorable impression and the more conspicuous parasitized caterpillars would easily appear to outnumber the healthy. This amount of parasitism would certainly be inefficient in America unless it were supplemented by a much larger amount of parasitism by other species.

On this account it has been necessary to depend very largely upon the study of the parasite material imported from abroad as a source for information of this sort. From the beginning accurate notes have been kept of the many thousands of boxes of eggs, caterpillars, and pupæ of the gipsy moth and brown-tail moth, in which are recorded the locality from which each lot came, its condition on receipt, and the number and variety of parasites reared in each instance. The records are necessarily based in most instances upon such information as may be gained through a study of the condition of the material on receipt and the parasites reared, but in a few careful dissection work has been carried on to determine the true conditions, and thus to check up the results of the rearing work. It has been found that the amount of dependence which can be placed upon the rearing records is relatively small, and that nothing more than a general idea of conditions actually prevailing can be gleaned from them. Nearly always some of the caterpillars or pupæ are dead or dying upon receipt as a result of the ordeal through which they have passed. On the average, taking the gipsy-moth material from all localities, not more than 25 per cent has arrived in good condition (when the shipment of eggs is excepted). The brown-tail moth material has averaged very much better, and probably 75 per cent has been in good condition on receipt.

It has been found that sometimes a larger percentage of parasites than of caterpillars or pupæ died en route, while at other times these conditions are entirely reversed, and since dissections can not be made in every instance, it has been necessary to consider the parasitism indicated by the notes upon two different bases, i. e., that of the number of hosts originally involved and that of the number successfully completing their transformations. If from a lot of 1,000 brown-tail caterpillars 250 individuals of *Parexorista cheloniæ* and 250 moths are reared, it is perfectly safe to assume that parasitism by Parexorista amounts to more than 25 per cent and less than 75 per cent. Further than this nothing absolutely definite may be said. Exactly the same is true of the determination of prevailing rates of parasitism of native insects through rearing work.

On account of the inadequacy of these methods when it comes to the point of securing absolutely authentic information, not nearly so much is known of the parasitism of the gipsy moth or of the brown-tail moth abroad as is needed to carry on the work to its best advantage. This much, however, can be said definitely, that in some instances existing parasitism is sufficient to answer the requirements of the situation in America; in others it is obviously insufficient; in most the results of the study of imported material are not sufficiently reliable to support either contention.

Here, again, was food for serious consideration when it came to the point of making definite recommendations concerning the continuation of the work. Would the foreign parasites certainly meet the demands which would be made upon them in America?

This has been answered in the affirmative through its consideration from quite a variety of different viewpoints. For one thing, the lack of accurate information as to the conditions under which the parasite material was originally collected, has rendered the results of its study in America of difficult analysis. No one, for example, would seriously question the statement that the white-marked tussock moth is under well-nigh perfect control in America except in cities. Nevertheless, if it was desired to transport caterpillars or pupæ of this insect to Europe in order that its parasites might be reared, the agent intrusted with the collection of the material for exportation would certainly go to the city for it, and the person who received and studied it upon the other side would find so few parasites present as to justify exactly the same doubts concerning the parasitism of the tussock moth in America as have actually arisen concerning the parasitism of the gipsy moth and the brown-tail moth in Europe. The tussock moth is not often subjected to the full extent of parasitism necessary to effect its control in any locality from which caterpillars can be secured in quantity. It is reasonable to suppose that something of the same sort is true of the gipsy moth or of the brown-tail moth.

Furthermore, the study of the tussock moth has resulted in demonstrating another fact which is of peculiar interest in this connection, which is, that the parasites which assist in effecting its control in country districts where this control is perfect are sometimes entirely absent in the city. Something of the same sort may be true of the parasites which assist in effecting the control of the gipsy moth in many localities in Europe where it is so uncommon as to make collection of material for exportation in any quantity impossible. Some of the most interesting lots of caterpillars or pupæ which have been received were from such localities, and it may well be that there are parasites abroad which have not been received at the laboratory in Massachusetts in sufficient quantity for colonization, and which can never be received there until new methods for collecting and importing them are devised, but which at the same time are actually among the important species. This fact can only be determined definitely by careful study of the gipsy moth in localities where it was not sufficiently abundant to permit of its collection in large quantities. These studies, it is hoped, will be instituted in 1911, and so long as the gipsy moth continues to be a serious pest in America the investigation of its parasites abroad ought to be continued.

The ramifications of the parasite work have been so many and so diverse and have led so far afield, both literally and metaphorically speaking, as to make it practically impossible to report upon it as a whole as fully as would be desirable and practicable were it less extensive and varied. A chapter might be written upon the parasitism of the gipsy moth and another upon that of the brown-tail moth in each of the several countries in Europe from which the parasite material has been imported, but it is wholly impracticable to do so. At the same time, now that a new phase of the work is being entered upon, it will not be out of place to review in some slight detail the results of the work which has been carried on in a few of the localities which, for one reason or another, may be selected as of more than general interest in this immediate connection, but which are at the same time and in another sense typical.

PARASITISM OF THE GIPSY MOTH IN JAPAN.

From the viewpoint of gipsy-moth parasitism Japan possesses a peculiar interest, because, if we are to judge from the reports of those who have been there and incidentally or critically studied the situation, the Japanese gipsy moth is pretty thoroughly controlled through natural agencies, and among these its parasites appear to rank very high. This is the more interesting and encouraging because the Japanese race is notably larger and at the same time more fecund than the European, judging from counts as made at the laboratory of the number of eggs in a mass.

In 1908, after several unsuccessful attempts which had been made to import its parasites had served to demonstrate the futility of any less radical course, Prof. Trevor Kincaid, of the University of Washington, Seattle, was delegated to spend the summer there in the interests of the work. As a result numerous large shipments of parasite cocoons and puparia, as well as of caterpillars in various stages and of pupæ were received at the laboratory. The condition of the material on receipt compared more than favorably with the average of similar shipments from Europe, and for the first time opportunity was afforded for the actual first-hand investigation of the parasitic fauna of the gipsy moth in Japan.

Similar shipments were made in 1909 and 1910, with even better results in so far as the condition of the material on receipt was concerned, and several of the more important parasites have now been liberated in the field in America under conditions which are apparently ideal and which ought to encompass their introduction and establishment, if such a thing is possible.

TABLE I.—*Sequence of gipsy-moth parasites in Japan.*

PARASITES.	EGG.		LARVAL STAGES.							PUPAL STAGES.			ADULT.
	FRESH 10 DAYS.	OLD 280 DAYS.	FIRST 7 DAYS.	SECOND 7 DAYS.	THIRD 7 DAYS.	FOURTH 7 DAYS.	FIFTH 7 DAYS.	SIXTH 7 DAYS.	SEVENTH 7 DAYS.	PRE-PUPA 2 DAYS.	FRESH 3 DAYS.	OLD 7 DAYS.	
ANASTATUS BIFASCIATUS	••••												
SCHEDIUS KUVANAE	••••	••••											
APANTELES FULVIPES					FIRST GENERATION ••••••••••								
							SECOND GENERATION ••••••••••••						
*LIMNERIUM DISPARIS					••••••••••••								
*METEORUS JAPONICUS					••••••••••••••••								
CROSSOCOSMIA SERICARIAE							••••••••••••••						
TACHINA JAPONICA							••••••••••••						
*THERONIA JAPONICA										••••••••••			
*PIMPLA PLUTO										••••••••••			
*PIMPLA DISPARIS										••••••••••			
*PIMPLA PORTHETRIAE										••••••••			
CHALCIS OBSCURATA										••••			

* SPECIES NOT CONSIDERED TO BE OF MUCH IMPORTANCE ECONOMICALLY.

A total of 14 species of parasites has been reared from the imported material, of which 7 were present in sufficient abundance to indicate that they were of real importance in effecting the control of the moth. Two species are of such doubtful host relationship as to have been omitted from Table I.

Specimens of one species, *Meteorus japonicus*, the importance of which is not indicated by the examination of the imported material, have been sent to us by Mr. Kuwana with the statement that it is sometimes, locally at least, a common parasite, but none for colonization has been received. Still another is of possible importance, judging from the very limited opportunity which we have had for its investigation, but none of the others is of proved worth. Since nothing is actually known of the conditions under which particular lots of parasite material were collected, it can not be stated as confidently as the circumstances render desirable that some among the others are not incidentally of value in keeping reduced the numbers of the moth in localities where it is too rare to permit of collection of material for shipment to America.

Prof. Kincaid's reports upon the effectiveness of the parasites, even when taken with more than the prescribed grain of conservatism, have been so consistently optimistic as to leave no room to doubt that the parasitism to which the moth is subjected in Japan, even in localities where it is more than normally prevalent, is sufficient to meet and overcome the rate of increase of the gipsy moth in America.

How these parasites work together in bringing about the control of the moth in Japan is indicated in Table I, which, with its explanation, was published in a somewhat abbreviated form in the popular bulletin by the junior author which was issued through the office of the State forester of Massachusetts a year ago.

The addition of the names of the species marked with an asterisk makes the list complete, so far as it may be completed through the information now available. The species so designated are those which have never been received in sufficient abundance to make their colonization possible, and among them are some which are doubtless of wholly insignificant importance from an economic standpoint, while others may, upon investigation, prove to be of more than sufficient importance to justify an attempt to secure their introduction into America.

Opposite the name of each parasite, extending across a certain number of the vertical columns, is a dotted line. The vertical columns indicate different stages in the development and transformations of the gipsy moth, as the egg, the caterpillar, and the pupa, and these are still further divided into caterpillars of different sizes and eggs and pupæ of different ages and conditions. At the head of each column is stated the approximate number of days during which the individual gipsy moth remains in that particular stage.

The dotted line following the name of the parasite indicates those stages in the life of the gipsy moth during which the latter is likely to be attacked by the parasite in question, and it will be seen that in a number of instances, as, for example, Chalcis and Theronia, this period is exceedingly short. The solid line indicates the stages in the life of the gipsy moth during which it is likely to contain the parasite in its body. This, it may also be noted, varies considerably. Crossocosmia, for example, gains lodgment in the active caterpillar while it is only about half grown, and the extension of the solid line across all of the columns which stand for the later caterpillar stages, as well as for all of the pupal stages, indicates that the larvæ of this parasite do not leave the host caterpillar until after it has transformed to a pupa, and until the moth would naturally have emerged had the pupa remained healthy and unparasitized.

It will be noted that the parasites not designated by the asterisks, and which are therefore to be considered as of some importance in effecting the control of the moth, form, when taken together, a perfect sequence, and that every stage of the moth from the newly deposited egg to the pupa is subjected to attack. It is furthermore of interest to note in this connection that, so far as may be determined from the scanty information available, all of these parasites are present in more or less efficient abundance within a limited area in the vicinity of Tokyo, from which a part, and presumably the greater part, of the material was collected for exportation.

PARASITISM OF THE GIPSY MOTH IN RUSSIA.

The earliest first-hand knowledge of the gipsy moth and its parasites in Russia was secured as the result of the visit paid to that country by the senior author in the spring of 1907. Through his instrumentality several of the Russian entomologists were interested in the parasite-introduction work to such a practical extent as to collect or cause to be collected and forwarded to America several small and a few large shipments of the eggs, caterpillars, and pupæ of both the gipsy moth and the brown-tail moth. The difficulties attending the importation of material from Russia proved to be considerably more real and less easily surmountable than those which were so successfully overcome in the instance of the Japanese shipments, and for the most part the Russian material was of more interest from a technical than from an economic standpoint by the time it arrived at the laboratory.

From a technical standpoint it was exceedingly interesting and valuable, since there were found to be present in the boxes of young gipsy-moth caterpillars the cocoons of several species of hymenopterous parasites which had either not been received from other sources or which were not known to be sufficiently abundant in any other part of Europe to make possible their collection in large quantities. Prof. Kincaid's successful prosecution of the Japanese work encouraged his selection as the best and most experienced agent available for the decidedly more difficult proposition of visiting Russia and attempting to secure an adequate supply of the several species of parasites which could only be secured in that country to advantage, so far as could be determined from the information then at hand.

The manner in which he was impressed by the gipsy-moth situation which he encountered there is best described in the following extracts from his letters, in the course of which occasional comparisons are made between Russian and Japanese conditions.

BENDERY, BESSARABIA, RUSSIA, *June 11, 1909.*

The season here is in full swing, but the situation causes me considerable anxiety, as the whole business is so utterly different from my experience in Japan. The damage wrought by *dispar* in the forests and orchards of Bessarabia this season is enormous and parasite control seems to be most inefficient in checking the depredations of the caterpillars. When I think of the masterly and well-ordered attack of the Japanese parasites and the splendid fashion in which they wiped out the caterpillars in large areas before depredation took place I am surprised by what I see here. When I left Gauchesty on May 31 the forests of that district were thoroughly riddled over thousands of acres, and yet I had seen no sign of insects, fungous, or bacterial attack except a couple of clusters of Apanteles cocoons. * * * *Dispar* seemed to be having its own way as fully as in America, so far as could be seen on the surface. * * * Here at Bendery the season is slightly more advanced, owing to the lower altitude, and the prospects of securing parasites of *dispar* seem better. Even here, however, the situation seems to me quite remarkable. I have available three extensive and very different collecting grounds. The great forest of Gerbofsky about 6 versts from Bendery is composed almost exclusively of oaks. An immense area is covered by trees of this species, forming magnificent groves of fine trees 80 to 100 years old. Thousands of these great trees are completely defoliated, so that no sign of foliage remains. In the same forest are groves of young trees of the same kind, also greatly damaged. Examining the myriads of caterpillars in the field, I have found no sign of parasite attack or of fungous disease except the work of Apanteles. * * * The percentage of attack by Apanteles is so small that rearing in trays is of little practical importance. One would have to have billions of caterpillars to do any good. Collecting in the field is very difficult, as the caterpillars creep into crevices or suspend themselves to branches at some height from the ground, where they are hard to reach. No sign of bacterial disease has appeared in this area, nor have I seen any evidence of tachinid attack in the field or in my rearing trays. Another great forest of about 500 acres is at Kitzkany, about 7 versts from Bendery on the banks of the Dniester, a low damp situation. This forest has no oaks but is much mixed. The principal tree is *Populus nigra*, but there are many other trees, as Ulmus, Acer, Salix, etc. Here again the damage is tremendous, with almost no sign of parasite attack. Prolonged search yielded a few cocoons of Apanteles. On the other hand, thousands of the older caterpillars were found in the pendulous condition so characteristic of bacterial attack. The third condition I found in the numerous orchards adjacent to both of these forested areas. These orchards have been almost overwhelmed by *dispar*. The more progressive peasants have protected their trees by rings of axle grease or by strips of cotton wool, but others have done nothing and the trees are quite stripped. * * * The sight of a tree covered with hundreds of dead caterpillars bearing clusters of Apanteles cocoons such as I saw in Japan seems not to be hoped for.

KIEF, *June 26, 1909.*

From what I can see in the field and from what I can gather from Prof. Pospielow, *dispar* was almost exterminated in this district last year through the activity of the parasites. Only a few isolated colonies seem to have survived, the most important of these being at Mishighari, a small place on the river about two hours by steamer from Kief. In this place, which is perhaps 100 acres in extent, the trees are plastered with cocoons of *Apanteles fulvipes*. The attack of the parasite was so thorough that the first generation seems to have been sufficient to wipe out the caterpillars, as I can find no large caterpillars about the place, and a few days will doubtless witness the complete wiping out of *dispar*. * * * Tachinids also appear to be very active, as I find many eggs, but as these are laid upon caterpillars suffering from the attacks of Apanteles it would seem as if the emaciated caterpillars could not supply sufficient nourishment to bring the tachinids to maturity. * * * From the standpoint of parasite control the situation at Kief is most inspiring, but as a field in which to gather a quantity of material it is evidently not very hopeful. The whole situation is in violent contrast to what I found at Bendery and Gauchesty, where *dispar* is vastly more abundant this year than last, with little sign of the multiplication of parasites.

BENDERY, RUSSIA, *July 10, 1909*.

In the forest of Kitzkany where *dispar* caterpillars prevailed to an incredible extent three weeks ago, not a single caterpillar or pupa is to be found. An epidemic of a bacterial nature swept them away in millions. In the forest of Gerbofsky, among the great oak trees, the number of caterpillars that have formed pupæ is surprisingly small. Vast numbers of caterpillars swarmed over the trees, completely stripping them of leaves. Deserting the trees, the caterpillars swarmed over the ground in search of other food and vast numbers died of starvation and disease. These trees are now putting forth new leaves which promise to sustain the life of the forest.

After the close of the "caterpillar season" in 1910 the junior author took a vacation trip to Europe and, thanks to an extension of leave for the purpose and still more to the kindness of Mr. N. Kourdumoff, entomologist of the experiment station in Poltava, was enabled to spend about 10 days in the field in Kief and Kharkof Provinces. In Kief the forest at Mishighari, which is mentioned by Prof. Kincaid as the one locality where he found the parasites in control, was visited, as well as several other localities in that province. This portion of Kief Province, topographically, meteorologically, and otherwise, is radically different from Massachusetts, and much more like portions of Minnesota than any other part of the United States with which the visitor is at all familiar. The forests, which are limited in extent as compared with those of Massachusetts, are less diversified. For the most part they are of pine, mingled with a small quantity of oak, wild pear, birch, and occasionally other trees. Everywhere the gipsy moth was rare or at least uncommon, and everywhere the cocoon masses of *Apanteles fulvipes* were at least as abundant as the egg masses of their host.

At Mishighari the conditions remained much as described by Prof. Kincaid, except that the cocoon masses of Apanteles were even more abundant than his letters would indicate. Upon some trees they were litterally matted together by the thousands in such semiprotected situations as are selected by the caterpillars at the time of molting. The forest in this particular locality was varied to an extent not noticed elsewhere. In addition to the generally distributed oak, birch, and poplar were quantities of beech, alder, Carpinus, maple, elm, and other species, while the shrubs were equally varied and abundant. The forest was situated upon the steep bluffs overhanging the Dnieper, running down on one side to its banks, where great willows bore evidence of the high water which sometimes covered their trunks, and on the other extending back some little distance until it was met by a wide stretch of treeless prairie. Here, at least, parasite control of the gipsy moth appeared to be pretty thoroughly effective, since there were in evidence a vastly larger number of old cocoon masses of the Apanteles than there were of old egg masses of its host, and the new egg masses were enormously outnumbered by the old. According to Prof. Pospielow, who was a member of the party, the forest in this locality was almost completely defoliated in 1908, but there was no indication of damage to any of the trees composing it, and every indication that the parasites alone were responsible for the disappearance of the moth.

In several other localities along the banks of the Dnieper the conditions encountered were essentially the same, differing principally in the lesser abundance of both egg masses and parasite cocoons. In one locality quite near to Kief, fresh egg masses were more common than at Mishighari, and cocoon masses of the season of 1910 were also more common. It seemed to offer opportunities for the collection of a sufficiently large quantity of this parasite to make an experiment in importation and colonization possible and practicable in 1911, which it is hoped may be carried out.

In Kharkof conditions, both as regards the gipsy moth, its parasites, and the country at large, were essentially different from those in Kief. Numerous localities from 5 to 20 miles out of the city in different directions were visited, and everywhere indications of the recent presence of the gipsy moth were found in abundance. Old egg masses were massed around the base of the trees in a manner exceedingly suggestive of uncared for woodland in Massachusetts, and mingled with them were a very few fresh masses; so few, relatively, as to indicate most conclusively that the moth had encountered very adverse conditions during the season of 1910, with the result that its abundance had been most materially reduced.

In every locality the conditions were the same, although the character of the forest varied to a material extent. For the most part the province of Kharkof is devoid of forest, and quite suggestive of parts of North Dakota in appearance. Such forest as does occur is mostly confined to the valleys in the neighborhood of streams, and though it may be fairly extensive, it is rarely very diverse. No pine was seen. Oak predominates very largely and, with the exception of some birch, forms practically pure forests away from the lowlands, except in the best watered localities.

Everywhere, irrespective of the character of the forest, the gipsy moth was found under the circumstances recounted above. Everywhere there had been an abundance of eggs in the spring, everywhere there had been an abundance of caterpillars, a considerable proportion of which had gone through to pupation, and everywhere the number of fresh egg masses was very much smaller than was that of the old. Nowhere was there evidence of parasitism by *Apanteles fulvipes* to anything like the extent which prevailed in the vicinity of Kief. Cocoon masses were occasionally found, nearly always old, sometimes very old and so discolored as to be with difficulty distinguished from the bark to which they were attached. In Kief the number of cocoon masses was everywhere in considerable excess over the number of egg masses. Here the number of egg masses was enormously in excess of the number of cocoon masses.

Examination of the pupal shells for evidences of parasitism was unavailing. It could be said with assurance that pupal parasites were certainly not common and that the death of the pupæ (for proportionately very few of them hatched) was not due to any of the pupal parasites which were known from western European localities. The earth beneath the cocoon masses was examined for evidences of tachinid puparia. For a time none was found, but search was finally rewarded by the discovery of *Blepharipa scutellata* in most extraordinary abundance in a single one among the numerous localities visited. This particular forest, which was very near to the village of Rhijhof and about 8 miles from Kharkof, was unique among the others visited in the variety of its trees. The soil was rich, the trees were larger, and the undergrowth was more abundant and varied, but at the same time there was less diversity than was encountered in the forest at Mishighari. Unfortunately, the presence of the puparia could not be considered as of much significance, because they were practically all hatched and obviously dated back more than one year. The parasite had surely not been responsible for the reduction in numbers of the gipsy moth which had taken place in the season of 1910, and neither had it prevented the moth from increasing to such numbers as to bring about partial defoliation of the forest in 1910 before disaster in one form or another had overtaken it.

Of other tachinids there were practically none, and it is certain that they would have been found had they been present. *Compsilura concinnata* is even now so abundant as a parasite of the gipsy moth in Massachusetts as to bring about an appreciable percentage of destruction in 1910, and its puparia are recovered from the field with ease. Had it been one-tenth as common in Russia it could not have failed of detection. The same is true of Tachina, which, although it effects a parasitism of less than 1 per cent in Massachusetts, is not difficult of detection, and it is safe to say that not much if any more than this amount of parasitism prevailed in Kharkof. All told, not enough parasites were found to indicate that they had played any important part in the reduction of the moth from a serious menace to the wellbeing of the forest to such small numbers as to require several years at least before it would be possible for such conditions to recur.

The investigations having been conducted in September, some time after the death of all of the caterpillars and pupæ, it was no longer possible to determine with assurance the cause for the peculiar conditions, but everything conspired to indicate that nothing less than an epidemic of disease had been responsible. The condition of the pupal shells which hung upon the trees in countless thousands was in every respect identical with the condition of the pupal shells which are to be found in Massachusetts in every locality where the disease has prevailed to a destructive extent the season before. Among the old egg masses which plastered the extreme base of nearly every tree in most of the localities visited were found a variable, and sometimes a very large, proportion which had hatched only in part or not at all. The appearance of these unhatched masses was identical in every respect with the appearance of similarly large numbers which are frequently found in Massachusetts. The reason for the nonhatching of the eggs is not yet plain, but it is the consensus of opinion that this is probably associated with the "wilt" disease. It is known that affected caterpillars may pupate before death, and it seems not illogical to suppose that slightly affected caterpillars may pupate and produce moths which are able to deposit their eggs, but that these eggs fail to hatch as the direct result of the taint in the blood of their parent.

These Russian experiences seem, on the whole, to indicate that in that country the gipsy moth is not controlled by its parasites to an extent which serves to remove it from the ranks of a destructive pest. But as one day after another in the field at Kharkof served more and more indelibly to deepen this conviction, it served equally, first to create, and finally in retrospect to confirm, the observer in another, which was, in effect, that if this was the best that could be expected of disease as a factor in the control of the gipsy moth in its native home then something better than disease must be found to control it in America. Just so long as conditions similar to those seen in Kharkof or pictured in the letters of Prof. Kincaid are allowed to prevail in Massachusetts just so long will the incentive remain to see the parasite-introduction experiment carried on until success is either achieved or proved impossible. Conditions similar to those prevailing in Russia emphatically do not prevail in western Europe, nor, according to all accounts, in Japan. Natural conditions in western Europe and in Japan are in many respects more like those of our own Eastern States than are those of Kharkof Province. Conditions in Kief Province, even, are much more like those of Massachusetts than are those of Kharkof, and in Kief parasite control seemed to be an accomplished fact, although of course there is no assurance that it is continuous and perfect.

The final outcome of the Russian experience was, therefore, the opposite of what might have been expected, and it resulted in a firmer determination than ever to carry the work through to its end.

PARASITISM OF THE GIPSY MOTH IN SOUTHERN FRANCE.

Following the 10 days in Russia a shorter period was spent by the junior author in somewhat similar field work in southern France, where, with the aid of M. Dillon, he was enabled to visit the localities from which the largest, and in that respect the most satisfactory, shipments of parasite material ever received at the laboratory were collected. As the direct result of the senior author's visit to Europe in 1909 some thousands of boxes containing hundreds of thousands of gipsy-moth caterpillars had been collected in the vicinity of Hyères, about 50 miles to the eastward of Marseilles. These caterpillars were largely living upon receipt, and in the winter of 1909–10 Mr. W. B. Thompson dissected several hundred preserved specimens and the actual percentage of parasitism was thus determined. Some few pupæ which had also been received from the same locality made possible a fair understanding of the extent to which the pupæ were parasitized.

The results of these investigations, taken in connection with the actual rearing work, were disappointing. It was evident that the moth was fairly common in the region from which the material was collected—as common, perhaps, as it would need to be in Massachusetts to provide for an increase of sixfold annually. Nevertheless, the amount of parasitism which was indicated by this, the most thorough study of parasitism of the gipsy moth abroad which was ever undertaken in the laboratory, was less than enough to offset a twofold, much less a sixfold, increase.

For this reason much curiosity was felt as to the conditions which prevailed in a country where parasitism of such comparatively insignificant proportions was sufficient.

No sooner was the character of the country districts in this portion of France seen than the wonder which had been felt at the small percentage of parasitism which was sufficient to hold it in check was replaced by a much greater astonishment that the gipsy moth should exist under such conditions at all. It was a country of olive orchards and vineyards, with a strip along the littoral which was so nearly frostless as to permit the culture of citrus fruits, and even of date palms. The hills were semiarid, with the soil exceedingly scanty and often covered by loose stones. The principal forests consisted largely of cork oak and pine, except in the low and well-watered valleys and bottom lands where other trees in considerable variety occurred.

The slopes of the higher mountains were fairly well forested and a larger variety of trees and shrubs thrived than on the lower elevations. Over much of the country the soil was either too dry or too scanty or both to permit of cultivation, even in a land so densely populated by so thrifty a race, and here were found occasional thickets of scrub oak, apparently of a deciduous species, which sometimes reached the dignity of a small tree. For the most part such country was covered with a scanty growth which would be called chapparal in some of our States, composed of a variety of uninviting looking shrubs, judging them from the probable viewpoint of a gipsy-moth caterpillar in search of food. Taken altogether, the country may more aptly be compared with southern California than with any other part of the United States.

It seemed to the visitor that if the gipsy moth were to be found in any portion of this region it would most likely be within the rich and well-watered bottom lands, where occasional hedges, or rows and groups of large trees in considerable variety, seemed to offer fairly acceptable conditions for its existence. But to his surprise and amazement he was assured by M. Dillon that it was from the chapparal covered, arid, and uncultivated elevations that most of the enormous quantities of caterpillars had been collected. In support of this assertion, after the visitor had searched in vain in what would be the most likely situations in Massachusetts for the concealment of egg masses, pupal shells, or molted skins, M. Dillon proceeded to turn over a few loose stones among those which fairly covered the ground, and thereby disclosed sufficient indication of the presence of the moth in fair abundance to convince the most skeptical. In this particular locality in the vicinity of the little provençal town of Meoun, in a thicket of deciduous oak surrounding and concealing the ruins of an ancient chapel, there were sufficient egg masses of the moth to represent a fair degree of infestation, but eggs, pupal shells, and molted larval skins were all so completely hidden as to evade completely the eyes of one who had been trained to look for first evidences in sheltered places on the bark or in the knot holes and hollow trunks of trees.

As a matter of fact, as was abundantly evidenced by that day's experiences, as well as of the several days which followed, the gipsy moth departed most materially from its characteristic habits in the cooler, better watered and forested localities in which it is present as a pest in America. Instead of being a typically arboreal insect, it is rather terrestrial, and thereby becomes subjected to a variety of natural enemies to which it is practically immune so long as it remains arboreal. In the course of the several years past a variety of species of the larger European Carabidæ has been studied at the laboratory for the purpose of determining their availability and probable worth as enemies of the gipsy moth. Of them all, not one refused to attack and devour the caterpillars and pupæ of the gipsy moth with business-like dispatch, once given an opportunity, but with one or two exceptions none has shown a disposition to climb trees in search of its prey. Being essentially terrestrial in habit, they were essentially unfitted to prey upon an essentially arboreal insect.

We know little of the predatory beetles which are to be found in that part of France which was visited upon this occasion, nor does this lack of knowledge vitiate the strength of the argument to any great extent. The fact was that if present (and undoubtedly some species are to be found) any of the numerous forms which have been studied at the laboratory and discarded as unfit for the purposes desired in Massachusetts would immediately assume high rank as enemies of the gipsy moth. In other words, the conditions under which the gipsy moth exists in southern France are wholly incomparable with those under which it exists in New England, and the agencies which are effective in accomplishing its control are likewise incomparable. The unimportant rôle obviously played by the parasites immediately loses its significance. Those species of true parasites which assist in this control are practically the same as those which assist in other localities, but the demand upon them and their opportunities for multiplication are insignificant compared to those existing in Massachusetts, if they are ever established there. True to their character as agencies in facultative control, they do not increase in efficiency to an extent which would practically mean the extinction of their host.

The results of the rearing and dissection work carried on at the laboratory indicated that a parasitism varying from 25 per cent to something in excess of 40 per cent prevailed in this locality. After seeing the conditions under which the gipsy moth struggled for existence, real wonder was felt that it should be able to survive, and the trip resulted in a firmer conviction than ever in the efficacy of parasitism, and the validity of the theory upon which the parasite-introduction work was conceived.

SEQUENCE OF PARASITES OF THE GIPSY MOTH IN EUROPE.

The parasitic fauna of the gipsy moth varies considerably in various faunal divisions of Europe, and no attempt has been made to prepare separate lists of the parasites peculiar to those regions which have been represented in the material imported. In Table II, which is constructed in accordance with that representing the sequence of parasites in Japan, as explained on page 122, all of the various species reared from the European material are listed. As in the table of Japanese parasites those species which are of no consequence in the control of the moth (so far as known) are marked with an asterisk.

THE BROWN-TAIL MOTH AND ITS PARASITES IN EUROPE.

Reference has already been made to the fact that in those sections of Massachusetts in which both the gipsy moth and the brown-tail moth occur, the latter is considered as the lesser pest of the two. This opinion, as held by those who are thoroughly familiar with the comparative noxiousness of the two, speaks quite plainly of the character of the gipsy moth as a pest, in view of the very considerable agitation which has come about on account of the brown-tail moth in

TABLE II.—*Sequence of gipsy-moth parasites in Europe.*

PARASITES.	EGG.		LARVAL STAGES.							PUPAL STAGES.			ADULT.
	FRESH 10 DAYS.	OLD 280 DAYS.	FIRST 7 DAYS.	SEC-OND 7 DAYS.	THIRD 7 DAYS.	FOURTH 7 DAYS.	FIFTH 7 DAYS.	SIXTH 7 DAYS.	SEV-ENTH 7 DAYS.	PRE-PUPA 2 DAYS.	FRESH 3 DAYS.	OLD 7 DAYS.	
ANASTATUS BIFASCIATUS	•••••												
*APANTELES SOLITARIUS				••	•••••	•••••	•••••	•••••					
APANTELES FULVIPES				FIRST GENERATION ••••• •••••		SECOND GENERATION ••••• •••••							
*METEORUS VERSICOLOR					•••••	•••••	•••••	•••••					
*METEORUS PULCHRICORNIS					•••••	•••••	•••••	•••••					
*LIMNERIUM DISPARIS				•••••	•••••	•••••							
*LIMNERIUM (ANILASTUS) TRICOLORIPES				•••••	•••••	•••••							
BLEPHARIPA SCUTELLATA						•••••	•••••	•••••					
*CROSSOCOSMIA FLAVOSCUTELLATA					•••••	•••••	•••••						
COMPSILURA CONCINNATA					•••••	•••••	•••••						
*DEXODES NIGRIPES					•••••	•••••							
ZYGOBOTHRIA GILVA						•••••	•••••						
CARCELIA GNAVA						•••••	•••••						
*PALES PAVIDA						•••••	•••••						
PARASETIGENA SEGREGATA						•••••	•••••						
TRICHOLYGA GRANDIS						•••••	•••••						
TACHINA LARVARUM						•••••	•••••						
*ICHNEUMON DISPARIS											•••••	•••••	
*THERONIA ATALANTAE											•••••	•••••	
*PIMPLA EXAMINATRIX											•••••	•••••	
*PIMPLA INSTIGATRIX											•••••	•••••	
*PIMPLA BRASSICARIAE											•••••	•••••	
CHALCIS FLAVIPES												•••••	
MONODONTOMERUS AEREUS												•••••	
CALOSOMA SYCOPHANTA				•••••	•••••	•••••	•••••			•••••	•••••	•••••	•••••

* SPECIES NOT CONSIDERED TO BE OF MUCH IMPORTANCE ECONOMICALLY.

localities into which it has preceded the gipsy moth or where the latter has not as yet reached a state of destructive abundance.

On account of the lesser interest aroused in the brown-tail moth in Massachusetts, its parasites have not been given quite the consideration, in some respects, that has been given to those of the gipsy moth, but this lack of consideration has had entirely to do with the question of the future policy of the laboratory, and has not extended to the actual handling of the parasites themselves. In every respect other than as a basis for calculations as to future policies of the laboratory, they have received as much and as careful consideration as have the parasites of the more dangerous pests.

So far as known the brown-tail moth does not occur in Japan, and in consequence no determined efforts have been made to secure, from Japanese sources, parasites likely to attack it. It has an ally and congener there in *Euproctis conspersa* Butl., which is attacked by a variety of parasites, some of which may be expected to attack the brown-tail moth if given an opportunity. A few of them have been collected and forwarded to the laboratory through the great kindness of Mr. Kuwana, but unfortunately have arrived in such condition, or at such time of the year, as to make their colonization impossible. It is intended in the near if not in the immediate future to devote some time to the investigation of the Japanese parasites likely to be of service in this respect, and, if any can be found of promise, to attempt their importation into America.

In Europe the brown-tail moth appears to be the more common of the two insects under consideration and, taken all in all, it is probably the more injurious as well. Neither in Europe nor in America does it bring about the wholesale defoliation characteristic of an invasion of the gipsy moth, but its injury is of a more insidious character and more evenly distributed throughout the years. In Russia, in the fall of 1910, the junior author was astounded at the tremendous abundance of its nests in many localities, notably on the irregular hedgerows planted as a windbreak alongside the railroad in the midst of an otherwise open prairie. Occasionally small scrubs of Crataegus, or wild pear, completely isolated by what seemed to be miles of open prairie, would be fairly covered with the nests.

In gardens in the vicinity of Kief pear and apple trees were frequently injured to a considerable extent by its caterpillars, and sometimes to a greater extent by the caterpillars of *Aporia crataegi* L., which are similar in their habit, and were constructing their own hibernating nests side by side with those of the brown-tail moth. In the forests round about it was common, but except occasionally not quite so common as in southeastern New England. On one occasion in excess of 50 nests were noted upon a small hawthorn which stood at the edge of an oak forest. This was just a little worse than anything which has been seen in America.

In southern France the circumstances under which it occurred were as surprising as those under which the gipsy moth was encountered, in respect to their departure from that which past experience led the visitor to consider as the normal. M. Dillon, who had collected and forwarded to the laboratory a considerable quantity of the winter nests, undertook to guide the visitor to the locality where they were collected. The way led through a rich and fertile valley, with many sorts of trees, including apple and pear, as well as hawthorn and oak, every one of which is a favored host plant in other regions, but not a single brown-tail nest was seen and, according to M. Dillon, it was never found upon these trees. Farther on an elevated plain was passed, with occasional ridges of uncultivated land upon which were growths of a deciduous scrub oak. Gipsy-moth eggs, pupal shells, etc., could usually be found by a little search under the stones on these ridges, but the brown-tail moth was conspicuous by its total absence.

The next day the route selected passed through an extensive forest of cork oak, mingled with pine, and finally up the sides of the mountain, until great plantations of aged chestnut trees were indicative of a change in climatic conditions brought about by the considerable altitude. Various shrubs and a few trees unknown or rare in the lower elevations became a feature of the forest, and among them the arbusier (*Arbutus* sp.), closely resembling in its growth, in the appearance of its evergreen foliage, and in its habitat the mountain laurel of our own southern mountains. It is very beautiful and unusual in its appearance, partly on account of its flowers (which are suggestive of Oxydendron) but more particularly because of its fruit. This was globular, about the size of a marble, and hung pendant on long stems in more or less profusion, and in all stages of ripening. In the course of this process it passed from green through a sequence of vivid yellows to orange, and finally intense scarlet. It was at once recognized as the host plant of the hundreds of nests which had been collected and shipped by M. Dillon. Although it was occasionally met with sufficiently far down the mountain side to mingle with orchards and hawthorn hedges, according to M. Dillon the brown-tail moth invariably seeks it out, even there. The selection of a food plant representing a totally different order from any selected in other parts of Europe or in America, and this in spite of the fact that what are ordinarily its most favored hosts were frequently much the more abundant, was considered to be quite as remarkable as the assumption of terrestrial habits by the gipsy moth.[1]

In central and western Europe generally the brown-tail moth finds a stronghold in the dense Cratægus hedges which are commonly planted in many localities, and upon them as well as upon oak and fruit trees it is frequently abundant. In these regions, also, not only the food plants, but the seasonal and feeding habits are quite like those in New England. Occasionally an apple tree or an oak will be found carrying an abundance of nests and, as noted by the senior author in northwestern France in 1909, the moths are sometimes so

---

[1] It has since been learned that in the warmer parts of the region visited, the brown-tail moth caterpillars not only remain active but feed to some extent during the winter. In the middle of January, 1911, the nests were found commonly, always upon Arbutus, in parts of the coast regions near Hyères, and in nearly every instance the caterpillars were active and in most they were feeding. In this particular locality the nests were very different from those typical of the caterpillars in northern localities, being loosely woven, and not at all designed for hibernation in its stricter sense.

numerous as to lay their eggs in quantities on growing nursery seedlings and low-growing plants.

Among the very many lots of caterpillars and cocoons which have been received at the laboratory there is occasionally one in which a fungous disease is present. Usually, when it is present at all, the majority of the caterpillars received from that particular locality will be found dead and "shooting" the ascidiospores upon receipt. According to Dr. Roland Thaxter, to whom specimens have several times been submitted, it is specifically identical with the fungus which is so effective in America as to have largely assisted in reducing the moth from the preeminent place which it would otherwise have occupied as a pest. Its presence under these conditions, as it was, for example, in 1909, in practically every box out of a large number which were forwarded to the laboratory from lower Austria, is strongly indicative of the importance of this disease.

Looked at from one standpoint, the brown-tail moth situation in America is less satisfactory than is the gipsy-moth situation. In numerous localities throughout western Europe as well as in eastern Europe it frequently increases to such an extent as to become a pest. It hardly seems as though more could be expected of the European parasites in America than is accomplished by them in Europe, but if even this much can be secured it will aid materially in reducing the frequency of the outbreaks. At the same time, it must be admitted that from nearly every point of view the prospects of unqualified success with the gipsy-moth parasites are better than with the parasites of the brown-tail moth.

SEQUENCE OF PARASITES OF THE BROWN-TAIL MOTH IN EUROPE.

The accompanying table (Table III), in which are listed all of the parasites of the brown-tail moth which have been definitely associated with that host in the course of the studies of imported European material, is constructed in the same manner as the tables of parasites of the gipsy moth in Japan and in Europe (see pp. 121, 132). It will be noted that the number and variety are slightly larger than of European gipsy-moth parasites, and that the species which are or which appear to be promising as subjects for attempted importation are also slightly more numerous. Very rarely, however, does any one among them become as relatively important as any one of several among the gipsy-moth parasites which might be mentioned. Neither has any lot of brown-tail material produced so many parasites of all species (as high a percentage of parasitism) as have several lots of gipsy-moth material.

PARASITISM OF THE GIPSY MOTH IN AMERICA.

Although the gipsy moth is attacked by a considerable variety of American parasites the aggregate effectiveness of all the species together is wholly insignificant, so far as has been determined by the rearing work which has been conducted on an extensive scale at the laboratory. Actual effectiveness may be greater than indicated, however, because it is possible that the caterpillars or pupæ may be attacked by parasites, the larvæ of which are unable to com-

TABLE III.—*Sequence of brown-tail moth parasites in Europe.*

PARASITES.	EGG.	FALL STAGES. FIRST	SECOND	THIRD	WINTER STAGE	SPRING STAGES. FIRST	SECOND	THIRD	FOURTH	PUPAL STAGES. PRE-PUPA	FRESH	OLD	ADULT.
*TRICHOGRAMMA SP.	•••••												
*TRICHOGRAMMA PRETIOSA-LIKE	•••••												
*TELENOMOUS PHALAENARUM	•••••												
APANTELES LACTEICOLOR			••••• FALL GENERATION	•••••			••••• SPRING GENERATION	•••••	•••••				
METEORUS VERSICOLOR													
ZYGOBOTHRIA NIDICOLA		•••••											
*PTEROMALUS EGREGIUS			•••••										
*LIMNERIUM DISPARIS						•••••	•••••	•••••					
PAREXORISTA CHELONIAE						•••••	•••••	•••••					
DEXODES NIGRIPES						•••••	•••••	•••••					
COMPSILURA CONCINNATA[1]			..1..			•••••	•••••	•••••					
*BLEPHARIDEA VULGARIS						•••••	•••••	•••••					
*CYCLOTOPHRYS ANSER						•••••	•••••	•••••					
*MASICERA SYLVATICA						•••••	•••••						
EUDOROMYIA MAGNICORNIS						•••••	•••••						
ZENILLIA LIBATRIX						•••••	•••••						
PALES PAVIDA						•••••	•••••						
TACHINA LARVARUM						•••••	•••••	•••••					
*TRICHOLYGA GRANDIS						•••••	•••••	•••••					
*PIMPLA BRASSICARIAE										•••••	•••••		
*PIMPLA INSTIGATRIX										•••••	•••••		
*PIMPLA EXAMINATRIX										•••••	•••••		
*THERONIA ATALANTAE										•••••	•••••		
MONODONTOMERUS AEREUS[2]						2					•••••		
*DIGLOCHIS OMNIVORA											•••••		
*PTEROMALUS SP.											•••••		

1, ATTACKS YOUNG CATERPILLARS BEFORE HIBERNATION, BUT LARVAE APPARENTLY FAIL TO MATURE.
2, ADULT FEMALES HIBERNATE IN WINTER NESTS.
*, SPECIES NOT CONSIDERED TO BE OF MUCH IMPORTANCE ECONOMICALLY.

plete their transformations under the conditions in which they find themselves. This is known to be true in the instance of what would otherwise be a very important parasite, *Tachina mella.*

In such instances the host usually remains unaffected and the parasite perishes. At other times, as proved through a series of experiments carried on by Mr. P. H. Timberlake, of the Gipsy Moth Parasite Laboratory, in the spring of 1910, the host may perish without exhibiting any external symptoms of its condition. No serious attempt to determine whether this actually happens in the field has been made, but undoubtedly it does occasionally result when the parasite larva finds itself under unnatural surroundings. It is thus well within the bounds of possibility that effective parasitism should pass unnoticed in the course of investigations in which reliance is placed entirely upon the results of rearing work.

As will be shown in another place, death of the host through superparasitism by a species fitted to attack it may similarly occur without the true cause becoming apparent.

A sufficiently large quantity of the native caterpillars of the gipsy moth has been dissected at the laboratory to indicate that such concealed parasitism, if it is ever a factor in the control of this insect, is of rare occurrence, or else of insignificant proportions. This can not be said of the pupæ of the moth in America, which have not been studied sufficiently well as yet.

The following native parasites have been reared from the gipsy moth in Massachusetts:

Theronia fulvescens Cress.

This, the most common American parasite completing its transformations upon the gipsy moth, was mentioned by Forbush and Fernald in their comprehensive report upon "The Gypsy Moth" under the name of *Theronia melanocephala* Brullé. The true *T. melanocephala* appears not to have been reared from this host. The importance of *T. fulvescens* as a gipsy-moth parasite is indicated by the summarized results of the rearing work conducted in 1910.

In his account of the parasites of the forest tent caterpillar (*Malacosoma disstria* Hübn.) in New Hampshire by the junior author it was credited as being a secondary parasite of *Pimpla conquisitor* Say, and was not recognized as a primary parasite. Investigations at the laboratory have served to throw considerable light upon its life and habits, and it is now known to be a true primary parasite, but one which, like *Pimpla conquisitor* itself, is able to complete its transformations under a variety of circumstances. The supposed secondary parasitism, in this instance, is to be classified rather as "superparasitism" and is believed to result through the circumstance that the primary host chances to contain the larva of Pimpla, rather than through the deliberate searching out by the parent Theronia of pupæ thus parasitized. In its relations to the gipsy moth, which is not successfully attacked by Pimpla at all frequently, Theronia has always been a primary parasite so far as known.

Pimpla pedalis Cress.

One or two specimens have been reared from the pupæ of the gipsy moth collected in the field, but it is of extremely rare occurrence as a parasite of this host, so far as recent rearing work indicates. It was mentioned as one of the more common parasites by Forbush and Fernald, but it is possible that the next following species is intended.

PIMPLA CONQUISITOR SAY.

Judging from observations made from time to time in the field the pupæ of the gipsy moth are frequently attacked by this species, but, unfortunately, the young larvæ of the Pimpla appear not to thrive upon this host and rarely complete their transformation. It is safe to say that more female Pimplas will be found attacking the gipsy-moth pupæ in the course of a day's observations in the field at the proper season of the year than would be reared if that day were to be spent in collecting pupæ instead. It is believed that the affected host usually dies, but the subject has not received the attention which it deserves. If it is true, *Pimpla conquisitor* may prove to be of some assistance in the control of the moth.

PIMPLA TENUICORNIS CRESS.

Recorded as a parasite by Forbush and Fernald, but never reared at the laboratory. Possibly *P. conquisitor* was actually the species reared.

DIGLOCHIS OMNIVORA WALK.

Mentioned by Forbush and Fernald as of some consequence as having been reared from this host, but during late years it has been so rare that only a single pupa has been found in which it has completed its transformations.

ANISOCYRTA SP.

Mentioned by Forbush and Fernald, but the record has not been confirmed by later rearing work.

LIMNERIUM SP.

A single cocoon, which was directly associated with the remains of the host caterpillar, was collected by Mr. R. L. Webster in 1906 during his association with the laboratory. It was very likely that of *L. fugitiva* Say, but the fact will never be known, because a specimen of *Hemiteles utilis* Norton, a hyperparasite, actually issued.

APANTELES SP.

In 1910 a colony of the caterpillars of the white-marked tussock moth was established upon some shrubbery in a locality where the gipsy moth was fairly common. The young caterpillars were sparingly attacked by a species of Apanteles, possibly *A. delicatus* How., although the fact was not determined. At the same time and place a young gipsy-moth caterpillar was found from which an Apanteles larva had issued, and spun a cocoon identical in appearance with that of the species from the tussock moth. This is the only known instance of the parasitism of the gipsy moth by an American Apanteles, and it is believed that it resulted through the fact that the parasites were first attracted, and subsequently excited into oviposition, by the tussock caterpillars.

A considerable number of a minute black and yellow elachertine secondary parasite was reared from this cocoon, so that the specific identity of the Apanteles originally constructing it will forever remain in doubt.

SYNTOMOSPHYRUM ESURUS RILEY.

In July, 1906, Mr. R. L. Webster, who was at that time associated with the parasite laboratory, found a pupa of the brown-tail moth from which he reared a number of Syntomosphyrum, probably *S. esurus* Riley. On the same date, July 18, the pupa of a gipsy moth was found to contain the early stages of a chalcidid parasite, presumed to be the same as that reared in connection with the brown-tail moth.

At about the same time several chalcidids, apparently of Syntomosphyrum, were found ovipositing in pupæ of the gipsy moth, but in no instance was the oviposition successful, so far as the notes indicate.

TACHINA MELLA WALK.

In their report on the gipsy moth Forbush and Fernald speak of having collected no less than 300 caterpillars of the gipsy moth bearing Tachina eggs which were reared through in the laboratory. The most of these produced moths and the remainder died. No parasites were reared.

In 1907 and subsequently large numbers of caterpillars have been found in the open, bearing tachinid eggs, and many hundreds have been kept under observation in confinement with results substantially the same as those above mentioned. In one or two instances, however, the tachinids have completed their transformations and in each instance the species was *Tachina mella*. It is believed, therefore, that this is the species which deposits its eggs so freely and injudiciously.

The fact that effective parasitism failed to result was attributed by Forbush and Fernald to the molting off of the eggs before they had hatched, and this doubtless does occasionally happen. Mr. C. H. T. Townsend reinvestigated the subject and came to the conclusion that the explanation was to be found in the inability of the newly hatched larvæ to penetrate the tough integument of the caterpillars, since he actually observed such failure in one instance, and found many caterpillars upon which the eggs had actually hatched, but from which no parasites were reared.

That this explanation may serve in part to elucidate the mystery is also true, but still later observations have shown conclusively that the parasite larvæ may gain entrance into their host and yet fail to mature. Two explanations have grown where one was deemed sufficient, as the result of certain technical studies which have been made at the laboratory during the past year. Mr. P. H. Timberlake and, later, Mr. W. R. Thompson have thoroughly demonstrated the fact that a parasite larva gaining lodgment in an unsuitable host may die, and its body may be in great part absorbed through action of the phagocytes without causing the host obvious inconvenience. This very likely takes place with *Tachina mella* in its relations with the gipsy moth and is probably a better explanation of its failure to become an effective parasite than any other which has yet been put forward.

Mr. Thompson also discovered another most remarkable and peculiar phenomenon in connection with parasitism by those tachinids the larvæ of which inhabit integumental funnels similar in character to those formed by Tachina. These funnels appear to be formed as a direct result of the tendency of the skin to grow over and heal the wound caused by the entrance of the tachinid maggot into the body of the host. This wound is kept open by the larva itself, and as a result the growing integument takes the form of an inverted funnel, more or less completely surrounding the parasitic maggot, which continues to breathe through the minute orifice in its apex. When the caterpillar molts the old skin is usually torn away from around this opening, leaving the maggot in situ and unaffected, but occasionally its attachment to the funnel may remain so strong as to result disastrously for the maggot, and the whole funnel, maggot included, may be withdrawn. Thus, not merely the eggs may be molted off, but the internal feeding maggots which have hatched from the eggs may be molted out and perish.

ACHÆTONEURA FRENCHII WILL.

A very few specimens of this species have been reared from time to time in the course of the work at the laboratory. It is probable that the species is synonymous with that mentioned by Forbush and Fernald under the name of *Achætoneura fernaldi*.

EXORISTA BLANDA O. S.

Occasionally reared as a parasite of the gipsy moth.

UNDETERMINED TACHINIDS.

Dr. S. W. Williston, in reporting upon a collection of Diptera reared from the gipsy moth in Massachusetts and sent to him in 1891 by Dr. Fernald, stated that there were present two species of Exorista and four of Phorocera. Unfortunately these specimens appear to have been lost before being definitely determined. No such variety of tachinids has been reared from this host in Massachusetts during recent years, but several species as yet undetermined, or represented only by unfamiliar puparia, are in our collection.

SUMMARY OF REARING WORK CARRIED ON AT THE LABORATORY IN 1910.

In Tables IV and V are the condensed results of a part of an extensive series of rearing experiments primarily instituted for the purpose of determining the present status of the introduced parasites of the gipsy moth in America in 1910. They also serve excellently as an indication of the effectiveness of parasitism by native species. In this respect they are typical of the results secured from similar work in previous years.

TABLE IV.—*Results of rearing work in 1910 to determine progress of imported parasites and prevalence of parasitism by native parasites of gipsy-moth pupæ.*

Laboratory No.	Localities.	Number of dispar. pupæ.	Moths reared.		Parasites reared.				Scarcophagid puparia.
			Male.	Female.	Blepharipa scutellata.	Theronia fulvescens.	Compsilura concinnata.	Miscellaneous tachinids.	
2185	North Andover...	2,700	1,222	259	2				
2186	Melrose..........	500	185	159	1				
2187	North Andover...	1,800	683	299	2	3			2
2188	Stoneham........	1,400	482	461		5	1		
2189	Saugus...........	1,000	412	250		2	2		3
2190	Wakefield........	800	126	217					1
2190A	....do...........	[1]106	4	38	1		[1]12		
2191	Swampscott......	300	159	68					
2192A	North Andover...	300	106	92					
2192B	Woburn..........	282	69	115		1			
2192C	North Andover...	100	67	22					
2193	Saugus...........	1,050	274	172			2		1
2194	Melrose [2]......	1,700	312	211	1				1
2196	Stoneham........	1,047	379	127					
2197	Beverly..........	1,000	81	144					13
2199	....do...........	1,000	147	78		2			5
2280	Saugus...........	1,250	253	209		2	1	[3]1	10
2281	North Andover...	580	232	102		1			
2282	....do...........	1,700	442	218	8	17		[4]1	6
2283	Beverly..........	500	261	125		1			
2284	....do...........	400	152	133					
2285	Gloucester.......	1,700	354	346		3		[5]2	5
2286	Beverly..........	600	82	87					1
2287	....do...........	300	85	49					
2289	North Andover...	1,000	143	461		1			
2290	....do...........	1,100	118	323		3			1
2291	....do...........	900	198	215		4			2
2293	Beverly..........	500	229	102					
2294	....do...........	500	81	104		2			2
2295	Middleton........	1,100	193	156					
	Total........	27,215	7,531	5,342	15	47	18	4	53

[1] Prepupæ.
[2] Partly from another locality.
[3] Tachina-like species, puparium.
[4] *Exorista blanda.*
[5] One tachina-like and one unknown native species, puparia.

Table IV includes the results of rearing work in which caterpillars collected in the open were used. A large proportion of the collections, probably half or more, was from localities in which *Compsilura concinnata* is not as yet established, although it will probably have extended its range to include these by the end of another year. It will be observed, however, that the number of the Compsilura reared is very far in excess of that of any native tachinid or other parasites attacking this caterpillar as a host. This is especially encouraging in view of the fact that one year ago less than one one-hundredth of the proportionate number of Compsilura were reared under similar circumstances, and only from material collected over a much more restricted area.

TABLE V.—*Results of rearing work in 1910 to determine progress of imported parasites and prevalence of parasitism by native parasites of gipsy-moth caterpillars.*

Laboratory No.	Localities.	Number of caterpillars.	Number of pupæ reared.	Number of moths reared.		Number of parasites reared.					
				Male.	Female.	Compsilura concinnata.	"Tachina-like."[1]	Exorista blanda.	Apanteles fulvipes.	Theronia fulvescens.	Blepharipa scutellata.
4317	Wellesley	1,029	158	15	13						
4330	Stoneham	535	298	43	81	9					
4332	Wellesley	2,110	716	419	69		8				
4333	....do	3,000	962	371	115		4				
4334	Stoneham	2,000	599	59	177	108					
4337	....do	800	300	20	88	73					
4340	Wellesley	1,250	846	234	245		3				
4341	Swampscott	600	308	124	144	1					
4349	Wellesley	2,000	902	363	317		6		²1		
4351A	Medford	300	234	18	202	6					
4351B	Winchester	100	54	5	34						
4352	Danvers	1,500	566	30	199	6					
4353	Wilmington	125	69	7	25	2					
4354	Winchester	110	14	1	6	12		1			
4356	Wellesley	1,000	818	245	344		2		²3	³8	
4359	....do	1,125	934	150	301		1			³4	
4360	Stoneham (?)	500	207	11	125	59	1				1
4362	Beverly	230	56	4	30			1			
4366	Lexington	400	20	1	5	2					
4367	Salem	1,000	338	29	139	15					
4369	Marblehead	250	61	4	22	5					
4370	Salem	800	286	21	106	14				³1	
4372	Stoneham-Woburn	150	103	11	43	21					
4373	Wellesley	150	65	9	11					³1	
4377	Nahant	425	181	29	114	1					
4378	Winthrop	300	191	18	128	1					
4386	Burlington	100	37	2	22	1					
4387	Lexington	53	6		4	1					
4392	Manchester	341	281	21	42						
	Total	22,283	9,610			337	25	2	4	14	1

[1] The identity of this tachinid, the puparia of which resemble those of tachina, but which hibernates as a pupa, is wholly unknown. The adult has never been reared.
[2] From near site of colony of 1910. Recovery of no significance. Masses of cocoons counted.
[3] A few pupæ included in this collection.

In Table V are included results of rearing work in which collections of gipsy-moth pupæ were used. These were largely made in localities near the center of colonization of *Blepharipa scutellata* of the same year. The number of Blepharipa secured is higher in proportion to the number of pupæ collected than would be the case had these collections been made irrespective of the localities where the species was so recently colonized.

Parasitism by Theronia is somewhat less on the average than in some other years when similar studies have been made. At times it has amounted to as much as 2 per cent.

Sarcophagids are not considered as parasites, but rather as scavengers. Their true status is yet to be determined, however.

Compsilura concinnata is not commonly secured from the pupæ, and in one instance in which more than an insignificant number of this species was recovered the collection consisted of caterpillars, which had prepared for, but not undergone, pupation.

No Monodontomerus were reared from any of the collections of pupæ included in Table IV, nor has the species ever been recovered from counted lots of pupæ collected in the open. It was found in 1910, as in 1909, by the examination of unhatched pupæ after the most of those remaining healthy had produced the moths, and issued in unsatisfactory numbers from collections of pupæ made at the same time. These were not counted at the time of collection, and on that account were not included in the table.

PARASITISM OF THE BROWN-TAIL MOTH IN AMERICA.

The brown-tail moth in America is subjected to a considerably higher percentage of native parasitism than is the gipsy moth, but at the same time, as will appear in the summarized results of the rearing work in 1910, the aggregate is scarcely sufficient to be considered as consequential.

The following species have been reared, and doubtless the list will receive additions in the near future.

Trichogramma pretiosa Riley.

A very considerable percentage of the egg masses collected in the open is parasitized by this species, but because of the inability of the parasite to attack any but the more exposed eggs in a mass, the actual percentage of parasitism is insignificant.

Limnerium clisiocampæ Weed.

In 1907 a single specimen of this common parasite of the tent caterpillar was reared from a brown-tail caterpillar collected in Exeter, N. H. One or two other rather doubtful records have been made since. It is unquestionably not an important parasite of the brown-tail moth.

ANOMALON EXILE PROV.

Quite commonly reared as a parasite of the tent caterpillar and not infrequently as a parasite of the brown-tail moth, apparently attacking the caterpillars before pupation and probably while they are still very young. Its frequency as a parasite of the brown-tail is well indicated in Table VI (p. 147), which records the results of the summer's rearing work of 1910.

THERONIA FULVESCENS CRESS.

This is probably, as in the case of the gipsy moth, the most common native hymenopterous parasite. No attempt has been made to determine whether it is commonly primary or secondary in this connection, but it is presumably primary in the majority of instances.

PIMPLA CONQUISITOR SAY.

The pupæ of the brown-tail moth seem to afford much more suitable conditions for the development of the Pimpla larvæ than do the pupæ of the gipsy moth. In consequence this Pimpla is frequently reared and is probably about as important as a parasite of the brown-tail moth in America as are the European species, *Pimpla examinator* and *Pimpla instigator*, abroad.

PIMPLA PEDALIS CRESS.

This species is never so common as *Pimpla conquisitor* in its association with the brown-tail moth. It is apparently identical in habit with the more common species, but if results in studies in parasitism of other hosts are to be excepted, it is more apt to occur in forests and woodland than in open country.

DIGLOCHIS OMNIVORA WALK.

At times Diglochis is a common parasite of the brown-tail moth pupæ, but in 1910 it was unexpectedly scarce in Massachusetts, although it seemed to have been much more common in Maine, judging from the small amount of material which has been received from that State.

SYNTOMOSPHYRUM ESURUS RILEY.

Of irregular occurrence as a parasite of the brown-tail moth, but among the more effective of the native species in 1910. It was first reared in 1906 by Mr. R. L. Webster, while associated with the laboratory, and not again encountered until 1910, when large numbers issued from material collected in certain localities, as will be seen in Table VI which shortly follows.

Chalcis sp.

Upon several occasions specimens of Chalcis have issued from cocoons of the brown-tail moth collected in the open. The species has not been definitely determined nor compared with *Chalcis ovata*, because it is thought likely at the present time that two species may be confused under that name. One of them is believed to be a primary parasite of lepidopterous pupæ and the other to be essentially a secondary parasite attacking tachinid puparia.

Euphorocera claripennis Macq.

Several times reared from brown-tail caterpillars collected in the field, but always, apparently, rare in this connection.

Tachina mella Walk.

Never a common parasite in connection with this host. Usually about on a par with Euphorocera and without economic significance.

Phorocera saundersii Will.

A single specimen thus determined by Mr. Thompson was reared in 1910 under circumstances which quite conclusively indicate this host relationship.

Exorista boarmiæ Coq.

Like the above, only a single individual of this species has been reared from brown-tail caterpillar collections, but under circumstances which were not so decisive as in the last-mentioned instance. Mr. Thompson is authority for this determination also.

Undetermined Tachinid: "Native Parasite of chrysorrhœa."

One of the most remarkable instances of attack on an unsuitable host by a tachinid parasite is that of the species which in the laboratory notes is always referred to as above, upon the caterpillars of the brown-tail moth. The history of this species, the identity of which is very much in doubt, is in certain respects comparable to that of *Tachina mella* in its relations to the gipsy moth. As may be seen by the summarized results of the rearing work in 1910 (Table VI), it is not infrequently a parasite of some little consequence, and in all many hundreds of its larvæ have been secured from field collections of brown-tail caterpillars.

Invariably, however, these larvæ died without forming perfect puparia. For a long time it was thought that this was due to unfavorable surroundings at the time when pupation was attempted and that if the larvæ were allowed to enter damp earth as soon as they issued from the host, better results might be obtained. In order to provide for this, the style of rearing cage which is shortly to be described and figured, was devised. With its aid we were enabled to secure a large proportion of the parasite maggots within a few moments after they had finally separated themselves from the cocoon which the host invariably spins before dying, and these were given every advantage which could be afforded to assist in the successful completion of their transformations. The results, however, were always the same and not one perfect puparium has been secured.

The reasons for this may not be far to seek, but the chances are that it will be a long time before an adequate explanation is afforded. When it was found that the larvæ failed to pupate under the most favorable conditions, they were carefully examined on the supposition that death might possibly accrue through the action of the poisonous spines into which they must necessarily come in direct contact in leaving the host cocoon. It was at once discovered that there was invariably a number of minute reddish spots scattered irregularly, and more or less abundantly, over the whole or a part of the body. It looked, at first, as though these spots might be the result of contact with the poisonous spines, but upon further examination it was found that they were of a character which could hardly be attributed to this cause. They are somewhat variable in size and seem to consist of a thickening of the epidermis which becomes slightly raised, shining, and brick-red in color. No attempt has been made as yet to determine whether they are present in the maggot before it leaves the body of its host, but little doubt is felt that they will be found when such examination is made.

That these spots are directly or indirectly responsible for the failure of the maggot to pupate is well indicated by the study of the numerous half-formed puparia which result from the attempt on the part of the larva of the parasite to do so. These are all more or less larviform, but occasionally one is found one end of which is smooth and rounded exactly as though pupation had successfully resulted, while the other is shrunken and withered, resembling a dead larva. Careful examination revealed that in such specimens the reddish spots were absent from the perfect portion and present in the withered.

SUMMARY OF REARING WORK IN 1910.

The accompanying tabulated results (Table VI) of an extensive series of rearing experiments for the purpose of determining the progress of the imported parasites of the brown-tail moth also indicate the extent to which that host is attacked by native parasites.

TABLE VI.—*Results of rearing work in 1910, to determine progress of imported parasites and prevalence of parasitism by native parasites of the brown-tail moth.*

Laboratory No.	Localities	Estimated number of chrysorrhoea	Moths reared.		Imported parasites reared.							Native parasites reared.							
			Male.	Female.	Compsilura concinnata.	Zygobothria nidicola.	Meteorus versicolor.	Monodontomerus aereus.	Undetermined Tachinid "N.P.C."	Pimpla conquisitor.	Pimpla pedalis.	Theronia fulvescens.	Anomalon exile.	Syntomosphyrum esurus.	Diglochis omnivora.	Chalcis.	Phorocera saundersii.	Exorista boarmiae.	Tachina mella.
2100	Saugus	500	80	9	7												1		
2101	Melrose	500	73	127	6														
2102	Andover	500	2	9					1										
2104	Saugus	500	29	69	2				2										
2105	Lynnfield	550	10	20	2	1												1	
2109	Stoneham	300		3	21			5											
2110	Andover	200	10	49			1		1										
2111	Tewksbury	1,000	93	15					10	1									1
2112	Reading	350	14	65					4	1									
2113	Bedford	500	67	35						1									
2114	Billerica	500	50	17					3										
2115	Woburn	500	10	14					7										
2116	Lexington	500	18	17					1	1									
2122	Saugus	300	12	63	13														
2123	Waltham	500	47	13					6	1									
2125	Arlington	500	9	19				1	2										
2126	Chelmsford	500	10	8					1										
2127	Drayeut	500	4	177	1				12	1									
2131	Melrose	450	198	24					4										
2132	Manchester	150	8	70	2	3			1										
2133	Saugus	500	221	26	3		4										2		
2134	Melrose	100	16	59					19	1			1						
2135	Reading	500	88	48					2										
2136	Concord	500	26	18															
2137	Lincoln	300	12	45															
2138	Stoneham	250	68	24	9	1		20	3	3	1	2	4	136					
2139	Gloucester	350	21	2															
2140	Saugus	200		85															
2141	Beverly	200	116	71															
2142	Merrimac	500	82	71					6	2									1
2143	Middleton	500	76	93				9		3									
2144	Groveland	500	104	15						1									
2145	Littleton	1,000	17	59					7		2	7		59					1
2146	Westford	200	50						54										

148 PARASITES OF GIPSY AND BROWN-TAIL MOTHS.

TABLE VI.—*Results of rearing work in 1910, to determine progress of imported parasites and prevalence of parasitism by native parasites of the brown-tail moth*—Continued.

Laboratory No.	Localities.	Estimated number of chrysorrhoea.	Moths reared.		Imported parasites reared.					Native parasites reared.									
			Male.	Female.	Compsilura concinnata.	Zygobothria nidicola.	Meteorus versicolor.	Monodontomerus aereus.	Undetermined Tachinid. "N.P.C."	Pimpla conquisitor.	Pimpla pedalis.	Theronia fulvescens.	Anomalon exile.	Syntomosphyrum esurus.	Diglochis omnivora.	Chalcis.	Phorocera saundersii.	Exorista boardmiae.	Tachina mella.
2147	Beverly	250	22	16															
2149	Methuen	250	63	67															
2150	Andover	250	32	12															
2151	do	250	18	22															
2152A	do	150	19	21															
2152C	do	100	35	38															
2152B	do	100	14	10															
2152D	do	150	15	23								1	1						
2152E	do	150	11	23					9	1		2							
2156	Metheun	200	30	32					15										
2157	do	150	31	64				3	9										
2158	do	50	16	3															
2159	do	150	46	42					10	2		1							
2160	Andover	200	72	52				7	12	1		1		7					
2161A	Melrose	90	47	34			5	14				1	2						
2161B	do	160	74	70			1					2	1						
2161C	do	150	49	50			3	6			1	6	12						
2161D	do	250	96	62															
2162	Andover	500	9	15	1				5			1							
2163	Boxford	500	87	91															
2164	Salisbury	500	121	119				3	2	1		1	1						
2166	Middleton	150	46	42								1							
2167	Lynnfield	175	42	63		1			2					38	53				
2168	Peabody	230	70	74	1			2		3		2	1	69					
2169	Reading	180	22	33								1	7						
2170	Saugus	400	89	141								2	1						
2171	Reading	200	27	27								1							
2172	Lynnfield	200	37	39				12						40					
2173	Salem	150	64	29															
2174	Melrose	1,000	304	314	4	1	5	57	2	1		10	1	52					
2175	Concord	200	45	57					17	7		1							
2176	Bedford	200	38	63						1		1							
2177	do	150	13	17				24											
2178	Danvers	175	24	54				19							2				

2179	Reading	200	56	48															
2180	Lynnfield	175	68	63					2										
2181	Melrose	325	217	240					1		2			1					1
2182	Swampscott	325	79	87							1								
2183	...do	225	96	96							4				54				
2184	Lynn	240	64	59							1								
	Total	24,350	3,955	3,851	72	7	19	313	276	43	4	58	32	452	55	2	1	1	4

As may be noted by reference to the table, the imported parasites are beginning to become sufficiently abundant so that parasitism by them will compare favorably with that by American species, but are not as yet so abundant as to exceed the American species in relative effectiveness. The table as presented does not indicate at all accurately the actual status of the several species of parasites mentioned, on account of the difference in the condition of the material at the time of collection.

Compsilura, for example, is much more apt to leave the caterpillars before they spin for pupation, and the same is true of Meteorus. Monodontomerus, Pimpla, Theronia, etc., never attack caterpillars before spinning, and Monodontomerus and Theronia frequently reserve attack until some little time after the host has pupated. As it stands, parasitism by Monodontomerus is about equal to that of Theronia and in excess of that by Pimpla or Anomalon. Parasitism by Compsilura is distinctly more effective than that by all of the other native tachinid parasites of the caterpillars. Meteorus is much more common than indicated in the limited territory over which it is now known to exist, and the specimens reared represented the second generation of adults to develop upon the brown-tail moth in 1910.

Apanteles lacteicolor Vier. is not represented in these collections, since it does not attack caterpillars so large as those involved.

In carrying on this work several styles of rearing cages were used, of which one was devised for the special purpose of securing the tachinid parasites with the minimum of exposure to the effects of the irritating hairs of the brown-tail caterpillar. This worked very satisfactorily, and since it may possibly be found of service in conducting similar work elsewhere, the following description is presented:

The basis of this cage (see fig. 10) consisted of a box of stiff pasteboard 8 inches square and 12 inches high. About 4 inches from the top a stiff paper funnel (a) was fitted and held in position by the cleats (b), which, in turn, were fastened to the sides of the box by broadheaded upholsterer's tacks driven in from the outside. These cleats served to support the tray (c), which just fitted into the cage. The bottom of this tray was covered, in some instances with coarse mosquito netting, and in others with a wire screen of ¼-inch mesh. Two holes in the side of the tray corresponded with two 1-inch holes in the side of the box, and these in turn with similar holes in a wooden strip (d), which was fastened on the outside. When the tray was in position, paper cones (h) and large glass tubes (g) were inserted in these holes.

The stiff paper funnel (a) had its apex inserted into another hole bored diagonally in a similar wooden strip which was fitted in the bottom of the cage. Inside of this hole a stiff paper cone (formed like h by rolling up a section of a strip of paper cut to a circular shape) was held in position by a tack which passed through it into the wooden strip. The end of this cone, passing through the bottom of the cage, permitted a third glass tube (*f*) similar to the two above mentioned, to be held in position. No further support to this tube was needed than that afforded by the cone itself.

In using this cage a mass of cocoons of the brown-tail moth was placed in the tray, and the cover was put on with the several tubes in position. Tachinid maggots issuing from the prepupal caterpillars, or pupæ contained in the cocoon mass, in attempting to seek the earth would pass through the bottom of the tray and be conducted by the stiff paper funnel into the lower tube, where they were quickly noticed and easily removed. All other parasites, as well as the brown-tail moths themselves, when they emerged, were attracted by the light into the two upper tubes, and could be similarly removed with little difficulty. (See Pl. V, fig. 1.)

By the aid of this contrivance we were enabled to secure a quantity of the larvæ of the unknown tachinid, already mentioned, within a few minutes after they had issued from the host, and thereby determined that the failure of this species to pupate was in no way due to the unnatural surroundings. Sometimes the tubes were partly filled with damp earth, in order that these larvæ might immediately come in contact with it, and at other times the larvæ were removed as soon as they dropped and placed upon earth similar to that which they would naturally have encountered had they issued from cocoons in the field under wholly natural conditions.

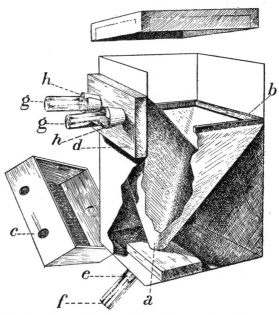

Fig. 10.—Rearing cage for tachinid parasites of the brown-tail moth: *a*, Paper funnel; *b*, cleats holding paper funnel in position; *c*, tray; *d*, wooden strip on outside of cage; *e*, paper cone connecting paper funnel *a* and glass tube *f*; *h*, *h*, paper funnels supporting glass tubes *g*, *g*. (Original.)

The use of these cages also saved a large amount of exceedingly painful work which would otherwise have been necessary in determining whether or not *Parexorista cheloniæ* was present in any of the field collections.

IMPORTATION AND HANDLING OF PARASITE MATERIAL.

Since insects like the gipsy moth and the brown-tail moth are subjected to the attack of different species of parasites at different stages in their development, it has been necessary, in order to secure all of these, to import the host insects in as many different stages as possible and practicable. If the present experiment in parasite introduction is brought to a successful conclusion, it will undoubtedly encourage the undertaking of other experiments in which similarly imported pests are involved. Even should it fail, from a severely practical standpoint, and the complete automatic control of neither the gipsy moth nor the brown-tail moth should be effected, it seems to us that the technical results already achieved are sufficient to give encouragement rather than the opposite to similar undertakings in the future. It is therefore desirable to describe in some detail the various methods employed for the importation and subsequent handling of the parasite material.

With very few exceptions the methods first employed proved more or less unsuitable. Sometimes they were entirely discarded; usually they were modified to suit the exigencies of the occasion. Sometimes these modifications were in comparatively unimportant particulars which would scarcely be pertinent to any other insect than the gipsy moth or the brown-tail moth, and realizing this there will be no attempt in such cases to enter into lengthy descriptions. At other times radical modifications have been found necessary on account of unforeseen difficulties which would be likely to occur in pretty nearly any other undertaking along anything like similar lines.

EGG MASSES OF THE GIPSY MOTH.

The importation of egg masses of the gipsy moth (see Pl. VI) from European sources has been attended with no difficulty whatever, beyond that of securing the collection of these eggs in sufficiently large quantities. Any style of package, provided that it were sufficiently tight to prevent loose eggs from sifting out, was as good as another, and any one of the established means of transportation served the purpose.

In the case of shipments from Japan serious difficulties were encountered. One of the parasites peculiar to that country and unknown in Europe invariably issued en route and died without reproducing. Various attempts to overcome this difficulty without having recourse to cold storage failed and it was only after cold-storage facilities were perfected and used that living parasites of this species were secured in numbers.

As in the instance of similar shipments from Europe, no special form of package was required, but at the same time a word of appre-

FIG. 1.—VIEW OF INTERIOR OF ONE OF THE LABORATORY STRUCTURES, SHOWING REARING CAGES FOR BROWN-TAIL PARASITES. (ORIGINAL.)

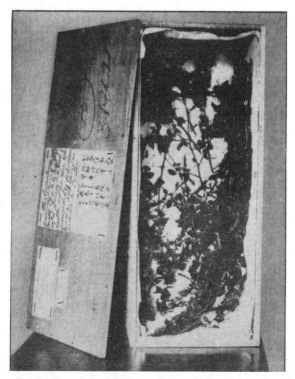

FIG. 2.—BOX USED IN SHIPPING IMMATURE CATERPILLARS OF THE GIPSY MOTH FROM JAPAN. (ORIGINAL.)

ciation must be said for the wonderful care with which the Japanese entomologists packed the egg masses for shipment. Good-sized and wonderfully well-constructed wooden boxes were used and each mass was wrapped separately in a small square of soft rice paper.

Considering the ease with which egg masses of the gipsy moth ought, theoretically, to be obtained and shipped to the laboratory, the number received in response to the requests which were made for their collection and shipment was astonishingly small during the first two winters. Up to that time only a very few dead parasites of an undescribed genus and species had been received from Japan, and none at all had issued from any of the few European importations.

In 1908 the several lots of eggs were placed in small tube cages of the ordinary type and the caterpillars killed as they issued. Some time after the eggs had hatched a few parasites began to appear simultaneously from the European and Japanese material, which proved upon examination to be *Anastatus bifasciatus* in each instance. Later a few *Tyndarichus navæ* were reared from the Japanese eggs and supposed, rightly enough, to be secondary, although probably not, as at first supposed, upon Anastatus. This hyperparasitism was by no means certain, and it was resolved to determine the fact definitely the following fall and winter, provided additional importations could be secured. The desired material was imported and an exhaustive study of the parasites which were present was made, with the result that the five species were reared and their host relations as well as their relations one to another definitely determined. The execution of this project proved to be much more tedious than was expected, and was, in fact, the feature of that winter's work. Further mention of the investigation will appear in the discussion of *Schedius kuvanæ*.

GIPSY-MOTH CATERPILLARS, FIRST STAGE.

In the spring of 1907 an attempt was made to import the caterpillars of the gipsy moth in their first instar, and a considerable number was received from several different localities. The experiment was not a success and was not repeated. The mortality was heavy en route and only a small proportion of the caterpillars would feed after receipt. Some few were carried through to maturity, but no parasites were reared.

It is very probable that if recourse were had to cold storage, caterpillars could very successfully be transported in this stage, but the importation of slightly larger caterpillars indicates that the percentage of parasitism would average to be very small at the best, and it is probable that the best would rarely be achieved.

GIPSY-MOTH CATERPILLARS, SECOND TO FIFTH STAGES.

EUROPEAN IMPORTATIONS.

The first importations of gipsy-moth caterpillars in the second to fifth stages were made in 1907. Small wooden boxes, each with a capacity of about 40 cubic inches, were used for the purpose, and all shipments were by mail. The caterpillars, usually to the number of 100, were inclosed in these boxes, together with several twigs bearing fresh foliage.

The method was of doubtful utility, and at the same time no improvement upon it could be devised short of cold storage en route. On receipt the twigs would usually be stripped bare of foliage. Some of the caterpillars were invariably dead—whether from starvation or from injuries received at the time of collection or subsequently could not be determined. The remaining caterpillars were in all stages of emaciation and many of them, though still living, were too weak to recuperate.

Parasites in considerable variety but always in very small numbers issued, for the most part en route, but occasionally from the caterpillars after receipt. Nothing could be decided as a result of these importations and their repetition was resolved upon.

It was planned to import much larger numbers in 1908 without modifying the methods employed the year before. In this respect success was not achieved, principally, it would appear, on account of the difficulty of collecting these small caterpillars in numbers, especially in localities where the gipsy moth was not very abundant. Furthermore, it became increasingly evident that the percentage of parasitism (so far as it could be determined by the actual number of parasites secured) was so insignificant as to make the task of importing sufficiently large numbers of any one parasite for the purpose of colonization wholly impracticable. Many of the lots of caterpillars which were received in the best condition produced no parasites at all. It was therefore evident that if extensive operations in any locality should be determined upon, complete failure might result through the absence of the parasites in that particular locality during that particular season. Nothing less than an improvement in the service of several thousand per cent over that of 1907 or 1908 would answer, and this was altogether out of the question, except at an expenditure which even the generous funds appropriated by the State and Federal Governments could not cover. Further importations from Europe were regretfully decided to be impracticable.

JAPANESE IMPORTATIONS.

It has already been told how Prof. Kincaid spent the summer of 1908 in Japan in the interests of the parasite work. While there, in cooperation with the Japanese entomologists, he evolved a wholly new method for the transportation of the immature caterpillars of the gipsy moth, which would have been applicable in the case of European importations if it had seemed to be worth while to continue these importations in 1909. Large oblong wooden boxes having a capacity of about 1½ cubic feet were used. Like all the boxes received from Japan, they were most excellently constructed of a sort of wood which was less affected by dampness than most. The success of the work was very largely dependent upon both the character of the wood and the excellence of construction. It is certain that ordinary packing boxes would have warped to such an extent as to permit the escape of the small caterpillars.

These boxes (see Pl. V, fig. 2) were first lined with several thicknesses of absorbent paper, which was then thoroughly dampened. Small branches of a species of Alnus were attached to the sides, so that the interior was a mass of green foliage; the caterpillars to the number of several hundred were introduced, the cover tightly attached, and the whole sent in cold storage from Yokohama to Boston with scarcely an interruption en route. Sometimes the ends of the branches were thrust into a piece of succulent root (radish or potato), but this proved unnecessary, and rather a detriment than otherwise.

The condition of these boxes on receipt was usually good, and in some instances surprising. In some of the best of them scarcely a leaf was withered or even discolored, and in one in particular it seemed almost as though the branches had been freshly collected, with the early morning dew still clinging to the leaves. This illusion was almost instantly destroyed, for within an hour practically every leaf had dropped from the stem and was already beginning to blacken, as though struck by a sudden blight.

There was a good deal of difference in the condition of the caterpillars. Those which had been shipped in the second and third stages almost invariably arrived in the best condition. There was scarcely any mortality en route, and physically they were all in perfect health and ready to feed voraciously. Larger caterpillars did not survive their journey so well, and among those that had reached the fifth stage there was always a heavy mortality, and the survivors were never very healthy and would mostly die without feeding. It would appear that they were so heavy as to be thrown to the bottom of the box while dormant through cold, and thus become injured.

While technically a success, these attempts were practically failures. No parasites were secured in anything more than the most insignificant numbers, which could not be secured much more easily in other ways, and no further importations were attempted in 1909.

Relatively such small quantities of this class of material have been received as to make unnecessary any specially devised methods for their economical handling. With very few exceptions the boxes were opened immediately upon receipt and most carefully sorted for parasites and living caterpillars. A few of the large Japanese boxes were not opened immediately, but holes were bored in the end, cones and tubes inserted, and living insects of all sorts thus attracted to the light and removed. The living caterpillars were placed in cages or trays and fed, and occasionally a few parasites were thus secured in addition to those present in the boxes upon receipt.

It is very much to be regretted that the dead and dying caterpillars were not preserved for subsequent examination and dissection, but it was only in 1909, after the shipments of this sort of material had been discontinued, that the wholesale dissection of caterpillars was attempted for the purpose of ascertaining the proportion of parasitized individuals. At the best, even after long experience, it is a tedious process, especially in the case of material which has been killed and preserved.

A few caterpillars, accidentally imported in their early stages with Apanteles cocoons in 1910, were saved and dissected with good results from a technical standpoint.

GIPSY-MOTH CATERPILLARS, FULL-FED AND PUPATING.

Importations of large caterpillars (Pl. VI) ready or nearly ready to pupate were first made in 1905, and it was demonstrated during that year that they could be brought to America with a fair degree of success, and that at least a proportion of the parasites with which they were infested could be reared.

Ever since 1905 we have been attempting to improve upon the methods first used during that year and have experimented with scores of modifications of the most successful, some of which were intentional while others were incidental to the fact that there have been many different collectors, each of whom has displayed some individuality in his methods of collecting and packing. It would be tedious and is probably unnecessary to go into detailed descriptions of even a part of these various intentional or accidental experiments.

The most successful method yet devised involves the use of rather shallow wooden boxes having a capacity of from 40 to 70 cubic inches. (See Pl. VIII, fig. 3.) Quite a large number of shipments has been made in much larger boxes, but their condition on receipt has almost invariably been very bad. The boxes *must* be tight to prevent the escape of tachinid larvæ, which can apparently pass through any

DIFFERENT STAGES OF THE GIPSY MOTH (PORTHETRIA DISPAR).

Egg mass on center of twig; female moth ovipositing just below; female moth below, at left, enlarged; male moth, somewhat reduced, immediately above; female moth immediately above, somewhat reduced; male moth with wings folded in upper left; male chrysalis at right of this; female chrysalis again at right; larva at center. (Original.)

opening large enough to accommodate the head. No special provision for ventilation is necessary, but it is necessary to construct the boxes of soft and absorbent wood in order to secure best results. This will not only prevent too rapid evaporation, but superfluous moisture will first be absorbed and subsequently will evaporate. Tin boxes are wholly unsuitable and paper or pasteboard have never been at all satisfactory.

Twigs with foliage should be included in each box, and these must be long enough to remain firmly braced in case the caterpillars eat the foliage. Some very bad results have followed the use of loose foliage, a practice which certain collectors have been persistent in following.

The fewer the caterpillars included in each box the better the results. The number has gradually been reduced from 100 at first to 20 during the past few years. Undoubtedly 10 would be better yet, but not enough better to make the added expense an economy.

The more nearly the caterpillars are ready to pupate when packed the better. If collected just a few days before pupation, they usually arrive in good shape, provided conditions otherwise are as they should be.

Shipments by mail have generally been successful when the boxes were not smashed, as has sometimes happened, or when something else was not wrong. Shipments by express without cold storage have been equally successful when the boxes have been properly packed. As has been said, there is no need to provide for the ventilation of the interior of the box, but the exterior must be exposed to the air on at least one side to permit the evaporation of the moisture absorbed by the wood. Otherwise, as nearly always happens, when a part of the caterpillars or pupæ die, they decompose, and as a result of their presence a similar fate usually overtakes the remainder. Some very large shipments were a complete loss in 1907, merely because a European agent, prevented by newly enforced postal regulations from making shipments by mail, packed the boxes tightly in large packing cases and forwarded them by express. When the lids were removed from these cases the sides of the boxes were found to be thoroughly damp, and the whole exhaled an ammoniacal odor so strong that it would seem of itself alone sufficient to destroy any ordinary form of insect life.

Bundles of boxes wrapped in thick, glazed paper have almost invariably been received in bad condition. If the paper is soft and absorbent it is generally satisfactory. One collector wrapped several packages in a thick fabric composed of tarred paper strengthened by muslin, and the contents rotted.

Cold storage in the case of shipments of this character would never have been either necessary or even desirable had it not been for the difficulties experienced in the importation of *Blepharipa scutellata*. All other species of parasites could be secured equally as well or better from shipments under normal conditions, but because Blepharipa differed from all the others in this apparently minor characteristic, it was found necessary to make use of cold storage for practically all of the very large shipments made during 1909 and 1910.

No large shipments of full-fed caterpillars have been made from Japan. They were rendered unnecessary in the first place on account of the very excellent and intelligent service rendered by the Japanese entomologists. There are but three important parasites to be secured from these large caterpillars in Japan, and the cocoons and puparia of these have been reared and forwarded to us in specially devised packages, with almost uniformly good results.

There have been a few very valuable lots of material of this character shipped otherwise than as above described. It is by no means certain that if sufficient time and experimentation were to be devoted to the subject some of these occasional and successful modifications might not be developed into something better than has yet been tested. Any deviation is apt to prove disastrous, however, as witnessed in 1910, when failure resulted because the quality of the paper used for wrapping the bundles of boxes was changed in several instances from that employed at any time previously. It is very difficult, and in practice impossible, to foresee such minor contingencies and provide against them. The really serious phase of the situation lies in the fact that such a slight modification may result not only in the complete loss of the shipment itself, but in a year's delay before it can be remedied. By the time the first shipments are received and the trouble recognized, it is apt to be too late to apply a remedy that year, even by the use of the cable.

The laboratory methods in use for the handling of the parasite material of this sort have been modified in various ways, more especially for the express purpose of overcoming the difficulty in hibernating the puparia of *Blepharipa scutellata*. Such of these modifications as have been primarily made for this special purpose will be discussed in the account of Blepharipa which will be found elsewhere.

In general it has been the practice to open the boxes immediately upon their receipt, and to sort the contents in accordance with their character. The tachinid puparia were always carefully counted, and of late years they have been sorted to a certain extent into species.

In 1907 all the puparia were placed in jars, without sorting, with a little very slightly dampened earth which was kept from drying by the use of a wet sponge. In 1908 they were sorted to species, so far as this was practicable, and all were kept dry. In 1909 the Blepharipa puparia were sorted out and placed in earth as soon thereafter as possible. The remainder were placed in small tube cages, and taken to the field where they were to be liberated. An attendant counted the number of flies issuing, and watched for secondary parasites. Since no secondary parasites issued in the summer from puparia secured in this manner, in 1910 the puparia were merely placed in cages which were taken to the colony site and left unattended until the flies had ceased to issue.

After the adults of the summer-issuing forms have all ceased to emerge, the sound puparia are more or less carefully sorted. Those supposed to be *Parasetigena segregata*, indistinguishable externally from dead Tricholyga or Tachina, are buried in damp earth for the winter. Mr. J. D. Tothill, one of the assistants at the laboratory, devised an ingenious method for separating the puparia containing the healthy pupæ of Parasetigena from those containing dead Tachina or Tricholyga, by holding them so that they were viewed against a narrow beam of very strong light. The method was not infallible, but served its purpose fairly well, and was the first of many which had been experimented with which was at all successful.

The living caterpillars removed from the boxes have been placed in cages or trays and fed, but only an insignificant proportion of them has ever lived long enough to be killed by the parasites which many of them have contained. Large numbers of them have been dissected for the purpose of determining the proportions parasitized.

The dead caterpillars not infrequently contain the puparia of Tricholyga, when this parasite happens to be common in the locality from which the shipment originated. Under such circumstances they are placed in tube cages for the emergence of the flies. If Tricholyga is not present in the boxes in the form of free puparia the dead caterpillars may as well be discarded, since it is only very rarely that any other tachinid pupates in this manner.

Pupæ, both living and dead, nearly always contain a considerable number of the larvæ of *Blepharipa scutellata*. They are, therefore, placed over damp earth in order that the larvæ may pupate under natural conditions. Other tachinids may occur in the pupa, but never in anything but insignificant numbers.

GIPSY-MOTH PUPÆ.

It would seem as though it ought to be an easy matter to import the pupæ (Pl. VI) of the gipsy moth in good condition, but for reasons which are not altogether clear in every instance the vast majority of the importations of pupæ have been worthless, or worse than worthless, since the handling of worthless material involves an additional waste of labor. Too often the cause of failure is directly and obviously the result of careless packing, and the number of lots of pupæ which have been received packed with the care which is essential to success is very small. The most successful shipments ever received were carefully packed in slightly dampened sphagnum moss, so arranged that the individual pupæ rarely touched each other. One or two successful importations thus received in 1908 were used as the basis for future instructions to collectors, and in every instance in which the directions were carefully followed in 1909 the results were equally good. In 1910 additional material apparently packed with the same care was received from the same source and via the same route. For no apparent reason whatever it was worthless when received.

In consequence of this another method which has occasionally been followed, will be recommended for shipments in 1911. This has been employed successfully upon various occasions, although without anything like uniform success, and is in effect the same as that used for the shipment of full-fed and pupating caterpillars and the same precautions must be used. It will probably be better to place a somewhat larger quantity of foliage in the box to prevent the pupæ from being thrown about too much in transit.

So very few shipments of gipsy-moth pupæ have been received at the laboratory as to have rendered unnecessary any special devices for their handling after receipt. Each year a few of the lots of caterpillars have contained a few individuals which were collected as prepupæ or as pupæ, and from such, an occasional parasite of one or another of the species peculiar to the pupæ has been reared. These have been so few, however, as to be of entirely inconsequential value, except from a technical standpoint.

The actual shipments of pupæ collected as such have been handled exactly as though they consisted of active caterpillars which pupate en route, with the one difference that the pupæ received have usually been inclosed in darkened cages with tubes attached, in order more easily to remove the parasites as they issued.

In 1908 pupæ were received in satisfactory condition for the first time since the preliminary shipments were made in 1905, and it was not until then that the host relations of several of the parasites, notably *Chalcis* spp. and *Monodontomerus æreus*, were finally determined. These lots were studied with the greatest care, each individual pupa being opened on receipt and for the most part isolated in a small vial, in order that it might be dissected after the contained parasite had issued.

BROWN-TAIL MOTH EGG MASSES.

No difficulty has ever been experienced in the importation of the egg-masses of the brown-tail moth (Pl. VII), except that when cold storage is not used a portion of the parasites are apt to hatch, and either escape or die en route. Wooden boxes of various sizes and

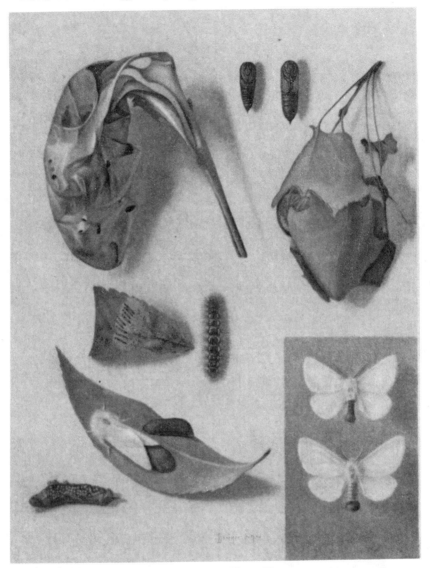

DIFFERENT STAGES OF THE BROWN-TAIL MOTH (EUPROCTIS CHRYSORRHŒA).

Winter nest at upper left; male and female adults, lower right; another winter nest, upper right; male and female chrysalides above, male at left; full-grown larva in center, somewhat reduced; young larvæ at its left; egg mass, the eggs hatching, at lower left; female ovipositing on leaf; egg mass also on same leaf. (Original.)

construction have been employed with uniformly good results, and nearly all shipments have been made by mail. All of the parasites have been found amenable to methods of laboratory control, and their reproduction has been undertaken as an economic venture in each instance. Under such circumstances there is no need to import large quantities except for the purpose of discovering other forms of parasites, should they exist.

HIBERNATING NESTS OF THE BROWN-TAIL MOTH.

The importation of hibernating nests (Pl. VII) of the brown-tail moth has been attended with very good success, as a rule, but by no means invariably. If they are sent too early in the winter and subjected to long continued high temperature before shipment, or while in transit, the caterpillars will die instead of resuming activity in the spring.

If sent too late in the spring, exposure to abnormally high temperatures en route results in premature activity of the caterpillars and they will arrive in bad and sometimes in worthless condition.

If sent in the middle of the winter they will be very nearly ready to resume activity on receipt, but if again exposed to cold they will become dormant and remain so until about the time when they would normally have become active. This seems not too prejudicial to them, if one is to judge by their activities during the first few weeks after the resumption of activity, but in some subtle manner a change has been wrought, and they do not commonly go through to successful pupation. This phenomenon, previously observed with other insects, is discussed at some length in the account of the tachinid *Zygobothria nidicola*, which hibernates within the caterpillars, but which does not destroy its host until after it has spun for pupation.

The failure to rear these caterpillars beyond a certain stage in the spring was at first attributed to some fault in the methods employed, and when it was finally apparent that the fault lay elsewhere there was no longer need to seek to remedy it. The one parasite desired was found to be already introduced and apparently well established as the result of a colonization some three years before.

The methods for handling the imported nests have varied from season to season in accordance with the habits of the parasites which it was desired to rear from them. These methods will be more fully described in the discussions of the several parasites involved: **Pteromalus egregius, Apanteles lacteicolor, and Zygobothria nidicola.**

IMMATURE CATERPILLARS OF THE BROWN-TAIL MOTH.

Taken all together, importations of active brown-tail caterpillars (Pl. VII) in the second to fourth spring stages aggregate a considerable number. These importations were undertaken on the supposition that there were parasites which attacked them as soon as they resumed activity in the spring, and left them before they reached their last stage. There are, indeed, several species which have such habits, but there is none amongst them which may not be secured equally well from either the hibernating caterpillars or from the full-fed and pupating caterpillars, or from both. In consequence the importation of partly grown caterpillars in the spring has never been attempted except experimentally.

Of the considerable number which has been received, nearly all have been packed in the same manner as are the full-fed caterpillars. They are nearly always alive on receipt, and usually feed voraciously when given an opportunity. Undoubtedly a considerable percentage contains parasites of those species which only emerge after the cocoon is spun, but every attempt to rear these parasites by feeding the caterpillars has resulted in failure. They will feed once, at least, but usually not more than a few times, and then die sooner than would have been the case had they not fed at all. Some among them will live for a long time, feeding a little but scarcely growing at all, and sometimes a very small percentage will complete growth and pupate. The percentage is so very small, however, and the labor and pain of handling the caterpillars is so great, as to render the work of feeding them of much more than doubtful economy, in every instance in which it has been attempted.

FULL-FED AND PUPATING CATERPILLARS OF THE BROWN-TAIL MOTH.

The temptation is strong to use the present opportunity for the purpose of giving vent to certain poorly suppressed and heartfelt expressions of opinion concerning the infliction known euphemistically and very inadequately as the brown-tail "rash." It is a very living subject of discussion during most of the year at the laboratory, but never more so than while the boxes of full-fed and pupating brown-tail caterpillars are being received from Europe.

Aside from the fact that the handling of this sort of parasite material has been productive of most acute physical anguish, it has been altogether the most uniformly satisfactory of any received. No modifications in the methods of packing have been suggested during the past five years, other than a slight modification in the form of the box in order to alleviate the trouble to which reference is made above.

Rather shallow boxes, having a capacity of about 50 cubic inches, are used for the purpose. (See Pl. VIII, figs. 1, 2.) The caterpillars are collected, preferably just before they spin their cocoons, and 100 are placed in each box, together with a few twigs with foliage attached, which serve less as food than as a support for the cocoons. When collected at the proper time practically all will spin and pupate en route, and the cocoons are so strong as to prevent the pupæ from becoming injured.

There has never been any trouble experienced through these boxes sweating en route, as has so frequently happened when boxes containing gipsy-moth caterpillars have been too closely confined in box or bundle, but at the same time those which have been exposed to free circulation exteriorly are noticeably in better condition than others.

Much the larger proportion of the material of this sort has been received in perfect condition at the laboratory, and large quantities of parasites have been reared from nearly every lot thus received. Occasionally boxes have been used which were not sufficiently tight to prevent the escape of the tachinid larvæ, and some loss has accrued in consequence. In a number of instances the boxes have been infected with fungous disease, and all or nearly all of the caterpillars or freshly formed pupæ have died in consequence. In rather an unnecessary number of instances, or so it would seem, caterpillars have been collected too young, and have failed to pupate en route. Such shipments properly fall in the class last mentioned, and are worthless for the purposes desired.

Through a misunderstanding nearly all of this class of material was sent in cold storage in 1909, with the result that the caterpillars failed to pupate en route, as would have been the result otherwise, and a good many of them failed to pupate after receipt. Considerable loss resulted on this account before the collectors could be notified to return to the original method of shipping by mail.

In handling the boxes of caterpillars and pupæ, a variety of methods has been employed, of which the most satisfactory appears to be simply boring a hole in the one end, and introducing a paper cone and tube. Even the removal of the covers from a dozen or more boxes without protection is accompanied by painful results, and owing to the difficulty of boring the holes without splitting the wood, after the box has been received at the laboratory, collectors are now instructed to prepare the boxes for the reception of the tube at the time of their manufacture.

In 1906 and 1907, when the first shipments of this character were received, and when little was known of the character of the parasites which were likely to be reared from them, it was thought necessary not only to open each box, but to sort it over, and remove the tachinid puparia which were always present in a larger or smaller number. It was known that there were always present certain species of tachinids which would only complete their final transformations successfully when their puparia were kept more or less moist, and it was expected that among the parasites of the brown-tail moth would be some possessing this characteristic. Opening the boxes without some sort of protection was utterly impossible. Automobile goggles were used to protect the eyes, various forms of respirators to prevent the inhalation of the spines, the hands were protected by rubber gloves, and the neck and face were swathed in accordance with the fancy of the operator.

Two ingenious types of headdress (Pl. IX, fig. 1) were devised by Mr. E. S. G. Titus in the hope that they would solve the difficulty, but it was found that they were not only unbearably hot, but that the glass fronts would quickly become covered with moisture which could not be removed.

In 1907 a much larger quantity of this sort of material was received than during the previous summer, and it was practically a necessity that some method be devised which would do away with at least part of the trouble. After some little experimentation the arrangement shown in the illustration was the result. (Pl. IX, fig. 2.) It consisted of an ordinary show case, with sides and top of glass and with a wooden slide in the back. The two ends were removed and replaced with boards in which armholes had been cut. Thick canvas sleeves were attached to these, through which the gloved hands of the operator were thrust, and it was found that the work could be done with what was, comparatively speaking, a minimum of discomfort and danger.

In 1908, for several reasons which need not be entered into here, it was thought desirable to discontinue, temporarily, the importation of large quantities of the pupating caterpillars, and it was also demonstrated that all of the parasites which were secured from them would complete their transformations without being kept moist. The work of sorting over the boxes of parasite material was thus demonstrated to be unnecessary, and, consequently, in 1909, when large importations were resumed, the covers were simply removed from the boxes, which were then stacked up in the large wooden tube cages (Pl. X, fig. 1), which had originally been constructed for the rearing of parasites from the imported hibernating nests.

BROWN-TAIL MOTH PUPÆ.

Several attempts have been made to ship the pupæ (Pl. VII) of the brown-tail moth, packed in moss, as was at one time recommended for the shipment of the pupæ of the gipsy moth. Such attempts have usually been more or less satisfactory, but never as satisfactory as when the cocoons were collected in the field and placed loose in the boxes together with the active caterpillars. If only a small portion of the pupæ is collected in the field, the only sure method of detecting their presence is by the occurrence of the pupal parasites

FIG. 1.—BOXES USED IN 1910 FOR IMPORTATION OF BROWN-TAIL MOTH CATERPILLARS, WITH TUBES ATTACHED DIRECTLY TO BOXES. (ORIGINAL.)

FIG. 2.—INTERIOR OF BOXES IN WHICH BROWN-TAIL MOTH CATERPILLARS WERE IMPORTED, SHOWING CONDITION ON RECEIPT. (ORIGINAL.)

FIG. 3.—BOXES USED IN SHIPPING CATERPILLARS OF THE GIPSY AND BROWN-TAIL MOTHS BY MAIL. (ORIGINAL.)

Fig. 1.—Headgear Devised by Mr. E. S. G. Titus for Protection Against Brown-Tail Rash. (Original.)

Fig. 2.—Show Case Used When Opening Boxes of Brown-Tail Moth Caterpillars Received from Abroad. (Original.)

Bul. 91, Bureau of Entomology, U. S. Dept. of Agriculture.　　　　　　　　　PLATE X.

FIG. 1.—LARGE TUBE-CAGE FIRST USED FOR REARING PARASITES FROM IMPORTED BROWN-TAIL MOTH NESTS AND LATTERLY FOR VARIOUS PURPOSES. (ORIGINAL.)

FIG. 2.—METHOD OF PACKING CALOSOMA BEETLES FOR SHIPMEMT. (ORIGINAL.)

among those reared. A large number of the shipments has produced small or large percentages of these parasites.

Shipments of pupæ collected as such would preferably be made in cold storage. The most of the parasites, including those which are or which appear to be of the most importance, emerge coincidently or nearly so with the moths themselves, and if sent by ordinary mail they are apt to issue and die en route.

COCOONS OF HYMENOPTEROUS PARASITES.

There is only one hymenopterous parasite of demonstrated importance which attacks the gipsy moth, and which spins a cocoon outside of the host. This is *Apanteles fulvipes*, of Europe and Japan, and it is probable that the numbers of its cocoons imported as such have amounted to at least 1,000,000.

Little care is necessary in packing these for shipment, other than that they must not be crushed, nor yet too damp. A considerable degree of dampness has been sustained without injury, but upon one occasion in which they were packed between sheets of damp blotting paper, there was sufficient moisture present to thoroughly soak the cotton and some loss resulted.

The Japanese have displayed no little ingenuity in devising new methods for sending these, and with one exception, just noted, all have been good so far as packing was concerned. One method, which possessed a certain advantage over the others in permitting the adults which chanced to emerge en route a certain amount of very advantageous freedom, was used in a single shipment, which, partly on that account but principally on others, ranks as immeasurably the best ever received. The cocoons, to the number of about 1,000, were inclosed in a little wicker cage, which in turn was inclosed in an envelope of mosquito netting which prevented the cocoons from scattering out, but did not hinder the escape of the adults. This cage was supported in the very center of a large, otherwise empty wooden box by means of strings which were passed through screw eyes in the middle of each side and drawn taut. There was nothing loose in the box to crush the delicate parasites, no matter how roughly it was handled, and they were not only given ample space to expand and stretch their wings, but they were kept inactive by the perfect darkness (or at least were presumably so). It would be a simple matter to spray a portion of one side of the box with a very fine dew of honey, and if this were done the life of the adults would probably be considerably prolonged.

Cold storage is an absolute necessity if the cocoons of this parasite are to be as much as a week en route. The transformations are apt to be concluded in considerably less than one week after the spinning of the cocoon if the weather is hot, and ever under the best of conditions which can be devised for keeping them alive the mortality is heavy. Even in the ordinary temperature of a steamship's cold room development continues.

TACHINID PUPARIA.

The importation of tachinid puparia is by no means so simple as the importation of Apanteles cocoons, but at the same time it is easy as compared with the difficulties attending the importation of live gipsy-moth caterpillars from which to rear these puparia in America.

In all, quite large numbers have been received from both Europe and Japan. A variety of methods has been tested in the hope of hitting upon one that would be applicable for the purpose. Shipping in damp earth was early attempted, and seems to be the very first method which suggests itself to anyone wishing to ship a quantity of them, but of all ways it is very nearly the worst. It would probably be the best, if the larvæ could be allowed to enter the earth naturally and if they were left there wholly undisturbed throughout the time they were in transit, but mingled with damp earth and placed in a box to be sent by mail or express, disaster is pretty sure to result. Cotton has also been used several times, and it is usually as bad and sometimes worse. With the exception of excelsior, cotton is about the worst packing for living insects that has come under observation at the laboratory, although gritty moss, of a sort which dries brittle, is also bad. Presumably there are other worse substances, but they have not been discovered at first hand.

Probably the best packing material is slightly damp and preferably living sphagnum moss. The live moss retains its moisture in a manner wholly different from the moss which has been killed, dried, and subsequently dampened. Test shipments, which were sent to France and back without being opened, returned to the laboratory in good and almost unaltered condition, in the case of those which were inclosed in tight boxes. Even when fully exposed to the air the living moss seems to dry much more slowly and to hold its moisture more naturally. Sphagnum possesses the great additional advantage of being much softer when dry than most other kinds of moss.

One disadvantage attending the shipment of puparia, no matter how they are packed, is that of secondary parasites. A single colony of Dibrachys, issuing en route from one of a lot of puparia, will result in the parasitism of a large proportion of the remainder. This might possibly be prevented by packing in sand or earth, but this appears to be about the only advantage possessed by that method.

The puparia of certain tachinids must be kept damp, but this is not at all necessary in the case of all. Methods of packing and shipment will depend upon the characteristics of the particular species under consideration. So far as known, no tachinid which forms a free puparium outside of its host is injured by exposure to moisture.

The pupal period of the majority of the tachinid parasites of the gipsy moth and the brown-tail moth is quite short, usually lasting less than two weeks. It is therefore necessary to make use of cold storage en route, in order to make certain that the adults will not hatch before arrival. By far the larger part of the puparia which have been received at the laboratory have thus hatched, except when they were of species which naturally hibernated unless they were shipped in cold storage.

CALOSOMA AND OTHER PREDACEOUS BEETLES.

Quite a variety of the large carabid beetles has been imported from abroad for experimentation as to their serviceability as enemies of the gipsy moth, or for liberation in the field after this point had been demonstrated satisfactorily. At first some difficulty was experienced in accomplishing their importation successfully, but later it was found to be a simple matter if proper care was used in packing. The great majority of them have come in ordinary safety-match boxes (Pl. X, fig. 2), each box containing one beetle and a wisp of sphagnum moss. Usually one or two caterpillars or other sort of *succulent insect* have been included for the purpose of lunch en route, but the practice is of rather doubtful value in the case of those species which have been handled in the largest numbers in the laboratory. Should the beetle not fancy the quality of the sustenance provided, or refuse to eat for any other reason, death and decomposition of the victim may result disastrously and be prejudicial to the health of the beetle.

These small match boxes have been packed in larger wooden boxes and sent through the ordinary mails with little loss of life. Other small wooden or paper boxes have similarly been used with equal success.

Cold storage has occasionally been employed in a few minor shipments from Europe with very good results. In 1910 a large shipment of living beetles was received in cold storage from Japan in most excellent condition.

QUANTITY OF PARASITE MATERIAL IMPORTED.

In Mr. Kirkland's first report as superintendent for suppressing the gipsy moth and the brown-tail moth in Massachusetts, several pages were devoted to a detailed account of each shipment of parasite material received from abroad. After the first year no attempt to continue this practice was made, and if it were now attempted to treat each separate shipment with the same attention to detail, several hundred additional pages would be required.

Accordingly, in order that at least a rough idea of the quantity of material handled at the laboratory may be had, Table VII, which, without being absolutely accurate, is very approximately so, has been prepared by Mr. R. Wooldridge, an assistant at the laboratory.

TABLE VII.—*Table showing number of boxes received at the laboratory since beginning of work.*

	1905	1906	1907	1908	1909	1910
Porthetria dispar egg masses..................boxes..			1	.18	32	1
Porthetria dispar larvæ and pupæ.............do....	131	923	1,539	307	8,391	5,956
Euproctis chrysorrhœa egg masses.............do....		46	87	17	1	0
Euproctis chrysorrhœa webs..................webs..		117,259	55,082	32,830	29,295	29,696
Euproctis chrysorrhœa larvæ and pupæ.....boxes..		313	1,159	160	1,167	381
Apanteles fulvipes and Apanteles lacteicolor..do....				13	21	63

LOCALITIES FROM WHICH THE PARASITE MATERIAL HAS BEEN RECEIVED.

Mr. Wooldridge has also prepared the accompanying map (fig. 11) showing the various localities from which parasite material has been received each year from 1905 to 1910, inclusive. It will indicate the thoroughness with which the more accessible parts of the world have been searched for parasites of these pests.

THE EGG PARASITES OF THE GIPSY MOTH.

ANASTATUS BIFASCIATUS FONSC.

The first individuals of this species (fig. 12, female) were reared at the laboratory in the spring of 1908 from eggs imported the previous winter from Europe and Japan. The dissimilar sexes were not immediately recognized as of the same species, and for a few days there was some doubt as to whether one, two, or four were represented among the few scattering specimens emerging. The number was soon reduced to two through the obvious attraction between the sexes, and soon after to the one, when the senior author had an opportunity to examine and compare series from European and Japanese sources.

Their issuance had been anticipated a long time before, and a quantity of gipsy-moth eggs had been collected in the summer, before embryonic development had progressed beyond its initial stages, and placed in cold storage. It was thought possible that some species of parasite might be reared from imported eggs during the fall or winter which habitually and necessarily oviposited in undeveloped eggs, and it was hoped that those collected in the summer might be kept fresh enough to serve as host material for laboratory reproduction.

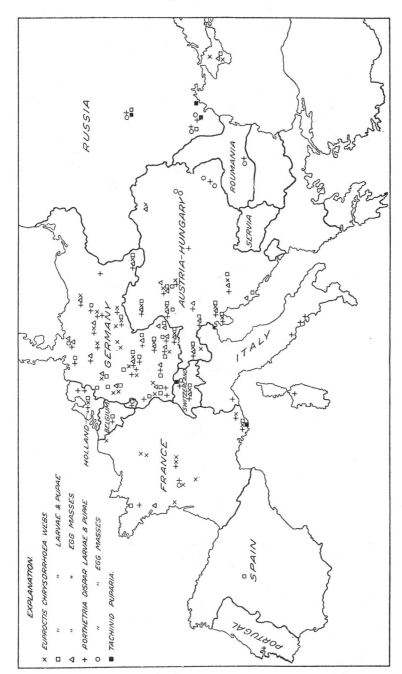

Fig. 11.—Map showing various localities in Europe from which parasite material has been received. (Original.)

As soon as females of Anastatus were secured, some of these eggs were removed from storage and found to be dead, with the contents partially decomposed. Nevertheless an attempt was made to use them, and the parasites were given their choice between them and others which contained embryonic caterpillars.

A few days after their emergence the females began to betray an interest in both sorts of eggs, and were several times observed in the act of oviposition or attempted oviposition. Apparently this was successfully accomplished, but without further results, for no second generation resulted. The experiment served one purpose, however, in indicating beyond reasonable doubt that the insect actually was a parasite upon the eggs of the gipsy moth, and not upon any chance form of insect life accidentally included.[1]

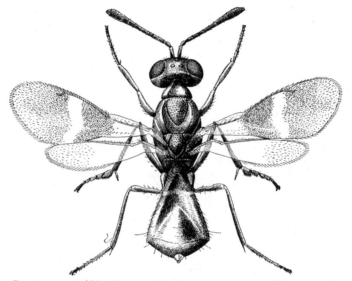

Fig. 12.—*Anastatus bifasciatus:* Adult female. Greatly enlarged. (From Howard.)

The exposure of the imported eggs to warmth for the purpose of hastening the emergence of any parasites which they chanced to con-

---

[1] How great is the likelihood of error when parasites are reared from unbroken egg masses has several times been demonstrated in the course of the investigations and rearing work at the laboratory. Upon several occasions small Lepidoptera have been reared from egg masses, and more than once their parasitized pupæ have been found. Eggs of other species of insects, and occasionally parasitized scale insects, have also been found attached to bits of bark to which egg masses were attached. Very frequently cocoons of *Apanteles fulvipes* are found, wholly or partially covered by the egg mass, and from them several species of hibernating secondaries have been reared. There is a record of a minute eulophid, allied to Entedon, having issued from a small lot of eggs which had been separated from nearly every trace of foreign matter. It was thought then and is still believed that these came from the eggs themselves, and that they were actually parasitic upon either Anastatus or Schedius, but when the material from which they issued was examined two or three cocoons of *Apanteles fulvipes* were found mingled with it, and what might otherwise have been a clear record was spoiled.

There is in Japan a limacodid moth—*Parasa sinica* Moore (*hilarula* Staud), as determined by Dr. H. G. Dyar—which appears habitually to seek out the gipsy-moth egg masses as a site for pupation. The larva buries itself in the mass before spinning its cocoon, and from outward appearances its presence is hardly noticeable. More than 25 of these moths have been reared under these circumstances from imported egg masses, or their cocoons have been found and destroyed.

tain in the hope that laboratory reproduction could be secured was soon recognized to be a mistake, and as the Anastatus continued to emerge considerably ahead of the time when they would obviously have issued under more natural conditions, it was resolved to remedy the evil, if possible, by placing the parasitized material in cold storage. This experiment was successful. The further transformations of the parasites were retarded without any apparent prejudicial effects upon their vitality, and in July some 500 were reared and colonized in the field.

FIG. 13.—*Anastatus bifasciatus:* Uterine egg. Greatly enlarged. (Original.)

Coincidently with the height of their emergence and subsequent to its close, a considerable number of a small black encyrtid, later described by the senior author as *Tyndarichus navæ*, issued, and all were destroyed on the supposition that they might be secondary. This was not by any means certain, and it was resolved to investigate their habits thoroughly so soon as opportunity should offer.

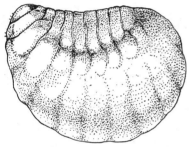

FIG. 14.—*Anastatus bifasciatus:* Hibernating larva. Greatly enlarged. (Original.)

Accordingly, in the fall of 1908, following the receipt of several considerable shipments of egg masses from Japan, an exhaustive investigation of the gipsy-moth egg parasites was inaugurated. These investigations were more intimately associated with the work upon Schedius, and more will be said of them in the discussion of that species. So far as Anastatus was concerned, its life and probable habits stood revealed from the start. Almost in the beginning its larvæ were found (fig. 14) and identified correctly, as was later proved. They were almost invariably found in eggs which had been destroyed before embryonic development had taken place, which showed conclusively that these eggs were attacked within a very short time after their deposition. It was known that the adults did not issue until after the caterpillars had hatched from healthy eggs in the spring, and the fact that the species was single brooded, with a life cycle that was correlated perfectly with that of the gipsy moth, was as certainly evident then as now, after two years' observation of its progress in the field has given ample confirmation.

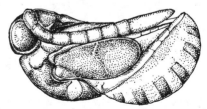

FIG. 15.—*Anastatus bifasciatus:* Pupa from gipsy-moth egg. Greatly enlarged. (Original.)

The egg of Anastatus has not been seen after deposition, but its appearance before is indicated by figure 13. The full-fed larva removed from host egg is well represented by figure 14, and the pupa by figure 15. This latter is very beautifully colored, the creamy ground color being set off by darker abdominal bands and wing covers, and by the delicately tinted reddish eyes.

It was soon demonstrated by a careful study of the European eggs that no other parasite and no secondaries were present. These eggs were therefore kept in confinement until after the caterpillars had all hatched in the spring. Then those which remained were examined, and the number which contained parasites carefully estimated, and found to be about 80,000, nearly all of which were contained in a very large shipment received during the winter through Prof. Jablonowski, and collected from various Hungarian localities.

The Japanese eggs, which contained numerous secondary parasites as well as Anastatus, were all carefully rubbed clear of their hairy covering, and those which contained Anastatus larvæ (Pl. XI, fig. 2) carefully and painstakingly picked out by hand, one by one. In this manner enough to make a total of nearly 90,000 of the parasites were secured.

One exceedingly important characteristic of the parasite was not considered with sufficient attention at a time when this might have been done. Several observations upon the activities of the females in the summer of 1908 had led the observer to question their ability to fly; but when several of them were placed upon a large sheet of paper and stirred into action, they disappeared with sufficient celerity to banish any doubts which may have been entertained. In considering these crude experiments in retrospect and in the light of subsequent developments, it would appear that their jumping abilities were rather underestimated, because it is now certain that they are either unwilling or else, like the female of their host, are unable to fly.

A most careful examination of egg masses in the vicinity of the locality where the colony of about 500 had been liberated the summer before had failed to discover the presence of parasitized eggs. It is now known that this was due to the accident of placing this colony in a locality where the gipsy-moth "wilt" disease proved later to be so destructive as to kill all the pupæ which were present when the first of the parasites were liberated, and which it was then thought would produce moths enough to deposit a sufficiency of eggs for attack. As a result of this extreme percentage of pupal mortality, there were practically no eggs within a radius of several hundred feet.

The cause of failure not being apparent, it was guessed that it might be due to the extremely rapid rate of dispersion rather than to the reverse, and to provide against loss through too rapid dispersion at first, very large colonies were decided upon as most advisable.

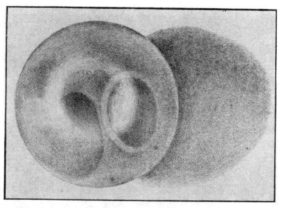

Fig. 1.—Egg of Gipsy Moth, containing developing caterpillar of the Gipsy Moth. Greatly enlarged. (Original.)

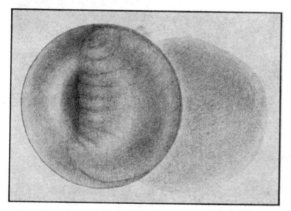

Fig. 2.—Egg of Gipsy Moth, containing larva of the parasite Anastatus bifasciatus. Greatly enlarged. (Original.)

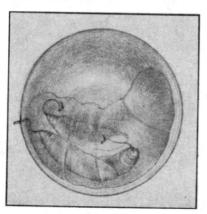

Fig. 3.—Egg of Gipsy Moth, containing hibernating larva of Anastatus bifasciatus, which in turn is parasitized by three second-stage larvæ of Schedius kuvanæ. Greatly enlarged. (Original.)

Bul. 91, Bureau of Entomology, U. S. Dept. of Agriculture. PLATE XII.

FIG. 2.—VIEWS OF CAGE PREPARED FOR USE IN COLONIZATION OF ANASTATUS BIFASCIATUS IN 1911. *A*, FRONT VIEW OF CAGE; *B*, BOTTOM OF CAGE. (ORIGINAL.)

FIG. 1.—VIEW OF CAGE USED FOR COLONIZATION OF ANASTATUS BIFASCIATUS IN 1910. (ORIGINAL.)

The 90,000 parasitized eggs were divided into five lots and placed in the field at the proper time in localities where an abundance of eggs was certain.

The parasites hatched in due course and were found attacking the egg masses in a businesslike manner that was quite encouraging, but only within a very short distance of the center of the colony. The species thus spreads slowly.

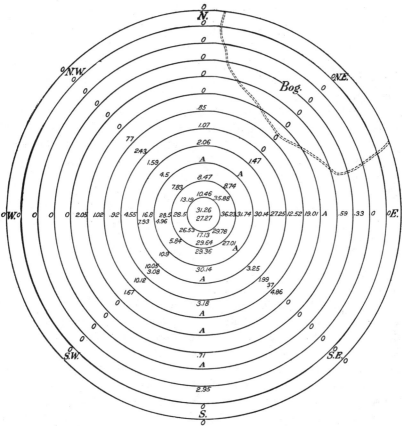

FIG. 16.—Diagram showing two years' dispersion of *Anastatus bifasciatus* from colony center. Each concentric circle represents a distance of 50 feet from the smaller or larger circle next it. *A* indicates parasitism of gipsy-moth egg-masses by *Anastatus bifasciatus* and *O* indicates absence of parasitism by Anastatus. The figures give percentages of parasitism. (Original.)

The accompanying diagram (fig. 16), which has been prepared by Mr. Wooldridge largely from the results of his own work, together with Table VIII, will serve as well as words to tell the story of the dispersion of this parasite in one of the 1909 colonies, and results of similar studies in various other colonies are substantially the same.

TABLE VIII.—*Average percentage of parasitism of gipsy-moth egg masses at different distances from center of colony.*

Distance from center.	Number of egg masses collected.	Percentage of parasitism.	Distance from center.	Number of egg masses collected.	Percentage of parasitism.
At center	20	29.26	350 feet	50	0.41
50 feet	66	24.68	400 feet	70	.18
100 feet	78	21.75	450 feet	60	.055
150 feet	80	14.43	500 feet	70	.42
200 feet	60	8.61	550 feet	70	.00
250 feet	100	3.59	600 feet	70	.00
300 feet	85	3.44			

When in the fall of 1909 it had become rather certain that the rate of dispersion of Anastatus was only going to be about 200 feet per year, plans for colonization along very different lines in 1910 were immediately put into execution. In four of the five colonies all of the egg masses which could be easily secured were collected and brought to the laboratory, where the eggs were separated from their hairy covering. This is best effected by gently rubbing them over a piece of cheesecloth stretched on a frame. (Pl. XX, fig. 2.) The hairs pass through and the eggs are left.

The number of parasitized eggs present was then estimated, and found to be very close to 90,000. In the spring, after all of the healthy eggs had hatched, those remaining, including all which were parasitized, were divided into 100 lots, each of which was supposed to contain approximately 900 parasite larvæ. An equal number of small, wire-screen cages was prepared (Pl. XII, fig. 1), and about the middle of June, when the male parasites began to issue, and when it was becoming possible to determine with some degree of assurance just where there were likely to be large numbers of gipsy-moth eggs a little later, the work of placing these cages in the open was begun. (See also Pl. XII, fig. 2, showing front and bottom of cage prepared for use in Anastatus colonization in 1911.) They were finally placed, each in a separate locality, and each, so far as has been determined by subsequent investigation, in localities where the parasites had an excellent opportunity to work to the best advantage as soon as they issued. Not all of these colonies have since been visited, and probably some of them never will be seen again, but all that have been examined have been found in the best of condition.

Early in the fall of 1910 the dispersion studies of 1909 were repeated, with results which have already been indicated in Mr. Wooldridge's diagram, and the egg collections were also repeated for the purpose of securing material for additional colonization work in 1911. With little difficulty some 270,000 parasitized eggs have been secured, and were it not for the fact that the proper care in placing the number of colonies thus provided for will probably tax all available resources at the time when they must be placed, if placed to advantage, more could easily be collected.

The rate of increase in the field, as indicated by the work which has been done, is not excessive, but probably amounts to something like sixfold per year. The extreme limit of dispersion discernible in 1909 was not quite half that of the extreme for two years, as indicated in the diagram. It is possible that it may become more rapid as time goes on, and it is rather expected that a high wind, at an opportune time, will assist materially in the dispersion of the species. Should it not, it will require a very long time for it to become generally established everywhere through the infested area. Even though there were a colony planted to each square mile, something like 16 years would elapse before all of them met and fused, unless the present rate of dispersion were accelerated.

It has been pretty definitely proved of Schedius that it can only attack the uppermost layer of eggs in each mass, and the same is equally well proved in the case of Anastatus. Since there are two layers of eggs, and usually three in all but the very smallest masses, it is evident that the usefulness of Anastatus is still further reduced through its physical limitations. The figures of percentages given in the diagrams probably represent about the maximum which can ever be expected. None the less, this means a distinct benefit, and with all its faults, Anastatus stands high in favor at the present time.

In its distribution abroad, Anastatus is, as might be expected, of quite local occurrence. It has been received from about half of the localities represented by the European importations, and in very variable abundance. The numbers found in five lots of what was estimated as 1,000 egg masses each, received from five different localities in Hungary through Prof. Jablonowski in the winter of 1908-9, is rather typical in this respect. As estimated through careful examination and counts, these numbers were as follows:

Laboratory Number.	Locality.	Number of Anastatus.
3017	Lippa (Temes)	34,000
3018	Bustyhaza (Maramoros)	0
3019	Huszt (Maramoros)	208
3020	Dorgos (Temes)	6,099
3021	Sistarobecz (Temes)	39,000
	Total	79,307

In Japan it is also unevenly distributed. The most which were received from that country were in a lot of eggs from Fukuoka Ken, received during the same winter as those above mentioned from Hungary. It is not at all common from the vicinity of Tokyo, and while it is present in nearly every lot of Japanese eggs which has been received, in every instance but one (the shipment above mentioned) the number present has not been sufficiently large to make the rearing of the parasite economically worth while. It is interesting and possibly significant that there was no Schedius in the one locality where Anastatus was sufficiently common to be considered as a parasite of consequence, while in the other localities, where Anastatus was rare, Schedius abounded. More than one instance has been observed in which parasites having similar habits alternate but rarely or never occur simultaneously in anything like equal abundance in one locality. Two fairly consistent examples of this sort will receive further mention later on, in which the tachinids *Dexodes nigripes* and *Compsilura concinnata*, and *Tachina larvarum* and *Tricholyga grandis* are respectively involved.

SCHEDIUS KUVANÆ HOW.

Only one species of gipsy-moth egg parasite has been received at the laboratory from Europe, but in Japan there are two, and, so far

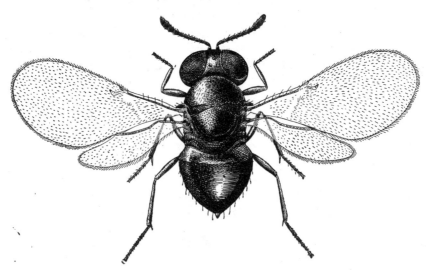

FIG. 17.—*Schedius kuvanæ*: Adult female. Greatly enlarged. (From Howard.)

as may be determined from their comparative abundance in the material from that country which has been studied, *Schedius kuvanæ* (fig. 17) is the more common and important as a factor in the control of its host. It resembles Anastatus in its choice of host, and in the fact that it is similarly limited through physical inability from attacking more than a limited percentage of the eggs in each mass. In every other respect the two species are widely different.

Anastatus is a true egg parasite, and rarely attacks successfully the eggs in which the young caterpillars have begun to form. Shedius, on the contrary, is strictly speaking an internal parasite of the unhatched caterpillar. Anastatus passes through but one generation annually, and its seasonal history is closely correlated with that of its host. Schedius, on the contrary, will pass through a generation per month, so long as the temperature is sufficiently high, and its seasonal history is in no way correlated to that of the gipsy moth. It appears not to hibernate in the gipsy-moth eggs, and it is quite probable that an alternate host is necessary to carry it through the summer months after the gipsy-moth eggs have hatched in the spring, and before the moths begin depositing eggs for a new generation.

At the time when the popular account of the parasite-introduction work was prepared for publication through the office of the Massachusetts State Forester it was considered to be much the more promising of the egg parasites, and its history in America was spoken of as one "of the most satisfactory episodes in the work of parasite introduction." The account of the first successful importation of living specimens as given at that time is included in the two following paragraphs, which are quoted verbatim.

As long ago as the spring of 1907 a few dead adults were secured in an importation of gipsy-moth egg masses received during the winter from Japan, but none was living on receipt. During the winter next following, large importations were made, and many thousands of eggs, from which some parasite had emerged, were found, but not a single living specimen was obtained. It was evident that it completed its transformations and issued in the fall, and that, if it hibernated in the eggs, it was warmed to activity while the packages were in transit to America, and the adult parasites either died or escaped en route.

In the fall, winter, and spring of 1908–9 a large quantity of eggs of the gipsy moth were received from Japan, the shipments beginning early in the fall and continuing until nearly time for the caterpillars to hatch in the spring. The first, received in September, contained hundreds, possibly thousands, of the parasites, which had issued from the eggs en route, and all of which, as usual, had died; not a single living individual was received. Specimens were referred to Dr. Howard, who found that they represented an entirely new and hitherto undescribed species, which he named after Prof. Kuwana, who collected and sent the eggs from which they had issued. A single pair of living specimens rewarded the careful attention which was lavished upon the importations received later in the fall and during the winter, and it was not until April, 1909, that a mated pair could be secured. During that month a total of 11 individuals issued from cages containing Japanese eggs recently received.

These 11 individuals served as the progenitors of a numerous and prolific race, but the story of the investigations which were made upon the various shipments of egg masses received at the laboratory from September, 1908, to April, 1909, which was not touched upon in an earlier account, is perhaps worthy of a place here.

LIFE OF SCHEDIUS AND ITS RELATIONS TO OTHER EGG PARASITES, PRIMARY AND SECONDARY.

Mention has already been made of the rearing of a small encyrtid parasite from Japanese eggs in company with Anastatus in the summer of 1908, of the doubts which were felt as to its true character, and of the resolve to investigate the matter thoroughly when the opportunity should arise. In accordance with this resolve an intensive study of the Japanese importations was begun in December, 1908. A large number of egg masses, which showed by the exit holes (of Schedius) that they had been freely attacked by some parasite which had issued in the fall, were selected, then "sifted," and the eggs from each mass were then carefully examined and sorted into three lots, composed, respectively, of the healthy eggs, the eggs from which parasites had issued, and the eggs which were neither one nor the other. Those falling in this third division were scrutinized again with still more care. Anastatus was quickly recognized, in most instances, and eggs containing its larvæ placed aside.. In the majority of the remainder there was evidently no life, but in a considerable number minute, white larvæ could more or less plainly be seen, surrounded and more than half concealed by the remains of the embryonic caterpillars which had been destroyed. These eggs were isolated in small vials, in order that there could be no question concerning the identity of the particular host egg from which any particular parasite issued.

Long before this work was completed the necessity for all the care that was being expended to secure accurate results was made manifest by the emergence of no less than three species of parasites from isolated or partially isolated eggs. The first of these to appear was a species of Pachyneuron (determined by the senior author as *P. gifuensis* Ashm.), and on account of known habits of other members of the genus was placed as probably secondary. Nevertheless it was given an opportunity to prove itself a primary if it would, and the specimens as they issued were confined in vials with gipsy-moth eggs, some of which contained the healthy caterpillars, while others harbored the larvæ of Anastatus. The Pachyneuron paid not the slightest attention to either, but invariably died without attempting oviposition.

The next species to issue was *Tyndarichus navæ* How., and it was with considerable surprise that it was recognized as different from Schedius. On account of the strong superficial resemblance between the two it had been supposed up to that time that they were one and the same.

The third was *Perissopterus javensis* How., of which a single specimen only was reared. To date this record is unique, and the species has previously been reared only from scale insects.

There was other and pressing work to be done with the parasites of the hibernating brown-tail caterpillars, and a realization of the difficulties which were likely to attend the prosecution of the egg-parasite investigations, thus complicated by the discovery that five and possibly more parasites were involved of which only one was definitely proved to be primary, was the prime argument which finally resulted in the detachment of Mr. H. S. Smith from the cotton boll weevil investigations and his transfer to the laboratory staff. By the time he was prepared to undertake his new work a large number of eggs from which Anastatus, Tyndarichus, and Pachyneuron were positively known to have issued were ready for dissection and study, and to these were soon added a number from which Schedius was similarly known to have come, secured in the manner about to be described.

The first Schedius which was ever reared in a living condition issued from an isolated egg in the laboratory in December, 1908. It was a male, and it died before it could be furnished with a mate. The next individual issued on January 8 from an egg which had been isolated on December 19. It was a female, and she was immediately transferred to a large vial containing an egg mass freshly collected from the field. Within a few days after being thus confined she was observed in the act of oviposition, and parthenogenetic reproduction ensued. Her progeny began to issue February 16, and up to February 25 no less than 28 males were reared.

The experiment was tried of confining her with several of her asexually-produced progeny in the hope that she might thus be fertilized and produce females. The experiment did not succeed at that time, apparently because she was not able to deposit any more eggs. She remained alive until March 2, but was dead on March 6, after at least eight weeks of active life.

FIG. 18.—*Schedius kuvanæ:* Egg. Greatly enlarged. (Original.)

The eggs from which these parthenogenetically-produced males issued were known beyond peradventure of a doubt to have produced Schedius, and never to have contained any other parasite, and together with those from which Anastatus, Tyndarichus, and Pachyneuron were known to have issued, made complete the series which was to be dissected.

The dissection work was mostly done by Mr. Smith, but he was not alone when it came to puzzling over the problems in parasite anatomy and parasitic interrelations which this work produced in abundance. The contents of the individual eggshells were scrutinized with the utmost care, and slowly the various anatomical remains found therein were associated with one parasite or another.

In the course of these studies it was discovered that Schedius deposits a large egg (fig. 18), which is supplied with a very long stalk. The egg is placed within the body of the unhatched but fully formed caterpillar, with the end of the stalk projecting outside.

Sometimes, and apparently usually, the end of this stalk passed through the shell of the egg as well as through the body of the caterpillar, as indicated in the figures (fig. 19, Pl. XI, fig. 3). When the egg hatches, the larva does not entirely leave the shell, but remains with its anal end thrust into it, and the stalk, which is hollow, becomes functional and acts like a lifeline attached to a submarine diver in supplying a connection with the outer air. As the larva grows the stalk increases in thickness, and the last anal segment of the larva becomes covered with a thick chitinized shield, which is unaffected by the action of strong caustic potash. There are two larval molts, and consequently three larval stages. During the entire course of both the first and second the young parasite remains quite firmly attached to its anal shield and lifeline and the cast skins are not entirely sloughed off, but are merely pushed backward. After the third ecdysis it retains this connection for awhile, and grows rapidly, but about the time when it reaches maturity the connection with the shield is broken, thus proving that it is not part and parcel of the integument. It would appear rather that this shield, including a tube within the egg-stalk (which, as stated, grows in thickness after the egg itself hatches), is actually part of the integument of the first-stage larva, and that the second and third stages merely continue to use what is in effect the skin of the first larval molt.

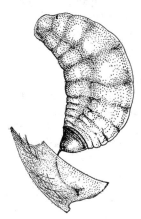

FIG. 19.—*Schedius kuvanæ:* Third-stage larva still retaining attachment to egg-stalk, and anal shield. Greatly enlarged. (Original.)

The host caterpillar is completely destroyed except for the harder chitinous parts, head, tarsal claws, hooks of the prolegs, etc., and the hair, which is left in a sort of hank, more or less completely surrounding and concealing the parasite larva. It is impossible to distinguish between the larvæ of Schedius and those of its secondaries from an external examination of the eggs.

FIG. 20.—*Schedius kuvanæ:* Pupa. Greatly enlarged. (Original.)

After the larva reaches its full growth and casts off its anal shield, it quickly pupates (fig. 20) and very shortly thereafter issues as an adult. There is no indication of a desire to hibernate during any part of the preliminary stages, in which respect Schedius differs from nearly every other chalcidid which has been studied at the laboratory.

Schedius is to all purposes, if not to all intents, a secondary parasite upon occasion. In the spring of 1909 a generation was carried through to maturity within the larvæ of Anastatus, and at that time there was no difficulty experienced in inducing the Schedius females to oviposit in such. In the course of later experiments which were designed to determine whether there was any preference shown between the eggs containing healthy caterpillars and those with the larvæ of Anastatus, only the healthy eggs were selected for oviposition by the parent females. What was more, although several later attempts were made to force Schedius to oviposit in eggs containing Anastatus larvæ, none but the first was successful.

Oftentimes two or more eggs are deposited in one host. Numerous instances have been found in which second-stage larvæ were feeding peaceably side by side as the result of such superparasitism, and still more have been observed in which the former presence of more than one individual was positively indicated by the presence of more than one egg-stalk and anal shield, but never, out of many thousands of examples under observation, has more than one adult parasite issued from one egg. What happens to the supernumerary individuals is not indicated further than that they disappear, and that their substance goes to nourish the sole survivor. Whether there is an actual struggle for supremacy in which victory comes to the strongest, or whether the struggle takes the form of a contest to determine which shall quickest consume the available food supply, the loser calmly surrendering his body to the winner by way of forfeit, has never been revealed.

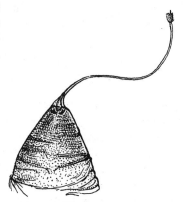

FIG. 21.—*Schedius kuvanæ:* Egg-stalk and anal shield of larva as found in host eggs of gipsy moth from which the adult Schedius has emerged, or in which the Schedius larva has been attacked by a secondary parasite. Greatly enlarged. (Original.)

FIG. 22.—*Schedius kuvanæ:* Larval mandibles. Greatly enlarged. (Original.)

FIG. 23.—*Tyndarichus navæ:* Larval mandibles. Greatly enlarged. (Original.)

The story of a triple tragedy is told in Plate XI, figure 3, which is drawn from a slide prepared by Mr. Smith. It represents a single gipsy-moth egg, which had been attacked by Anastatus before the embryonic caterpillar had developed sufficiently to leave perceptible remains. The Anastatus, after consuming the entire contents of the eggshell had reached the hibernating stage, and settled down to some 10 months of inactivity, when it was attacked by Schedius. No less than three Schedius eggs were deposited in fairly rapid succession (but probably by different parents) since the three larvæ, the outlines of which are shown, are practically equal in size. All are apparently about ready to molt for the second time, and after this molt, if they had been allowed to live, one would most certainly have gained the mastery and devoured the others.

FIG. 24.—*Pachyneuron gifuensis:* Egg. Greatly enlarged. (Original.)

But this conflict for supremacy, sanguinary as it is, is only the beginning of what might occur in the open in Japan. Tyndarichus and Pachyneuron are both habitually and essentially secondary parasites, and both prey not only upon Schedius, but upon each other with perfect impartiality. Either might attack the surviving Schedius, and be in turn the victim of the other, and there is no apparent reason why Schedius should not return to the fray and, by destroying its own secondary, start the battle all over again.

FIG. 25.—*Pachyneuron gifuensis:* Larval mandibles. Greatly enlarged. (Original.)

Such a long-drawn-out contest is hardly likely to occur very often, but in many instances tales scarcely less sanguinary have been told by the relics which strewed the field of battle. Among these relics the anal shield with egg stalk and the characteristic mandibles (figs. 21 and 22, respectively) have served as positive indication of the former presence of Schedius. Tyndarichus is betrayed by its mandibles (fig. 23), which, like those of Schedius, retain their characteristic form through all three stages. The former presence of Pachyneuron, curiously enough, is quite easily recognizable by its characteristic eggshell (fig. 24), which is of a substance which defies the action of hot concentrated caustic potash sufficently prolonged to result in the complete solution of the gipsy-moth eggshell. It may also be recognized by its mandibles (fig. 25), which are rather small and inconspicuous in any but the last stage. Anastatus, when its former presence can be proved at all, may be recognized by its mandibles also (fig. 26), but these are so small as to be very difficult to find, and it is altogether probable that there have been eggs dissected in which Anastatus was the original primary parasite, but of which fact no proof remained.

FIG. 26.—*Anastatus bifasciatus:* Larval mandibles. Greatly enlarged. (Original.)

In order that some idea may be had of the conditions which actually prevail in the open in Japan, results of the dissection of 43 eggs from Japanese importations are given below. Many other eggs were dissected, in some of which the tale was too complicated to be unraveled, and it is, of course, necessary to leave out of consideration here the results of those dissections which were made before the significance of that which was found was fully recognized.

In the formulæ which follow the symbols are to be read as follows: ×=Parasitized by; +=Superparasitized by.

Thus the conditions represented in the figure to which attention has already been drawn would be expressed:

   Porthetria dispar × Anastatus × Schedius.
                              + Schedius.
                              + Schedius.

The host relations revealed by dissections of eggs from which Pachyneuron emerged are similarly indicated as follows:

   Dispar × Schedius × Pachyneuron (20 times).
   Dispar × Schedius × Pachyneuron.
                      + Pachyneuron (1 time).
   Dispar × Schedius.
         + Schedius × Pachyneuron (3 times).
   Dispar × Schedius.
         + Schedius.
         + Schedius.
         + Schedius.
         + Schedius × Pachyneuron (1 time).
   Dispar × Anastatus × Pachyneuron (1 time).
   Dispar × Anastatus.
         + Schedius × Pachyneuron 1 time).

Dissections of eggs from which Tyndarichus emerged resulted as follows:

   Dispar × Schedius × Tyndarichus (11 times).
   Dispar × Schedius.
         + Schedius × Tyndarichus (2 times).
   Dispar × Schedius.
         + Schedius × Tyndarichus.
                   + Tyndarichus (1 time).
   Dispar × Schedius × Pachyneuron.
                   + Tyndarichus (1 time).
   Dispar × Anastatus × Tyndarichus (1 time).

Mention has already been made of the parthenogenesis of Schedius, and the fact that only males were produced in the first attempt of successful reproduction experiments in which only a single female was available. Numerous subsequent experiments have demonstrated beyond question that thelyotoky is the rule and that exceptions are rare if they ever occur.

In the course of the first unavoidable experiment in parthenogenesis the attempt was made to secure the fertilization of the female through union with her own asexually produced offspring, but, although she lived after they had completed their transformations, no results were secured. It seemed to be within the bounds of possibility that success would follow if the experiment were differently conducted, and accordingly in the fall of 1909 Mr. Smith repeated it, with this variation, that the females, after they had deposited a few eggs, were rendered dormant by exposure to moderate cold, awaiting the issuance of their progeny. This time no difficulty was experienced. The parthenogenetically produced males mated freely with their respective parents, and the subsequent progeny in each of several instances consisted of both sexes.

Females thus reared were mated with their brothers (which were at the same time their nephews), reproduced with the ordinary freedom, and their progeny were of both sexes in the usual proportions. Still another generation showed no signs of weakness or any sort of abnormality, and the experiment was discontinued.

In sexual reproduction the males appear always to be largely outnumbered by the females. Nothing like the diversity in this respect which has been noted in the case of other chalcidids has been observed in the case of Schedius.

REARING AND COLONIZATION.

When the first individuals of Schedius were secured from the imported Japanese egg masses in April, 1909, there was no difficulty in securing reproduction upon gipsy-moth eggs collected in the open, but by the time the second generation was secured those which had remained in the open were about to hatch, and would hatch almost immediately they were brought indoors. A large quantity of eggs had been placed in cold storage in anticipation of this, and it was found that these would hatch nearly as quickly when they were removed. Oviposition at any time within a few hours of the time when the eggs would otherwise hatch was generally successful, but when the eggs hatched within 36 hours after being exposed to the degree of warmth necessary to secure oviposition of the parasite, it soon became evident that not very much increase was to be expected. Accordingly, the experiment was made of killing the host eggs through exposure to just enough heat to bring this about. The parasites oviposited in these dead eggs with the same freedom that they would attack the living, and reproduction ensued. The progeny, however, were small and weak, and not as prolific as those secured earlier in the spring.

Thus, in one way and another the species was carried through the summer, and with the deposition of fresh gipsy-moth eggs early in July much better results were secured, and the parasites immediately began to increase rapidly in numbers with each succeeding generation. By August there were enough to make a small colony in the open possible without depleting the laboratory stock to a serious extent, and first one and later several small colonies were established in various localities in the moth-infested area.

At the same time reproduction work was continued on an ever-increasing scale at the laboratory, and by the first of the next year no less than 1,000,000 individuals, at a conservative estimate, were present in our rearing cages. Further attempts to increase this number were not successful, on account of the difficulties attending the handling of such an immense number at a time when the hatching of the host eggs followed too soon after their removal to high temperature.

The numbers in the laboratory suffered no decrease, however, and by the end of March colonization work on an extensive scale was begun. The parasitized eggs were divided into 100 lots, each of which contained approximately 10,000 of the parasite, and these were distributed to agents of the State forester's office, who placed them in the field in the hope and expectation that the parasites issuing from them would reproduce immediately upon the gipsy-moth eggs before the latter hatched.

There was also a large quantity of parasitized eggs remaining, and these were placed in cold storage in the hope that the emergence of the brood might be retarded until the fresh eggs of the gipsy moth should be available for attack in the latter part of the summer. This hope was not justified, because when the time came and the eggs were taken from cold storage not a single living parasite remained.

In Table IX are summarized the results of the reproduction work, as conducted in the laboratory from April, 1909, to the winter of 1909-10, and the dates when the first colonies were planted in the late summer and fall are therein indicated.

TABLE IX.—*Results of reproduction work with Schedius.*

Generation.	Number and source of parents.	Reproduction work begun.	Emergence of progeny.		Total number of progeny.	Colonized.
			Began.	Ended.		
First	11 from imported egg masses	Apr. 19	May 19	June 14	114	
Second	114 from first generation	May 19	June 23	July 16	645	
Third	645 from second generation	June 23	July 16	Aug. 10	1,350	
Fourth	1,350 from third generation	July 16	Aug. 16	Sept. 7	11,999	10,980
Fifth	1,019 from fourth generation	Aug. 16	Sept. 11	Sept. 29	6,286	3,280
Sixth	3,006 from fifth generation	Sept. 11	Oct. 5	Oct. 25	12,723	5,368
Seventh	7,355 from sixth generation	Sept. 30	Oct. 29	Nov. 15	35,423	5,639
Eighth	29,784 from seventh generation.	Oct. 25			[1] 219,627	20,115
Ninth	199,512 from eighth generation.	Nov. 21-Dec. 21			[1] 1,028,361	733,967
Tenth	294,400 from ninth generation.	Jan. 5			[1] 284,779	280,762

[1] Estimated.

The reproduction of the parasite in the field as a result of these early attempts at colonization was far in excess of expectations. The rate of reproduction in the laboratory (as indicated in the table) was greatly exceeded in the open, and hundreds of thousands of eggs in the immediate vicinity of the colony sites were known to be parasitized when the coming of cold weather put a stop to insect activity. In the one colony which was most closely watched, the parasitized eggs averaged some 30 to the mass (fig. 27), while everywhere within 50 yards of the center egg masses were so thick in spots as to hide the bark on the trees. Beyond the distance mentioned the number per mass fell off very rapidly, but some were found several hundred yards away from the point of liberation, in striking contrast to the results following the colonization of Anastatus.

In October adults of what appeared to be the second generation were not uncommon in the field, and on any warm day they could be found, apparently ovipositing for a third generation. At the same time larvæ and pupæ were in abundance, and only a few days' exposure to the warmth of the laboratory was needed to bring them out from eggs collected in the field. Collections of eggs were made from time to time during the fall, in order that assurance might thus be had of the continued wellbeing of the parasite, and until December nothing untoward occurred.

The first real winter weather came at the end of that month, and a few days later a lot of eggs was collected and brought in. Not a single parasite issued.

FIG. 27.—Gipsy-moth egg mass, showing exit holes of *Schedius kuvanæ*. Enlarged about four times. (Original.)

The experiment was repeated and with the same results, and although many hundreds of masses have since been collected (some of them in the spring, after the caterpillars had issued for the purpose of determining whether there might not be reproduction of hibernated adults at that time, and the rest of them in the fall to see if by any chance the parasite had escaped detection in the spring), no trace of its existence could be found. In every instance, so far as the above-mentioned colony was concerned, the results were the same, and there seems to be no doubt that, in this particular locality at least, the species has become extinct.

In the spring one large colony of the Schedius was planted coincidently with the distribution of the 100 lots of parasitized eggs for colonization by the State forester's agents, and for two months following weekly collections of eggs were made with the expectation that a partial spring generation would follow. None of these collected egg masses produced the parasite, and again it failed to come up to that which was expected of it.

In the fall, as has already been mentioned, very large collections of eggs made in the vicinity of that which was considered to be the best and most promising of the colonies of 1909 failed to produce Schedius, and at the same time numerous smaller collections were made in each of the other colonies of 1909, as well as in a considerable number of the spring colonies of 1910. In only one of the colonies of 1909 was the Schedius recovered, and this, curiously enough, from that in which every attempt had been made to secure evidence of spring reproduction. Here it was found in one direction from the center of the colony only, and over a rather limited area. In the immediate vicinity of the colony site (within 100 yards) none could be found.

The collections which were made in each of the other colonies of 1909 were followed by curiously similar results. The parasite was recovered in one of them, and in one only, and although collections of eggs were made in all directions from the center and at varying distances, parasitized egg masses were only found in a limited area to one side and some distance away.

It was pretty conclusively demonstrated that the larvæ and pupæ of Schedius could not survive the rigor of the winter, and it is very difficult to say whether the recovery of the parasite in this last-mentioned instance is indicative of its ability to survive the winter as an adult. In 1909 a quantity of the adults was placed in a small cage in the open before the beginning of severe weather, and, although mortality was heavy, some of them lived for a long time after all of the younger stages were destroyed. None of them lived through until spring, but there is nothing to prove that they would not have done so had they had their choice of situations in which to hibernate.

It may be that females successfully hibernated in the instance of this colony, which appears to have lived throughout one year in the open. It may also be that the recovery of the species under these conditions is the result of dispersion of the individuals from some of the many spring colonies, several of which were located within a not unreasonable distance of this spot. It will require another year to demonstrate the truth of the matter.

The recovery of Schedius under any conditions at all was considered as sufficient to justify the repetition of the rearing work of the winter before, and accordingly, using a few individuals secured from the field early in the fall, a series of generations has been reared in the laboratory until the number now on hand (Jan. 1, 1911) runs into the hundreds of thousands. In the spring it is planned to establish one or two exceedingly large new colonies, sufficiently far distant from any of the others to make the recovery of the parasite elsewhere a certain indication that it is able to pass the winter in New England and thereby justify the labors which have been expended in its behalf.

THE PARASITES OF THE GIPSY-MOTH CATERPILLARS.

APPARENTLY UNIMPORTANT HYMENOPTEROUS PARASITES.

It would be presumption to state without qualification that the parasites which are here brought together as unimportant are in reality that. It may well be that among them are some which will be of sufficient promise to make advisable the trouble and expense incident to an attempt to transplant them to America, and which will serve to fill in the gap in the sequence which the apparent failure of *Apanteles fulvipes* has left. To determine more definitely their relative importance abroad is one of the objects of the work for the season of 1911, as at present planned, and something more than is known now is certain to be known a year from now unless the plans for the season go wrong from the beginning.

The various species coming in this category are called unimportant because they have never been received in imported material in numbers sufficient to make colonization in America possible, and only upon very rare occasions and in the instance of a few amongst them only, in numbers sufficient to indicate that they were of any importance whatever in effecting the control of their host abroad.

The investigations into the parasites and parasitism of various native insects more or less similar in one respect or another to the gipsy moth have served to throw considerable light upon the status of such parasites as these. It has been shown, in the instance of the tussock moth, that a parasite may be entirely absent in localities where the host is abundant, or else very rare under such circumstances and yet be sufficiently common to effect an appreciable amount of control in localities where the host is very rare. It is thus possible that some among these species may play a very important rôle in keeping its host, when already reduced to relatively small numbers, from increasing sufficiently to become of economic importance, and that at the same time they may play no part at all in reducing that insect from a state of or approaching noxious abundance to within its ordinary limits.

On the other hand, studies with the parasites of native insects have revealed the existence of what may be called accidental or incidental parasites. These may be important parasites of one insect and of no importance whatever in connection with another, nearly allied. Sometimes this is due to the fact that the one species of host may excite in the mother parasite the desire to oviposit, which is not excited by the other, and occasionally, as has more than once been observed, the presence of the favored host in the immediate vicinity will induce the parasite to oviposit in another species which under otherwise identical circumstances would be entirely ignored. At other times an insect may be acceptable to the mother parasite, but for some reason unacceptable to her progeny, so that only a very few out of the many eggs which are deposited will go through to maturity, and the species will be of necessity considered as rare and unimportant.

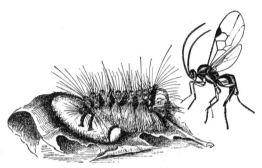

FIG. 28.—*Apanteles solitarius:* Adult female and cocoon. Enlarged. (Original.)

The fact that there are included in every list of the parasites of a given host a few species which are thus to be considered as incidental or accidental lends force to the contention that among the recorded parasites of the gipsy moth are several at least which come into the same category. Just which these are is not altogether plain at this time.

APANTELES SOLITARIUS RATZ.

Cocoons of a solitary species of Apanteles (fig. 28) which attacks the very young to half-grown caterpillars of the gipsy moth throughout the greater part if not the whole of Europe have occasionally been received in shipments in which the caterpillars were not all in the last or next to the last stage. In those shipments which consisted of caterpillars in the third, fourth, and fifth stages at the time of collection, the cocoons of this species have been the most common. In no instance has a sufficient number been received to make possible anything like a satisfactory colony of this species, and in all scarcely more than 100 have been received since the beginning of the work.

The parasite undoubtedly attacks the first-stage caterpillars as well as those of the later stages up to the fourth at least, and perhaps the fifth. The host probably molts at least once, subsequent to attack, and remains alive after the emergence of the parasite larva, clinging to and seeming to brood over the cocoon of its mortal enemy. Numerous experiments have been made with other, similarly living caterpillars from which parasites have emerged, in an attempt to make them feed, and invariably these attempts have been unsuccessful.

Of all of the gipsy-moth parasites in Europe of which there is no present prospect of introduction into America, this species is the most promising, and yet, if dependence is placed upon the results of the rearing records, it is so scarce as to be wholly inconsequential as a parasite of this host.

Meteorus versicolor Wesm.

Very occasionally cocoons of this common brown-tail moth parasite have been found in boxes of gipsy-moth caterpillars received from European sources, but never in any numbers. Altogether not nearly so many have been received as of the cocoons of *Apanteles solitarius*.

It is apparently an incidental parasite of no consequence, and were it not an enemy of the brown-tail moth as well, it is very improbable that any attempt would be made to introduce it into America.

As a parasite of the brown-tail moth it is of considerable promise, and as a brown-tail moth parasite it has been introduced and is apparently at this time thoroughly established over a considerable territory. Upon several occasions it has been reared from gipsy moth caterpillars collected in the field localities where it was particularly common as a parasite of the brown-tail moth, but, as in Europe, it expresses a strong preference for the last-mentioned host.

Meteorus pulchricornis Wesm.

Quite a number of this species has been reared from cocoons found in the boxes of gipsy-moth caterpillars received from southern France, and a very few have also been received from Italy. None of the Meteorus which have been reared from the brown-tail moth in any part of Europe have been anything else than *M. versicolor*, so far as known. It is, of course, possible that two species similar in appearance might easily have been confused, and no attempt has been made to determine the specific identity of every specimen which has been reared for liberation.

There is nothing to indicate that *M. pulchricornis* is ever of more consequence as a parasite of the gipsy moth than is *M. versicolor*, and until evidence to the contrary is forthcoming it will not be considered as of importance or promise.

Meteorus japonicus Ashm.

Specimens of this species were secured from boxes of young gipsy-moth caterpillars from Japan in very small numbers in 1908 and 1909, but so far as could be determined it was of no more importance in that country than were either of the two species already mentioned in Europe. In the winter of 1909–10 a few specimens were received from Mr. Kuwana, together with the statement that it was common as a parasite of the gipsy moth in Nagaoka, but not in Tokyo in 1908. Attempts to import it in 1910 were unsuccessful, and it is with the hope of confirming its importance, at least locally, and discovering some method of transplanting it to America, should such confirmation come about, that the investigations are undertaken in Japan in the year 1911.

LIMNERIUM DISPARIS VIER.

FIG. 29.—*Limnerium disparis:* Cocoon. Enlarged. (Original.)

This interesting parasite was first received in June, 1907, in a shipment of small gipsy-moth caterpillars from Kief, Russia. A total of 18 of its peculiar cocoons (fig. 29) was received in June and July of that year in boxes which contained only a relatively small number of caterpillars when sent, and its importance as a parasite appeared to be considerable. It seemed probable that the larvæ spinning them had issued from caterpillars in the fourth and fifth stages.

The cocoons usually approach more nearly the spherical than that used as the type for the drawing. The walls are thin, but so dense as not only to be impervious to moisture, but to prevent the drying of the meconial discharge for months.

FIG. 30.—*Limnerium disparis:* Adult male. Much enlarged. (Original.)

None of the cocoons was hatched on receipt. A single male adult (fig. 30) issued from one of these cocoons in August. No more adults appearing, some of the cocoons were opened from time to time during the fall and found to contain still living adults. Since it is known that several of the native species which spin similar cocoons actually do hibernate as adults within the cocoon, it is reasonable to suppose that the same is true of this.

No adults were reared, however, and their failure to emerge appears to be due to the drying during the winter of the semiliquid meconial discharge which effectually glued the adults to the sides of the cocoon and prevented their further movement.

The several native species spinning similar cocoons attack a variety of hosts, and one of them, *L. clisiocampæ*, is sometimes common and quite effective as a parasite of the host indicated by its specific name. The larvæ, after spinning the cocoon and before discharging their meconium, are very active for a period of about 24 hours, convulsively wriggling the body in such a manner as to make the spherical cocoon move about in an extraordinary manner. It is altogether probable that the gipsy-moth parasite has the same characteristic, and that the cocoons so spun in the trees are quickly dislodged, fall to the ground, and become hidden beneath leaves and débris.

Prof. Kincaid was especially instructed to seek for evidences of parasitism by this species in Russia on the occasion of his trip to that country in 1909. He did not find it at all abundant, however, and only secured three or four cocoons. In 1910 the junior author sought diligently for these cocoons in the forest about Kharkof, where the caterpillars had been very abundant the season before, but he was entirely unsuccessful and as a result thoroughly convinced that it was not an important parasite in any of the several forests visited. There was no opportunity at Kief to make a similar search, because the caterpillars had not been sufficiently abundant within recent years in any of the localities visited to make likely the discovery of these cocoons, even though the species had been of importance as an enemy of the gipsy moth.

In 1909 and again in 1910 it, or another practically indistinguishable species, was received in very small numbers from Japan, but at the same time under circumstances which were in a way as suggestive of the possible importance of the species as were those under which it was first received from Russia, as detailed above. As in the case of the Japanese Meteorus, it is hoped to be able to determine definitely whether it is to be considered as of more than technical interest in the connection in which it is here considered.

LIMNERIUM (ANILASTUS) TRICOLORIPES VIER.

From time to time several specimens of Limnerium cocoons, all of them oblong in shape, and most of them partially concealed by the skin of the host caterpillar, have been received from Europe. In no instance have they been in sufficiently large numbers to make the species appear promising as a parasite.

Less than a dozen specimens have been received, all told, and were it not for the fact that the remains of the host accompanied the cocoon, it would not be possible thus definitely to associate the parasite with its host.

APANTELES FULVIPES HAL.

The one among the hymenopterous parasites attacking the caterpillars of the gipsy moth which has ever been received under circumstances indicative of its unquestioned importance as an enemy of that host is at present known as *Apanteles fulvipes* (fig. 31). The name Glyptapanteles, as generically applied to it, has been regretfully dropped, the more so since this name has already become familiar to many whose interest in parasites begins and ends with those which are included among the enemies of the gipsy and brown-tail moths. It was accepted, in the first place, on account of the immediate distinction which it offered to Apanteles, as applied to *A. solitarius* and *A. lacteicolor* Vier., and because it seemed preferable to make the technical name the common name as well. Now, with an enforced change in the specific name vaguely in prospect, it would seem advisable to adopt an arbitrary common name rather than to attempt to popularize the technical name, and should it again become desirable to write of it in a popular way, this will probably be done.

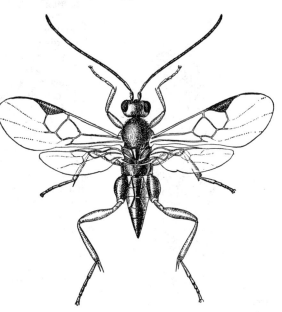

FIG. 31.—*Apanteles fulvipes*: Adult. Greatly enlarged. (From Howard.)

That a change in its specific designation will become necessary when it shall have been thoroughly well studied abroad seems probable, although there is no basis upon which to make such a change at the present time. If, as European taxonomists have agreed, it is synonymous with *A. nemorum*, described by Ratzeburg as a parasite of *Lasiocampa pini* L. and is at the same time specifically identical with the form so determined by Marshall as a common species in England, there seems to be no reason why it should not be introduced successfully into Massachusetts. A parasite with anything like the wide range of hosts accredited to this species abroad should find no difficulty in existing in America, and if the species which attacks the gipsy moth is proved to be identical with that which goes under the same name and attacks one or another of such a variety of hosts, no expense ought to be spared in attempting its introduction; always provided, of course, that the attempts already made prove not to be successful.

The story of these attempts, as told in the popular bulletin by the junior author, issued from the Massachusetts State forester's office in the spring of 1910, may well be quoted here, since there is little to be added to it.

Although this was almost the first parasite of the gipsy moth which attracted any attention in Massachusetts, and the first which it was attempted to import after the beginning of active work, it was one of the last to be liberated under satisfactory conditions, and its establishment in America is not yet certain. Extraordinary methods were necessary to bring it to America living and healthy, and it was not until Prof. Trevor Kincaid, who was selected by Dr. Howard as the best available man for the purpose, visited Japan and personally superintended the collection and shipment of the cocoons, that success was achieved. The story of Prof. Kincaid's experiences and of the difficulties which he met and overcame is interesting. He was accorded great and material assistance by the Japanese entomologists, and the work inaugurated by him in 1908 was continued with even greater success in 1909.

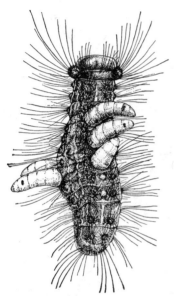

FIG. 32.—*Apanteles fulvipes*: Larvæ eaving gipsy-moth caterpillar. Enlarged. (Original.)

The adult parasite [fig. 31] deposits a number of eggs beneath the skin of the active caterpillars, and any stage, from the first to and possibly including the last, may be attacked. The larvæ, hatching from the eggs, become full grown in from two to three weeks, and then work their way out through the skin of the still living caterpillar [fig. 32], within the body of which they fed. Each spins for itself immediately afterward, for its better protection during its later stages, a small white cocoon. The number of parasites nourished by a single host varies in accordance with its size. There may be as few as 2 or 3 in very small caterpillars, or 100 or more in those which are nearly full grown.

The unfortunate victim of attack does not, as a rule, die immediately after the emergence of the parasite larvæ and the spinning of their cocoons, but it never voluntarily moves from the spot. Its appearance, both before and after death, surrounded by and seeming to brood over the cocoons, is peculiar and characteristic, and once seen can never be mistaken [fig. 33].

There is ample opportunity for two generations of the parasite annually upon the caterpillars of one generation of the gipsy moth. This is the rule in the countries to which it is native, and is to be expected in America.

The parasite was described from Europe more than seventy-five years ago, and has been known to be a parasite of the gipsy moth for a long time. Later it was described under a different name from Japan, and the Japanese parasite was for a time considered to be different from the European. Absolutely no differences in life and habit which can serve to separate the two are known, and, as the adults are also indistinguishable in appearance, they are considered to be identical.

It has been the subject of frequent mention under the name of Apanteles, as well as of Glyptapanteles, in the various reports of the superintendent of moth work, from the first to the fourth; and Dr. Howard, in the account of his first trip to Europe in the interests of parasite introduction, tells of its occurrence in the suburbs of Vienna. Largely on account of the fact that it is much more conspicuous than many of the other parasites, it has attracted more general attention. The Rev. H. A. Loomis, a missionary, and resident of Yokohama, was the first to call attention to its importance in Japan, and made several unsuccessful attempts to send it to America. Dr. G. P. Clinton, mycologist of the Connecticut Agricultural Experiment Station, who visited Japan in 1909, observed the parasite at work, and reported most favorably upon its efficiency as a check to the moth. Numerous other attempts on the part of European and Japanese entomologists, including one elaborate experiment which involved the shipment of a large wire-screened cage containing a living tree with gipsy caterpillars and the parasite, were made, but with uniformly ill success. Upon every occasion the parasites all emerged from their cocoons and died en route.

When every other means failed, Prof. Kincaid, as already stated, was deputed to visit Japan, and to make all necessary arrangements for the transportation of the parasite cocoons in cold storage to America. The arrangements which he perfected provided for continuous cold storage, not only en route across the Pacific, but during practically every moment from the time the cocoons were collected in the field in Japan until they were received at the laboratory in Melrose. Events justified the adoption of every precaution, and, with all the care, only a small part of the very large quantity of cocoons which he collected reached their destination in good condition. Hundreds of thousands were collected and shipped, and less than 50,000 were received alive—nearly all in one shipment in July.

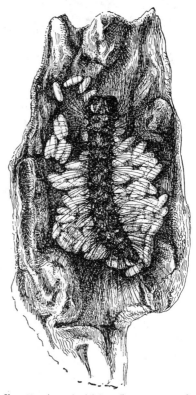

FIG. 33.—*Apanteles fulvipes:* Cocoons surrounding dead gipsy-moth caterpillar. Slightly enlarged. (Original.)

The season in Massachusetts was early, and nearly all of the gipsy caterpillars had pupated by that time, so that there was no opportunity for the parasite to increase in the field upon this host that season. In 1909 the sites of the colonies were frequently visited, but not a single parasitized caterpillar was found which could be traced to colonizations of the year before. Keen disappointment was at first felt, but later developments have tended to throw a more encouraging light upon the situation.

In 1909 importations were continued, through the magnificent efforts of Prof. S. I. Kuwana, of the Imperial Agricultural Experiment Station at Tokio, with much more satisfactory results. In 1908 the season in Japan was very late, and it was not practicable to send any of the cocoons of the parasite until June and July; while in America the season was early, and by that time all of the caterpillars, as has already been stated, had pupated. In 1909 the season was rather early in Japan and correspondingly late in America; and, besides, through special effort, Prof. Kuwana was enabled to send a few thousands of the cocoons of the first generation, which reached the laboratory early in June. About 1,000 adults emerged from these cocoons after receipt, and the most of them were placed in one colony in a cold situation on the North Shore, where the caterpillars were greatly retarded, and where there were still some in the first stage. The remainder were colonized in warmer localities, where the caterpillars were one stage farther advanced.

Immediate success followed the planting of these colonies. Within three weeks cocoons were found in each, and the number of parasitized caterpillars was gratifyingly large. A very careful investigation was conducted, to determine the proportion which was attacked by native secondary parasites; and, while this was so large in one instance as seriously to jeopardize the success of the experiment, it was not so large in the others.

There were several thousands of this first generation known to have developed in the open upon American soil, which issued from the cocoons some four or five weeks after the colonies were established, but in only that one on the North Shore, where the caterpillars were in the first and second stages when the parasites were liberated, was there a full second generation. Here the larger caterpillars were again attacked, and an abundant second generation of the parasite followed.

Meanwhile, additional shipments of cocoons of the second Japanese generation were received early enough to permit of a generation in the open upon the native caterpillars, and several other colonies were successfully established. It is known that there were many thousands of the parasite issuing in at least five different localities during August, but immediately thereafter they were completely lost to sight, and it is futile to hope to recover traces of them before another spring.

Until the late summer of 1909 nothing occurred to indicate that this parasite would be likely to fly for any great distance from the point of its liberation; and, as has been already stated, it was looked for in vain in the summer of 1909 in the immediate vicinity of the colonies of the year before. In July, 1909, a strong colony was planted in an isolated woodland colony of gipsy moths in the town of Milton. It was rather confidently expected that it would attack these caterpillars so extensively as to destroy the major portion; but it was the cause of some surprise, when the locality was visited after the parasites of the new generation had mostly issued from the affected caterpillars, to find a smaller number of cocoons than there were individuals liberated in the first place, and only about one-fourth, perhaps less, of the caterpillars attacked. The circumstance was as discouraging as anything which had gone before, and for a few days nothing happened to change its complexion. Then, to the intense surprise of the writer, Mr. Charles W. Minott, field agent of the central division, sent to the laboratory a *bona fide* example of the parasite, which had been collected in the Blue Hills reservation, upwards of a mile away. There was no possible source except the Milton colony, and a spread of upwards of a mile in something under a week was indicated beyond dispute. At almost the same time the brood of Monodontomerus was found for the first time in pupæ of the gipsy moth in the field; and when the history of this species is considered, in the connection which it bears toward the circumstances surrounding the recovery of the Glyptapanteles so far from the point where it was liberated, the whole situation is altered.

Granted that the parasite disperses at the rate of one mile in each week of activity, and that it is able to adapt its life and habits to the climate and conditions in America, the chances are, that, instead of looking for it in the immediate vicinity of the points of colonization, it is quite as likely to be found almost anywhere in the infested area within 25 miles of Boston. If it is thus generally distributed, very large numbers in the aggregate may exist, and it may increase at a rate as rapid as that of Monodontomerus, and at the same time escape detection until the summer of 1911 or 1912.[1]

There is not very much to add to the account given above, further than the statement that all attempts to recover the species in the field in 1910 from the vicinity of colonies of the year before failed. It hardly seems likely that so conspicuous an object as the cocoon mass of this parasite should escape the notice of the many field men who are familiar with its appearance, and who know of the great interest and importance which would attach to its discovery. In consequence the failure to recover the species is of more significance than the failure in the instance of any other parasite which could be mentioned.

At the same time all hope has not been given up, especially in consideration of the curious circumstances which will shortly be described, surrounding the recovery of *Pteromalus egregius* as a parasite of the brown-tail moth. If, as can no longer be doubted, a minute and to all appearances an inactive insect like Pteromalus has dispersed over a territory of approximately 10,000 square miles within five years as the extreme limit, and if during that period it remained so rare as to defy all of our efforts to recover it, it is not impossible that *Apanteles fulvipes* will do the same. Should this come about, the year 1911 or 1912 would probably witness its sudden and simultaneous appearance throughout the greater part of the territory infested by the gipsy moth.

It must be confessed, however, that hope rather than faith has dictated these last lines. It is believed, and not without some foundation, that the failure of *Apanteles fulvipes* to exist here is due to the absence of an absolutely necessary alternate host, and that further attempts to introduce it will be unavailing. That is the reason why the most will be made of every opportunity to determine the truth or fallacy of the European records which accredit it with attacking a variety of insects representing half a dozen families, and two or three times that number of genera, many of which are represented by closely allied and sometimes by the same species in America. If investigations uphold the truth of these records, no expense ought to be spared in further attempts to establish the parasite in America, because of all those which attack the gipsy moth it is the one which was not only the most promising at the beginning, but which remains the most desired at the present time.

---

[1] The occurrence of the cocoons in the near vicinity of the colony sites immediately following the liberation is most natural, and in perfect harmony with the wide dispersion. The female parasites as soon as they emerge are ready to deposit a small part of the eggs which they will eventually deposit if they live and have opportunity. After the deposition of this part, it is necessary for them to wait an appreciable time before they are ready to deposit any more.

In 1910 additional importations were made from Japan, and a large number of healthy adults was liberated sufficiently early in the season to allow for one generation upon the gipsy moth. As in 1909, cocoon masses were found in the vicinity of these colonies about three weeks after their establishment.

An attempt will be made in 1911 to import enough cocoons from Russia to make possible a strong colony of the European race. It is possible that it would succeed here when the Japanese would fail, and on the chance the experiment is undertaken.

SECONDARY PARASITES ATTACKING APANTELES FULVIPES.

It is safe to say that a better opportunity for an intensive study of the parasites of any one host which was itself a parasite has never been afforded than has come about at the laboratory in the case of the parasites of *Apanteles fulvipes*.

Hundreds of thousands of the cocoons of the primary parasite were collected in Japan after they had been exposed to attack by the secondaries, and, so far as can be judged, the latter stood the ordeal of the journey to America better than did the primary. Even in those shipments which were just a few days too long en route and in which the Apanteles themselves had all issued and died before their receipt the secondaries had hardly begun to issue. These, as well as the numerous shipments which were received in better condition, in so far as Apanteles was concerned, have produced many thousands of secondary parasites, which have all been carefully preserved, but not, as yet, carefully studied. It is not even known how many species are represented in the assortment, which includes a considerable number of undescribed forms, but apparently there are at least 30, and probably more, from Japan alone. Some are very rare, and are represented by but a few individuals among the thousands which have been reared. Others are common at times, and rare or absent at others. Some few are generally common, and practically always present.

The considerable shipments of cocoons which were collected in Russia by Prof. Kincaid and forwarded to the laboratory in 1909 were invariably so long en route as to permit the Apanteles to issue and die, but, as in the case of the Japanese shipments, the secondary parasites did not suffer. Not nearly so much material of this sort has been received from European sources, and probably on that account alone the variety of secondary parasites reared has not been so large. Nevertheless, more than 20 species have been recognized and probably at least 25 have been reared in varying abundance.

A good many of these secondary parasites have a very close resemblance to those which have been reared from the Japanese material.

In some instances they appear to be identical. In others it may be possible to find minor structural characters which, together with the difference in their habitat, will make it worth while to designate them by different names. Some very distinct species are peculiar to Europe or to Japan, and remain unrepresented by any nearly resembling them in the other country.

In 1909, as the immediate result of colonization work carried out under the happy auspices already described, it was possible to collect large numbers of *Apanteles fulvipes* cocoons under perfectly natural conditions in the open in America. This was accordingly done, with the result that no less than 18 additional hyperparasites were added to the list of those which attacked this host. Some of these were rare, others very common in this connection. A few appear to be undescribed.

The most interesting thing about them taken as a group, is the general resemblance which they bear to the similar groups of European and Japanese parasitic Hymenoptera having identical habits. Apparently there are about as many points in common between the American parasites of *Apanteles fulvipes* and the Japanese or the European as there are between the European and the Japanese.

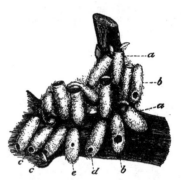

FIG. 34.—*Apanteles fulvipes:* Cocoons from which Apanteles and its secondaries have issued, as follows: *a*, *Apanteles fulvipes;* *b*, Hypopteromalus; *c*, *Hemiteles* sp.; *d*, Dibrachys; *e*, Asecodes. Enlarged. (Original.)

In the course of the work a total of 5,456 cocoons of *Apanteles fulvipes* was collected from several of the recently established colonies, but principally from two, representing the first among those planted in 1909 and in both of which a second generation occurred. Of this total, 1,531, or 28 per cent, had produced the Apanteles at the time of collection; 2,373, or 44 per cent, were attacked by secondaries (fig. 34); 634, or 12 per cent, were destroyed by various predatory insects, ants, etc.; and 918, or 17 per cent, remained unhatched in October, 1909. Among the unhatched cocoons was a considerable proportion which contained the hibernating larvæ of Asecodes, Elasmus, and Dimmockia. In more than one instance, too, hatching was prevented by superparasitism, and in others death probably resulted through the attack of predatory bugs. On at least one occasion *Podisus* sp. was found with its proboscis thrust through the wall of the cocoon and feeding upon the parasite larva or pupa within.

An idea of the variety of secondary parasites reared is conveyed by the tabulated list following.

List of secondary parasites reared from American cocoons of Apanteles in the order of relative abundance.

[In this list the number of individuals of Apanteles killed, not the gross number of the secondaries reared, is given. In case of tie, the species which was relatively the more important in the particular lot or lots from which it was reared is given preference.]

Hypopteromalus	1,276	Pezomachus No. 65	15
Dibrachys	583	Pteromalid No. 68	6
Asecodes	[1] 161	Hemiteles No. 63	5
Hemiteles No. 60	[2] 58	Pteromalid No. 70	2
Hemiteles No. 61	52	Eupelmus	2
Hemiteles No. 75	[2] 49	Hemiteles No. 66	2
Pezomachus	64	Anastatus	1
Eulophid	71	Total	2,288
Hemiteles No. 62	18		

Local conditions as affecting the control of this parasite through hyperparasites were well represented in 1909 by a comparison between the relative abundance of secondary parasites in cocoons from two colonies, the "Reading-Wilmington," and the "West Manchester," which were planted at about the same time. In both reproduction was abundant, and a large number of cocoons was collected from each. Only those which were left in the field until all of the Apanteles which remained healthy had issued are counted in the following:

	Reading-Wilmington colony.		West Manchester colony.	
	Cocoons.	Per cent.	Cocoons.	Per cent.
Apanteles	70	8	543	66.5
Hyperparasites	624	68	162	20
Predators	8	1	22	2.5
Unhatched Oct. 20	218	23	89	11
Total	920		816	

The West Manchester colony was located in rather dense forest, with a swamp, partly overgrown with brush and partly with thick forest on one side. The trees were large, and cocoon masses were frequently far beyond reach. Only those which could be reached from the ground were collected. There were more cocoons in this colony than in the other, but they were not quite so easily collected. It is of course possible that the larger number of cocoons explains in part the smaller percentage of hyperparasitism.

The increase in hyperparasitism in the cocoons of the second generation over the first can only be demonstrated in the case of the West Manchester colony, which was the only one where there was a second generation in sufficient abundance to permit of adequate field collections. In this it is or appears to be very striking, when the fact is taken into consideration that a considerable number of parasites hibernated in the cocoons of the second, while none were found in those of the first which failed to hatch after the 1st of September.

---

[1] Many Asecodes remaining unhatched within the cocoons will doubtless attempt to hibernate.

[2] Hemiteles No. 60 and Hemiteles No. 75 may possibly be one and the same species. It is possible, too, that further study will cause a change in the relative position of the two species.

Two lots of each generation were collected after the healthy parasites had issued, and the results follow:

	First generation.		Second generation.	
	Cocoons.	Per cent.	Cocoons.	Per cent.
Glyptapanteles	543	66.5	636	26
Hyperparasites	162	20	1,207	50
Predators	22	2.5	102	4
Unhatched	89	11	[1]469	[1]20
Total	816	100	2,414	100

[1] Oct. 20.

Other collections of cocoons from colonies where the Apanteles was liberated too late in the season to permit of two generations showed a high rate of hyperparasitism in the single generation, actually the first but corresponding to the second. Comparison in this instance is valueless, as local conditions enter in which can not be gauged.

This rather lengthy summary of a study in hyperparasitism has been prepared and is here presented with the object of illustrating the somewhat modified stand which it has been necessary to take concerning the subject in its relation to the project of parasite introduction. Were it within the bounds of possibility to introduce into America the parasites of the gipsy moth (*Apanteles fulvipes*, for example) without introducing the secondary parasites which preyed upon them abroad, it would unquestionably be possible to secure a greater meed of efficiency in America than that which the same parasites were capable of attaining in their native countries. This is on the supposition that the parasites themselves are no more likely to be attacked by the American hyperparasites than their hosts are likely to be attacked by the American primary parasites.

That the assumptions are fallacious, to a certain extent, is well proved by the results following the temporary establishment here of *Apanteles fulvipes*, as recounted above, and that the same results as those which followed the exposure of this parasite to American hyperparasites will result in the instance of others among the imported parasites is more than likely. In the case of *Compsilura concinnata* and *Apanteles lacteicolor* Vier. it is proved.

The truth of the matter is that the secondary parasites are very far from being as closely restricted to one or two species of hosts as are the primary parasites. This is in part due to the fact that they represent for the most part a much more degraded form of parasitism. Species like *Dibrachys boucheanus*, which is perhaps as generally abundant and omnivorous as any of the parasitic Hymenoptera, will attack anything which is dipterous or hymenopterous, provided it is physically suitable as food for its larvæ. *Apanteles fulvipes* and caterpillar parasites generally are governed in their host relations by physiological rather than by physical limitations, and the difference is as great as that which separates the true predator from the true parasite.

Did time permit, and were this the proper place, a lengthy digression might be made, in which several of the parasites typical of both groups, and which have been somewhat carefully studied at the laboratory, could be compared, the better to give strength to the statement just made. A little later attention will again be called to the matter.

At this time it is merely desired to define the modified stand which it has been necessary to take upon the question of hyperparasites. It is no longer possible, on account of the absence of their secondaries, to expect a much if any greater degree of efficiency from the imported parasites in America than the same species possess abroad. Since the foreign hyperparasites of the gipsy moth are generally the counterpart of the American species, which will become hyperparasitic upon the gipsy moth just so soon as there are any primary parasites, their introduction could not possibly do more than result in the existence in America of a somewhat greater variety of hyperparasites, which as a group would play exactly the same rôle as the lesser variety now existent here. Consequently the only secondary parasites which we have to fear are those which have no counterparts in America.

That such exist is beyond question; that they are in the minority is equally true. The only species which have been recognized as possibly or probably falling into that group are the hyperparasites reared from the gipsy-moth eggs from Japan, the Melittobia parasite of tachinids; the eulophid parasite of *Pteromalus egregius*, *Perilampus cuprinus*, and *Chalcis fiskei*; and, most unfortunately, two primary parasites already introduced, which are also secondary, *Pteromalus egregius* and *Monodontomerus æreus*. The two last mentioned are probably both beneficial rather than noxious in the final analysis, but nevertheless both are peculiarly adapted to act as secondary parasites of the brown-tail moth better than as secondary parasites of any other primary host.

It is not intended to ignore the secondary parasites in the future any more than in the past, but the same fears which have been expressed concerning their introduction are no longer felt in the same manner, and the benefits which were formerly expected to accrue through their exclusion are not so great as hoped.

TACHINID PARASITES OF THE GIPSY MOTH.

In proportion as one after another of the previously mentioned hymenopterous parasites of the gipsy moth have been eliminated from the lists as of no more than incidental or technical interest, and as the prospects for successfully introducing the one species which has been proved to be of preeminent importance abroad have grown less bright, the tachinid parasites have gained in the favor accorded to them, and from being considered as of secondary importance they have become of primary importance.

This change in attitude toward them would have come about in another way, even though it had not been forced through the comparative failure of the hymenopterous parasites to make good as yet. In nearly every instance in which the parasites of a native defoliating caterpillar have been studied, the tachinids have been found to play a part which was at least the equivalent of the part taken by the Hymenoptera, while in more than half the instances the tachinids have displayed superior efficiency. This is probably not true of the parasites of any other order than the Lepidoptera, and of only a portion of the larger representatives of that order.

For the most part the tachinids are restricted in their choice of hosts through purely physiological limitations, but to a material extent they are restricted through purely physical causes as well. The fall webworm offers a striking example of both. Literally thousands of tachinid parasites have been reared from it in the course of the past few years, and with the exception of an insignificant number of the imported *Compsilura concinnata*, only a single species has been found amongst them all. This species, at present known as *Varichæta aldrichi*, through its habit of depositing living larvæ upon the food plant instead of depositing eggs or larvæ upon the caterpillars, possesses a very distinct and powerful advantage in its attack upon this particular host. If the leaves or stems near a colony of young caterpillars are selected for larviposition, it is practically a certainty that the caterpillars will enlarge their nest to include these leaves; will thereby come in contact with the parasite larvæ, and thus complete the chain of circumstances through which parasitism comes about. A parasite having a similar habit would stand an infinitesimal show of providing for the future of its young if the webworm should suddenly change from a gregarious and nest building to a solitary and wandering insect. At the same time that the host escaped attack by Varichæta, it would lay itself open to attack by a variety of other species which are now only prevented from attacking it on account of the protection which its web affords.

But even though it were freely exposed to attack by all the species of tachinids which deposit eggs or larvæ directly upon or in their host, it would be immune to such attack by all but a small percentage of the species which might conceivably select it as a host. This is proved through the occasional occurrence of caterpillars bearing tachinid eggs, but with no evidences of internal parasitism showing on dissection. It is not merely necessary that the host be exposed to attack and acceptable to the instincts of the mother parasite; it is necessary that it possess certain physiological characteristics which force it to react in certain ways and no others to the stimulus of the parasite's presence. Unless the host does react in the manner to which the parasite is accustomed, the parasite which is unable to accommodate itself to circumstances beyond a certain extent will find itself in a position which would be comparable to that of a man suddenly thrust into a world where all the commonest laws of nature worked in an unfamiliar manner.

To say that many of the tachinids are physiologically restricted in their host relationships is equivalent to saying that they are restricted to a limited number of hosts, and this is true; probably more true than of the hymenopterous parasites taken as a whole, or of any large group of the hymenopterous parasites if the Microgasterinæ and a few similar groups of genera are excepted. It is probably true also that among those parasites which are the most closely restricted in their host relationships are to be found those which are the most effective in bringing about the control of their respective hosts. This is primarily due to the fact that a correlation usually exists between the life and seasonal history of such a parasite and some one or more hosts which it is particularly fitted to attack. The existence of a correlation between parasite and host of such intimate character makes possible the continued existence of the parasite independently of alternate hosts, and it is thus enabled to keep pace with the one species upon which it is peculiarly fitted to prey when other circumstances are favorable to its increase.

Some of the most interesting examples of correlation of this sort which have yet come to attention are to be found among the tachinid parasites of the gipsy moth or the brown-tail moth, and on this account as well as on a purely empirical basis they are now considered much more likely to become important enemies of these hosts than before their characteristics were so well understood.

THE REARING AND COLONIZATION OF TACHINID FLIES; LARGE CAGES VERSUS SMALL CAGES.

In more ways than can be recalled without taking up and discussing each species in turn has the necessity for a more complete knowledge of the tachinid parasites impressed itself upon those most concerned with their economical handling. The difficulties attending the successful hibernating of the puparia of Blepharipa and the mysterious disappearance of *Parexorista cheloniæ*, after it was considered to be thoroughly established in America, may be mentioned as conspicuous examples among the many oftentimes curious and sometimes apparently inexplicable problems which have come up for solution. Just at the present time there is pressing need of more and accurate

OUTDOOR PARASITE CAGE.

The tree is infested by gipsy-moth caterpillars, while the parasites are confined by the wire-gauze covering. Saugus, Mass., July, 1905. (From Kirkland.)

OUTDOOR CAGES COVERED WITH CLOTH AND INCLOSING INFESTED TREES; USED IN REARING PARASITES. (FROM KIRKLAND.)

VIEW OF LARGE CAGE USED IN 1908 FOR TACHINID REARING WORK. (FROM TOWNSEND.)

VIEW OF OUTDOOR INSECTARY USED FOR REARING PREDACEOUS BEETLES IN 1910. (ORIGINAL.)

Bul. 91, Bureau of Entomology, U. S. Dept. of Agriculture. PLATE XVII.

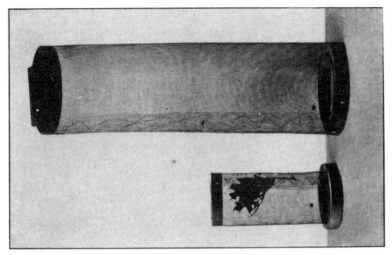

FIG. 2.—CYLINDRICAL WIRE-SCREEN CAGES USED IN TACHINID REPRODUCTION WORK IN 1910. (ORIGINAL.)

FIG. 1.—WIRE-SCREEN CAGES USED IN TACHINID REPRODUCTION WORK IN 1909. (ORIGINAL.)

information concerning the rapidity of dispersion of certain among these flies and concerning the host relations of certain others.

In an account of the methods used in conducting investigations into the lives and habits of the tachinids, published by Mr. C. H. T. Townsend three years ago, a good outline of the beginning of this work is given. It has been found necessary to modify to a certain extent the methods which seemed best at the time when this account was written, and in one particular at least it seems advisable to correct the statements therein made concerning the use of the large out-of-door rearing cage for tachinid reproduction work and investigation.

In the beginning the use of the large cages, consisting of a wooden frame covered with cloth or wire screen and inclosing a living tree, was attempted upon a considerable scale. Cages of this character had been so successfully employed in various somewhat similar lines of work as to justify their consideration in this, and accordingly a dozen or more were constructed and used for the confinement of all sorts of introduced enemies of the gipsy moth or the brown-tail moth, from *Pteromalus egregius* to *Calosoma sycophanta*, including the tachinid parasites.

The first of them, covered with wire gauze, was constructed in 1905 (Pl. XIII) and has been figured several times in various reports upon and accounts of the work, but it was never given a thorough test on account of the failure to secure parasites in any amount that first year. In 1906 cheesecloth coverings were substituted for wire and a number of cages, the general pattern of that figured herewith (Pl. XIV), was constructed and used that year and in 1907, but with pretty generally unfavorable results. It was found that only a very small number of caterpillars could be supported by the foliage of the inclosed trees or shrubs, and that it was necessary to feed them artificially exactly as was necessary in the smaller cages. The impossibility of keeping a variety of native insects out, as well as of keeping the foreign insects in, was another and only too apparent fault. In an experiment with tachinid reproduction in one of these cages in 1907, the number of flies introduced in the beginning grew steadily less day by day, with no adequate explanation for the disappearance of the missing individuals.

Another disadvantage accrued through the fact that when a caterpillar was in any way dislodged from the inclosed tree upon which it was expected to remain and feed, the chances were infinitely greater that it would find its way to the side, and then to the roof of the cage, than that it would, unassisted, regain its former position. The parasites, also, instead of staying about the tree where their business was supposed to demand their attention, would persistently remain in the uppermost recesses of the cage and refuse to come down.

All in all, the disadvantages were so many, in proportion to the advantages, and these latter were so largely imaginary in point of fact, as to result in the decision to discontinue the use of the large cages entirely in 1908.

The cage figured by Mr. Townsend (Pl. XV) was, however, an innovation in several respects. It was built independently of any tree which should serve as food for the inclosed caterpillars, but these caterpillars were confined within certain restricted limits and exposed to the attack of the tachinid flies at one and the same time by the use of the open "tanglefooted" tray. Here a most distinct advantage was gained. The floor of trodden earth (subsequently replaced by cement) effectually prevented the entrance of numerous insects which were formerly uninvited guests and thereby removed another serious disadvantage. An arrangement of double doors and wire-screened vestibule prevented the untimely liberation of the flies, and there were no longer so many inexplicable disappearances. The fact that the top of the cage was flat instead of being extended into the gable tended to keep the flies somewhere more nearly where they were wanted. In short, there were a great many advantages possessed by the new cage which were not possessed by the old, and there was some justification for considering it good.

In the meantime Mr. Burgess, who had taken over the Calosoma work in the fall of 1907, had developed the out-of-door cage along totally different lines, making it into nothing more than an out-of-door insectary (Pl. XVI), in which were conducted practically all of his numerous and varied investigations. It had seemed in 1907 as though the only one among the numerous imported insects which had done at all well in the out-of-door cages as then used had been the Calosoma, but the success attending their use for the rearing of this insect was so soon and so overwhelmingly eclipsed by the success which attended the use of small individual cages for single pairs of the beetles or individual larvæ as to render the advisability of their discontinuation for this purpose emphatic.

Some attempt was made to use the tachinid cage in 1909, but not to the extent to which it had been used the previous year. Late in the summer of 1909 reproduction experiments with small numbers of various species of tachinids were undertaken by Mr. W. R. Thompson, who used cages constructed after the familiar Riley type, but covered entirely with coarse fly screen. (Pl. XVII, fig. 1.) He succeeded in much of that which he undertook to do, and in 1910 continued the use of this type of cage, for a part of a quite extensive series of most interesting and successful experiments, but he also used a much smaller cage consisting of a wire-screen cylinder (Pl.

XVII, fig. 2, at right), about 8 inches in diameter and 12 inches high, with wooden top and bottom. His best results were secured through the use of this cylinder, and the reason appeared to be that the flies were less likely to fly and acquire sufficient momentum to injure themselves in small than in large cages.

In elaboration of the principle apparently involved, a still smaller cylinder (Pl. XVII, fig. 2, at left), scarcely 3 inches in diameter and shorter than that formerly used, was experimented with. Better results than ever before were secured upon the single occasion upon which this cage was used, and unless further experimentation results in additional modifications or in a reversal of the results first obtained, the cylinder cage figured herewith will be used almost exclusively in 1911.

As a basis for comparison of the utility of the large versus the small cages, the results attending the investigations into the biology of Blepharipa may be taken as an example.

Between 300 and 400 flies were used in an attempt to secure oviposition in the large cage in 1908, and no care that could be given them under these conditions was lacking. Not a single female completed her sexual development to the point at which she was capable of depositing fertile eggs, and no eggs of any sort were secured. Scores instead of hundreds of flies were used for the experiments in the spring of 1910, and many of the females lived throughout the period allotted for the incubation of their eggs and deposited them at the rate of several hundred daily, and abundant opportunity was thus afforded for the continuation of the studies into the lives and habits of the young larvæ under different conditions and in different hosts.

In short, after the most thorough tests, the use of the large out-of-door cages has been definitely abandoned for all phases of the work at the gipsy-moth parasite laboratory. It is not, however, intended to state thus dogmatically that similar large cages would not be adaptable to work with parasites of any other host.

HYPERPARASITES ATTACKING THE TACHINIDÆ.

Undoubtedly there is abroad an important group of secondary parasites of the gipsy moth and the brown-tail moth, included in which are some which attack the various species of tachinids to such an extent as indirectly to affect the welfare of the primary host. Very little is known of this hyperparasitic fauna, because practically all of the tachinids received have been from host caterpillars which were living at the time of collection. That it exists is well indicated by the tentative studies of the American parasites of *Compsilura concinnata*, which were made in 1910, and which will be the subject of mention at another place.

Occasionally, however, a few secondary parasites have been reared from puparia from abroad either because these puparia were collected in part in the open or because the parasites were of species which attacked the primary parasite during the life of the primary host. The number of secondary parasites having such habit is apparently very limited, and it has been definitely proved of but two genera, namely, Perilampus among the chalcidids and Mesochorus among the ichneumonids. The latter has never been reared as a parasite of any tachinid.

Because of the rather extraordinary precautions which were taken to avoid introducing into America the secondary, together with the primary, parasites of the gipsy moth and the brown-tail moth, the whole question of secondary parasitism is worthy of considerable attention in anything which purports to be a history, however abbreviated, of the operations conducted at the parasite laboratory. In the case of those attacking the tachinids it is better that they be briefly considered en masse, since there are very few among them with host relations restricted other than physically.

PERILAMPUS CUPRINUS FÖRST.

Actually, only a very little is known of this species from first-hand investigations further than that it is occasionally reared from puparia of any species of tachinid parasitic upon the brown-tail moth or gipsy moth in Europe, and under circumstamces which strongly indicate a habit of making its attack before the death of the primary host. At the same time it is felt that much is known of the probable habits of this species through analogy as the results of Mr. Smith's studies of the early history of the allied American species, Perilampus hyalinus Say, which attacks the parasites, both hymenopterous and dipterous, of the fall webworm. Presumably, like the American species, its minute first-stage larva, or "planidium," gains access to the host in some manner not quite clear, and after wandering about in its body for a time enters the bodies of such parasites as it chances to encounter.

That a secondary parasite having such habits might be expected to be peculiarly a parasite of the parasites of one particular host rather than of the same or similar parasites of another host, coupled with the fact that extraordinary precautions were obviously necessary to provide against its accidental importation, made Perilampus cuprinus appear peculiarly abhorrent, and for a time following the discovery of the early habits of P. hyalinus precautions against the importation of its congener were redoubled. In the course of time it was determined that it was never present in sufficient abundance to make it at all probable that it was a parasite of the gipsy moth or the brown-tail moth parasites to anything like the extent to which

P. hyalinus was thus peculiarly an enemy of fall webworm parasites, and thus a friend of the fall webworm. Neither, when it was present (which it was not, as a rule), was it ever known to emerge from infested puparia of the "summer issuing species" until long after the flies had ceased to emerge. From the puparia of species which hibernated as pupæ it never emerged until the spring and then appeared *before* the flies themselves. It was thus possible to provide against its escape with little trouble, and it is now considered as distinctly less menacing than the species which follows.

Melittobia acasta Walk.

Another most extraordinary parasite of tachinids in Europe is *Melittobia acasta*, according to a determination furnished some years ago by Dr. Ashmead. It is thought probable that a careful comparison between the parasite of the tachinids and *M. acasta* will reveal specific differences, but at the time of writing such comparison has not been made. Of all of the secondaries which have been imported with the parasite material this has proved the most annoying.

Its most annoying characteristic is its minuteness, which enables it to pass through 50-mesh wire screen at will, and this, coupled with an extreme hardiness and an insidious inquisitiveness which seems to know no bounds, has resulted upon two occasions in an infestation of the laboratory which is comparable to a similar infestation, which will receive further mention, by the mite Pediculoides.

No one knows where it came from upon either occasion or how it first succeeded in gaining a foothold in the laboratory. Its first appearance was in 1906, when Mr. Titus encountered it in several lots of puparia of different species of tachinids from several European localities. Mr. Titus evidently thought, judging from his notes, that it had been imported in each instance with the material from those localities. He studied its habits that first year, and found that it would oviposit freely in confinement and that such oviposition was successful. He did not give it full credit for its insidiousness, and as a result it succeeded in eluding his vigilance and gaining access to a number of the lots of hibernating puparia of Blepharipa, upon which it reproduced with great freedom.

In the spring of 1907 this circumstance became evident through its emergence in some numbers from several of the lots of hibernating puparia early in June, after most of the flies had issued. An examination of the remaining puparia was thereupon undertaken and a vast number of larvæ, pupæ, and unissued adults destroyed.

At that time it was supposed that each of the lots of puparia were infested at the time of their receipt, but when an even larger amount of similar material was received from an even greater number of localities in 1907, and a smaller, but still a considerable amount in the course of the year following, and no trace of Melittobia was encountered, it began to become apparent that the quite general infestation of the puparia in 1906 had taken place after their receipt at the laboratory.

Vigilance unrewarded during the two years slackened somewhat in 1909, and late in the summer a new infestation of Melittobia suddenly developed. Where it originated was and remains wholly a mystery. Possibly the first individuals were received in a large shipment of sarcophagid puparia which had been collected in Russia and forwarded to the laboratory by Mr. Kincaid, who considered them to be gipsy-moth parasites. This lot of several thousand puparia was thoroughly infested, and a very large proportion contained either the exit holes or the brood of Melittobia when their condition was discovered.

But the infestation did not stop here. Various small lots of puparia of various sorts, inclosed in small pasteboard boxes, in cloth-covered vials, or in other receptacles were found to have been attacked by the parasite. It seemed suddenly to have come from nowhere and to have attacked everything at once.

A very general cleaning up was immediately instituted, but again, it was felt, after the damage had been done. The sarcophagid puparia, which would otherwise have served as the basis for a very necessary and desirable series of investigations into the true character of these flies, had to be destroyed. A large percentage of them was attacked by the parasite, and the rearing of the healthy remainder involved the isolation of each and all of them in a series of tightly stoppered vials. The Melittobia were issuing daily and immediately attacking the healthy remainder and there was no method short of breaking open each puparium which sufficed to determine its condition.

After the cleaning up had been accomplished, Mr. Smith began a series of investigations into the life and habits of the parasite, the results of which he intended to have prepared for publication before leaving the laboratory. Since he did not do this, and since the species is one which is likely to become a cause of annoyance should similar work to the present be undertaken, the following brief summary of the results of his studies may be given.

The minute females, after having been fertilized by the still more minute, blind, and wingless males, issue from the puparium in which they have passed their early transformations and go in quest of others which they may attack. They will also attack hymenopterous cocoons, but with less success, apparently, than in the case of the more favored host. In the course of this search they will enter the damp earth for a distance of several inches in quest of puparia which have been buried therein, and since they can pass through well-nigh invisible cracks and are in possession of an acute maternal instinct, they are able to enter receptacles of all sorts by means of openings far too small to permit the passage of any other among the secondary parasites which have been studied, not excepting those from the gipsy-moth eggs.

Having located their prey, oviposition follows, the eggs are deposited upon the surface of the nymphs in an irregular circle surrounding a wound made by the ovipositor. They are very small but appear to swell somewhat before hatching, and if the puparium is broken open so that they are freely exposed to the air, they will not hatch at all. Contrary to expectations the larvæ and their mode of life presented nothing abnormal. The number of larvæ or pupæ which had been found in the hibernated Blepharipa in the spring of 1907 was so extraordinarily large in comparison to the size of the mother insect that it was considered likely that some form of polyembryony or pædogenesis would be found upon further study.

Becoming full fed, they will pupate immediately if the temperature is uniformly high, but will hibernate if it is allowed to fall below a point which was not determined. As soon as pupation has taken place the sexes are easily separable, through the absence of wings and eyes in the males. The male pupæ develop much more rapidly than the females and the adults issue in advance of their mates. They are invariably in the great minority, and their relative numerical strength is still further reduced through the terrific duels which follow their emergence. Notwithstanding their physical defects in the matter of sight and powers of flight, their seeming weakness otherwise, and their small size, even when compared to their mates, they possess a courage and a vigor that is most surprising. In the instance of a colony which had been removed, from the puparium in which it was reared through its early stages, to a small glass cell, the several males which issued well in advance of the females engaged forthwith in conflict, in the course of which a considerable number was killed. The survivors of this Lilliputian battle royal calmly awaited the issuance of the members of their harems and proceeded to mate with one and all with an ardor which seemed to know no limit.

Mr. Smith also conducted an experiment in parthenogenesis, the results of which were and remain unique in the annals of the laboratory. As in every other instance in which an attempt has been made to secure parthenogenetic reproduction with the hymenopterous parasites, it was successful, but in this case to a limited degree only, in that the females positively refused to deposit more eggs than they would normally have produced males had they been properly fertilized. Instead of depositing sufficient to provide for the complete consumption of the host, only four or five would be deposited at a time, and notwithstanding that after the depositing of what probably amounted to barely 5 per cent of those which filled their abdomens fairly to bursting, they ceased, and nothing short of impregnation served to arouse their maternal instincts again. As virgins they displayed a longevity lacking in the case of the fertilized individuals, and in those instances in which they were properly cared for easily outlived the time necessary for their scanty progeny to complete its transformation.

This progeny, as was expected, was exclusively of the male sex, which, when afforded opportunity, promptly united with their virgin mothers, who thereupon displayed the normal desire to deposit their eggs. As in the instance of Schedius, the fruit of such unnatural union consisted of both sexes.

Nothing approaching this characteristic of Melittobia has been encountered in any similar studies which have been made of the parthenogenetic reproduction of the parasitic Hymenoptera. In every instance either one sex or the other has been the result, and oviposition by virgin mothers, in so far as any observations to the contrary have been made, is perfectly normal and as free as by mated females.

It formed a strong argument in favor of the sex of the egg, in this particular species, having been determined before fertilization took place, a characteristic which is certainly not possessed by the majority of the parasites studied.

CHALCIS FISKEI CRAWF.

This large and fine representative of its genus has been received from Japan each year since the first large shipments came from that country in 1908 as a parasite of Crossocosmia and Tachina. It is of interest in that it is fairly common, and worthy of consideration on that account, but more on account of its having been reared under circumstances which tend to indicate that it somehow gains access to the tachinid larva before the latter leaves its host. This evidence is not sufficiently complete to justify an outright statement to the same effect, but it is sufficiently convincing to make its possibility worthy of mention. On this account the species acquires an importance which it would otherwise lack, and as a possible specific enemy of the parasites of the gipsy moth it is worthy of special endeavors looking toward its exclusion.

MONODONTOMERUS ÆREUS WALK.

As will be mentioned again under the discussion of this species as a primary parasite of the gipsy moth and the brown-tail moth, Monodontomerus is commonly reared as a secondary as well as a primary parasite. Its occurrence as a secondary is altogether too frequent and under such conditions as to make its recognition as such too plain to permit excuses in its behalf similar to those which have been put forward in the case of *Theronia fulvescens* Cress. It is therefore considered as a secondary just as much and as habitually as it is a primary and the question as to whether it is of enough more importance in one rôle than it is in the other to render it more than neutral remains to be decided. Apparently its value as a primary is sufficient to render void its noxiousness as a secondary, and to leave a considerable margin to its good, but this margin does not seem quite as wide now as it did a year ago, and it will require a year or two more to determine the true status of the parasite.

MISCELLANEOUS PARASITES.

There are quite a number of small chalcidids, the most of them being *Dibrachys boucheanus* Ratz, which are occasionally received with shipments of tachinids from abroad. None of them is of any importance whatever in this connection, from the point of view gained through the study of the material collected and sent under the conditions which have prevailed in the past. Sometimes when lots of loose puparia have been shipped as such, loosely packed, two or three among them have produced a colony of Dibrachys or some other parasite of similar size and habits, and these individuals have immediately set about the propagation of their species with such good effect as to bring about the destruction of the larger part of the remaining puparia.

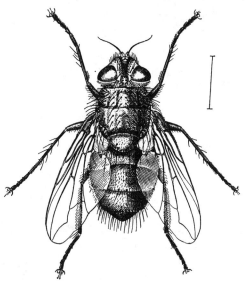

FIG. 35.—*Blepharipa scutellata:* Adult female. Enlarged. (Original.)

No serious effort has as yet been made to sort the Chalcididæ thus reared to species, much less to determine their specific identity.

BLEPHARIPA SCUTELLATA DESV.

Among the tachinid parasites of the gipsy moth caterpillars or the brown-tail moth caterpillars, *Blepharipa scutellata* (fig. 35) is the most conspicuous representative of the group characterized by the habit of depositing eggs (figs. 36 and 37) upon the foliage of trees or other plants frequented by its host with the deliberate intention that they shall be devoured. It is also an exceedingly close ally to the

Japanese *Crossocosmia sericariæ*, which was the subject of the original investigations by Dr. Sasaki through which this peculiar habit was discovered. The full life of the fly from the deposition of the eggs to the issuance of the adult, some 10 or 11 months later, has been the subject of a special series of investigations by Mr. W. R. Thompson, who, it is expected, will shortly publish the results of his studies.

FIG. 36.—*Blepharipa scutellata:* Eggs *in situ* on fragment of leaf. Enlarged. (Original.)

It is worthy of note that the results of Dr. Sasaki's observations have been abundantly confirmed in very nearly every respect in which there is not an actual difference between the habits of Blepharipa and those of Crossocosmia. Each female fly is capable of depositing several thousands of eggs upon the foliage of trees frequented by the caterpillars of the chosen host, but it is not known to what extent she employs discretionary powers in the selection of these trees. Presumably she is attracted to those upon which the host caterpillars are most abundant. Whether one sort of tree is more attractive to them than another is not known. The young larvæ hatching from the eggs which have escaped maceration by the mandibles of the caterpillars pass through the wall of the alimentary canal and immediately proceed to take full advantage of the physiological changes brought about in the host organism as the direct result of their presence. There are two larval ecdyses and three larval stages (as is the case with every other parasite of which the transformations are sufficiently well known to make any statement possible), and the manner of life undergoes a change with each ecdysis.

The first-stage larva embeds itself in the tissues of the host, which apparently react in a manner somewhat suggestive of the reaction which results in the growth of a vegetable gall following attack by a gall-making insect. The drawings of these gall-like bodies containing the larvæ (fig. 38), as well as the drawings of the egg and of the second-stage larval "funnel" were prepared under the direction of Mr. Thompson as illustrations for his forthcoming paper.

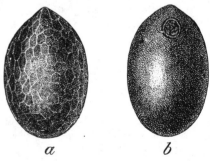

FIG. 37.—*a*, Egg of *Blepharipa scutellata*, showing characteristic sculpture and markings; *b*, egg of *Pales pavida*. Greatly enlarged. (Original.)

The second-stage larva undergoes a complete change in its manner of life, and its activities result in the formation of a tracheal "funnel," as illustrated in figures 39 and 40. In this stage the larva breathes through the spiracle of its host, to which the "funnel," which is apparently formed by the adventitious growth of a main branch of the trachea, is directly attached.

But few of the parasites, the early stages of which have been studied at the laboratory, exhibit a more clearly defined physiological relationship with their host than does Blepharipa. This relationship is comparable in many ways to that between the cynipid gall-makers and the oak tree which serves as their host. As is well known, many species of cynipids are closely restricted to one species of oak, or, at least, to several nearly allied species, and the same is to be expected of parasites like Blepharipa and others here spoken of as physiological, and thus limited in their host relationships. The gipsy moth itself is comparable to the parasites

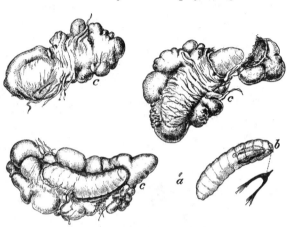

FIG. 38.—*Blepharipa scutellata.* First-stage larvæ: *a*, Natural size; *b*, greatly enlarged; *c, c, c*, greatly enlarged *in situ* in atrophied tissue of host. (Original.)

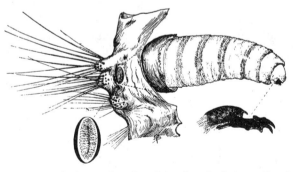

FIG. 39.—*Blepharipa scutellata:* Second-stage larva *in situ* in a portion of its tracheal "funnel." Greatly enlarged. (Original.)

in which the host relations are determined by physical rather than by physiological conditions. In its choice of food, although it prefers oak to almost any other of the native trees, it can and does attack all or nearly all varieties of deciduous trees, and even conifers and herbaceous plants when necessity demands.

The development of the Blepharipa is directly correlated to the development of the host, and as a parasite of the gipsy moth, its larva awaits the pupation of the host before assuming the aggressive, and destroying it (Pl. XVIII, fig. 1). Its own pupation is accomplished in the earth (Pl. XVIII, fig. 2), and the pupa develops adult characters in the fall. The space between the pupa or nymph and the shell of the puparium is filled by a small quantity of liquid, and the complete drying up of this liquid is very prejudical to the health of the individual, and is usually sufficient to prevent its emergence.

The difficulties which have stood in the way of a successful introduction of *Blepharipa scutellata* into America have differed in many respects from those which have accompanied the work with any of the other species, saving only the closely allied Crossocosmia. The first importations of full-grown caterpillars or freshly-formed pupæ of the gipsy moth in 1905 resulted in the securing of a considerable number of hibernating puparia. There were several hundred at least, but although they were kept under conditions which would be satisfactory in the case of most of the tachinids, not a single Blepharipa issued in the spring of 1906. The death of the insect did not take place until after the fly was fully formed and apparently nearly ready to issue from the puparium.

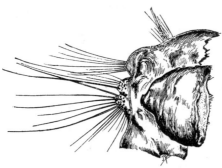

FIG. 40.—*Blepharipa scutellata:* Basal portion of tracheal "funnel." Greatly enlarged. (Original.)

A great many different methods of hibernating these puparia have been experimented with at the laboratory with variable, and until the winter of 1909 with poor, results. During the winter of 1907–8 the puparia were kept in moist earth and a 10 per cent emergence from a total of 5,000 was secured. The year before they were also hibernated in earth, but the emergence was less, amounting to only 3 per cent of the total, and the year following still less, being only about 1 per cent.

In 1909 for the first time since the inception of the work large numbers of living gipsy-moth pupæ containing the immature maggots of Blepharipa were received at the laboratory from Hyères, France, through the magnificent efforts of M. René Oberthür, of Rennes, and as a direct result of the senior author's trip earlier the same year. Some idea of the size of these shipments may be gained by reference to Plate XIX, figures 1 and 2, which show a small proportion of the total number of packages at the time of their receipt at the laboratory. For the first time it was possible to allow the formation of the puparia under natural conditions in the earth. During each of the preceding years the caterpillars and pupæ had been received from abroad by means of the ordinary methods of transportation and puparia had been formed in the boxes on receipt. They were often injured and always thoroughly dried when received. This year provision had

FIG. 1.—BLEPHARIPA SCUTELLATA: FULL-GROWN LARVA FROM GIPSY-MOTH PUPA. ENLARGED ABOUT SIX TIMES. (ORIGINAL.)

FIG. 2.—BLEPHARIPA SCUTELLATA: PUPARIA. SLIGHTLY ENLARGED. (ORIGINAL.)

Bul. 91, Bureau of Entomology, U. S. Dept. of Agriculture. PLATE XIX.

FIG. 1.—IMPORTATION OF GIPSY-MOTH CATERPILLARS FROM FRANCE IN 1909; EN ROUTE TO LABORATORY AT MELROSE HIGHLANDS, MASS. (ORIGINAL.)

FIG. 2.—IMPORTATION OF GIPSY-MOTH CATERPILLARS FROM FRANCE IN 1909; RECEIPT AT LABORATORY, MELROSE HIGHLANDS, MASS. (ORIGINAL.)

been made for cold storage in transit, with the results as mentioned above.

A very large number of the parasites were secured in this manner, and several thousands of the maggots were allowed to enter the earth in the open in forests infested by the gipsy moth. Others were allowed to pupate in a natural manner in forest soil or in a mixture of garden loam and sand in a variety of containers in the laboratory grounds.

An examination of these puparia was made from time to time during the winter and they were found to be uniformly in a much more satisfactory condition than the hibernating puparia had ever before been at that season of the year. So far as could be determined even up to within a few weeks before the emergence of the flies would naturally take place, there was no difference in the condition of the puparia hibernated in different kinds of soil or under slightly different environment.

Beginning quite early in the spring and continuing through a considerable period, flies emerged in very variable proportions from the different lots of puparia. The emergence in a few instances was well up toward 100 per cent. In others it was much lower, and in a few none of the flies completed their transformations. The reasons for these differences were not obvious in every instance, but it was obvious that unless conditions are practically identical with those which prevail in the open, the flies will fail to issue in the spring. Moisture is an essential, but is by no means the only essential to success. Nor can failure be attributed to unduly high or low temperatures, or unnatural and abrupt changes in the temperature during the period of hibernation.

The average percentage of emergence from all of the different lots of pupæ has not been as yet accurately calculated, but it was far in excess of any that was secured before, and three colonies which were considered to be satisfactorily large and strong were established in different parts of the infested area. It was not really expected that any of the new generation would be recovered from the field during the course of the first season, and it was therefore considered a particularly good omen when a few were recovered, without difficulty, and under conditions which indicated that dispersion at a quite rapid rate had accompanied a rapid rate of increase. The species has not yet been placed on the list of those considered as thoroughly established, since it is not certain that it will pass through the complete seasonal cycle in the field, but it is confidently expected that it will live through successfully and that it will be recovered in 1911 in larger numbers. If these expectations are realized there is every reason to believe that it will become a parasite of consequence within the next five years.

Curiously enough, among the imported gipsy-moth enemies that which most nearly resembles Blepharipa (if the practically identical Crossocosmia be excepted) in the part which it will probably take in the control of the gipsy moth is *Calosoma sycophanta*. No two of the imported enemies differ more radically in their method of attack than do these, the extent of their differences being fairly well exemplified by the fact that the gipsy moth eats Blepharipa, while the Calosoma eats the gipsy moth, which is literally true.

In one very important respect they are similar in that both are able to exist continuously upon the gipsy moth without being forced to have recourse to any other insect so long as the gipsy moth retains a certain degree of abundance. Both work to their best advantage and multiply most rapidly at the expense of the moth when the latter is superabundant.

It is only necessary to consider the powers of reproduction (potentially several thousandfold) possessed by the tachinid to see what an enormous rate of increase is likely to prevail in localities where practically complete defoliation occurs without becoming so complete as to bring about wholesale destruction of the gipsy moth through disease. Under such circumstances a very large proportion of the eggs deposited upon the foliage would perforce be eaten, as compared with the proportion eaten were the caterpillars present in small numbers. The percentage of parasitism would remain practically the same in both instances, but the gross number of parasites completing their transformations would be tremendously increased with a resulting increase in the percentage of parasitism the following generation, whenever the gipsy moth becomes unduly abundant. In like manner the Calosoma, which works at a disadvantage when the caterpillars are scarce, finds the conditions resulting through superabundance exceptionally favorable for its rapid increase.

Theoretically, therefore, Blepharipa ought to act as an agent in the reduction in the prevailing numbers of the gipsy moth whenever it exceeds a certain degree of abundance, and this is the rôle which it is expected to play. Theoretically, Calosoma will play practically the same rôle. Together their activities ought to result in the breaking up of dangerous colonies of the gipsy moth, and thereby render the work of the other parasites and of such native enemies as birds, predatory bugs. etc., doubly effective.

COMPSILURA CONCINNATA MEIG.

Quite a good many of the parasites of the gipsy moth attack the brown-tail moth also, but there is only one among them, *Compsilura concinnata* (fig. 41), which is equally important as a parasite of both. The remainder, if they attack both hosts, are more or less partial to one or the other.

In its method of attack Compsilura is the opposite of Blepharipa. Its eggs hatch in the uterus of the mother, and the tiny magots are deposited beneath the skin of the host caterpillar by means of a sharp, curved "larvipositor," which is situated beneath the abdomen. They usually seek the alimentary canal, in the walls of which they establish themselves during the first stage of their larval existence.

Growth is rapid, and in the summer is in no way correlated with the growth and development of the host. About two weeks are required for the complete development of the maggot, irrespective of the stage of the host at the time of attack, and at the end of that period it issues, and usually drops to the ground for pupation. The puparia from maggots which issue from caterpillars which have spun for pupation are not infrequently found in the cocoons in the case of the brown-tail moth; and even in the case of the gipsy moth, which does not spin cocoons worthy of the name, the puparia are often found immediately associated with the host remains.

It requires a surprisingly short time for the females to attain full sexual maturity after their emergence, three or four days apparently being sufficient. This, with two weeks for the

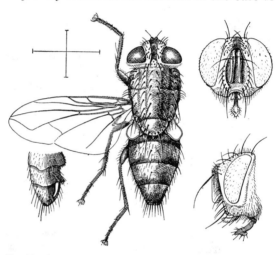

FIG. 41.—*Compsilura concinnata:* Adult female and details. Much enlarged. (Original.)

growth of the larva, and one week, or perhaps a little more, for the pupal period, makes possible a generation every four weeks during the warmer months of the year.

The position of the larva in the alimentary canal, together with certain structural characteristics, consisting of minute anal hooks, which are only known amongst other first-stage tachinids in the very similar genus Dexodes, makes possible the quite accurate determination of *Compsilura concinnata* from its first-stage larva alone, and only from observations which have been made upon these larvæ is it possible to say anything definite and at first hand concerning its habits of hibernation. Larvæ, which are almost certainly *Compsilura concinnata*, have been occasionally found in living brown-tail moth caterpillars during the winter months. It is presumed if these larvæ were able to mature under these circumstances, that they would have been reared before now from some among the hundreds of thousands of brown-tail caterpillars which have been carried through their first three or four spring stages in the laboratory. None having been reared under these circumstances, the only logical conclusion is that they start into activity so early and develop so rapidly as to cause the death of the host before they are sufficiently advanced to pupate successfully. This is not necessarily the true explanation of the failure to rear the species from hibernating brown-tail caterpillars fed in confinement, but it appears to be the best.

Ordinarily in the summer the larvæ do not pass over into the pupa of the host, but occasionally they do so. In the late summer and fall, when the host caterpillar is of a species which hibernates as a pupa, the parasite appears to be aware of that fact in some subtle manner, and likewise prepares for hibernation. Its larvæ (or what are without much doubt its larvæ) have several times been found in hibernating pupæ of several species. The adult has never yet been reared from pupæ under these circumstances, and the record is on that account open to some question.

The larger part of the Compsilura which were imported from 1906 to 1908, inclusive, issued from puparia (Pl. XX, fig. 1) found free in the boxes of brown-tail caterpillars from abroad. A companion species, *Dexodes nigripes*, which is indistinguishable from Compsilura in any of its preparatory stages, has also been reared under exactly similar circumstances, but curiously enough, if Compsilura was common in material from the same locality, Dexodes was apt to be rare, or vice versa. Some few were reared from gipsy-moth importations during this same period, but not in anything like the numbers which were secured from the brown-tail moth material, and it was not considered as of particular importance as a gipsy-moth parasite until 1909, when it was found to be very common among the tachinid parasites secured from shipments of gipsy-moth caterpillars from southern France.

The first colonies of Compsilura were planted in various localities within the gipsy-moth infested area in 1906, and in 1907, according to the records of the laboratory, a single fly was reared from gipsy-moth caterpillars collected in the immediate vicinity of one of these colonies. There is some reason to doubt the truth of this record, since every attempt at recovery made in 1908 failed.

In 1907 a much larger colony than any ever liberated before was located in the town of Saugus, in the near vicinity of one of those of the previous season. In 1908 none was colonized. In 1909 several very large and satisfactory colonies were planted in several places within the infested area, and for the first time it was felt that the species had been given a fair opportunity to prove its effectiveness as an enemy of the gipsy moth and brown-tail moth in America.

Fig. 2.—Tachina larvarum: Puparia. About Twice Natural Size. (Original.)

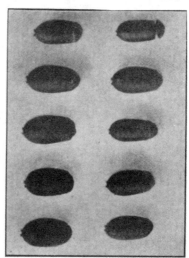

Fig. 4.—Parexorista cheloniæ: Puparia. About Twice Natural Size. (Original.)

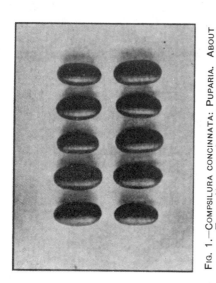

Fig. 1.—Compsilura concinnata: Puparia. About Twice Natural Size. (Original.)

Fig. 3.—Sarcophaga sp.: Puparia. Somewhat Enlarged. (Original.)

Hardly was the final establishment of what was for a few days considered to be the first satisfactory colony of *Compsilura concinnata* accomplished than the necessity for the expenditure of further labor on its account was obviated by the discovery that it could be recovered from the field in small but at the same time in very satisfactory numbers. Only an insignificant number was reared from the collections of gipsy-moth caterpillars made in 1909, but later in the fall of that year field men who were scouting for evidences of the spread of Calosoma and searching under burlap bands for its molted larval skins began to bring into the laboratory bona fide puparia of Compsilura found under the same circumstances. It was thus possible to delimit its range with some accuracy, and it was found to extend over a considerable territory, with the 1906–7 colony in Saugus much nearer to its center than any other more recently located colony. (See fig. 42.) There could be no doubt that the species was well established and spreading and multiplying at a rapid rate.

The results of the season of 1910 were awaited with very great interest, in expectation that they would confirm those of the year before. That these were confirmed, and most conclusively and satisfactorily, is evidenced by the results of the rearing work as summarized in Tables IV and V (pp. 141, 142), which give the results of rearing work for that year. The total number of the parasites reared or otherwise recovered from the field as indicated by these tables is very far short of the total secured.

Compsilura concinnata is recorded as a parasite of a large number of hosts in Europe, and will doubtless be found to attack an equally large number in America when it shall have become thoroughly established and abundant over a wide territory. Already some half dozen native hosts are known, and it would easily be possible to double or treble this list in the course of another season's work, should it be conducted with that end in view.

Few subjects for speculation are so overcrowded with possibilities as that of the effect which the importation of new parasites having a wide range of hosts will have upon native parasites and their hosts. The increasing abundance of Compsilura offers a most excellent opportunity to answer numerous questions which naturally arise when this subject is considered, and it is hoped that it may be made the most of. Already several highly significant observations have been made.

One of the most interesting of these resulted from a series of collections of tussock-moth caterpillars made by Mr. Wooldridge in the summer of 1910 for the purpose of determining the prevalence of parasitism in various localities and under slightly different conditions. All of these collections were of necessity made under urban conditions, since the tussock moth is rare in the country in eastern Massachusetts, and while it was expected that Compsilura would eventually be recovered as a parasite of this host, it was hardly expected that it would become of importance as a parasite so soon as 1910, or, for that matter, that it would become of importance as a parasite in cities at any time.

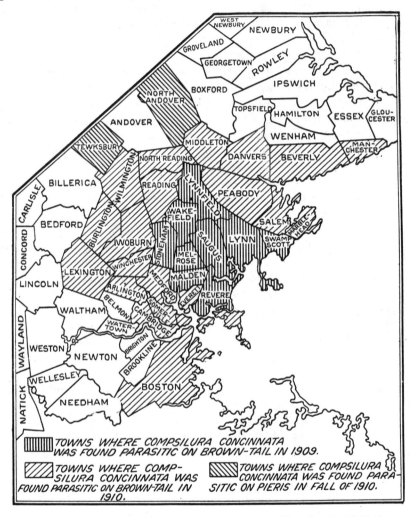

Fig. 42.—Map showing distribution of *Compsilura concinnata* in Massachusetts. (Original.)

The only one of the localities chosen for the tussock-moth collections which was within the limits of Compsilura's distribution so far as known when the work was instituted was in the city of Lynn, Mass., and from this a total of 110 caterpillars was collected on July 18, 1910. On July 29 the tray in which they were contained was carefully examined. Thirteen of the tussock-moth caterpillars had pupated and remained alive. The remainder had died, principally as the result of parasitism.

In all 96 tachinid puparia and 1 cocoon of Meteorus were found. Of these puparia 95 were *Compsilura concinnata* and the other apparently *Tachina mella*. The Meteorus, incidentally, proved to be of the introduced species, *Meteorus versicolor*.

Parasitism by native tachinids was probably considerably higher than would be indicated by the fact that only a single puparium was secured as against 95 of the imported species, but because the latter cómpletes its larval development much more rapidly than does *Tachina mella*, it would almost certainly be the victor in case of a conflict.

Later collections of tussock-moth caterpillars made for the express purpose of determining the limits to the distribution of Compsilura resulted in its discovery throughout practically all of greater Boston, and it may be that it will have some effect in reducing the importance of this insect as a pest in that city and its suburbs. With the end of experimenting further along this line the puparia secured from the Lynn collection, together with several hundred more from gipsy-moth caterpillars, were sent to Washington, where they were liberated upon the grounds of the Department of Agriculture, where the tussock moth is periodically a pest.

Another indication of good which may possibly result from the introduction of this tachinid resulted from an investigation, begun in September, 1910, by Mr. J. D. Tothill, into the parasites of the imported cabbage butterfly ([*Pieris*] *Pontia rapæ* L.). He found that in localities where Compsilura was known to be common the summer before, it was actually abundant as a parasite of this pest, and as high as 40 per cent had been attacked in some instances.

There is no native tachinid known to have quite the same habits as Compsilura, neither is there any with quite so varied a list of hosts. Both the cabbage butterfly and the tussock moth are commonly considered as pests, the one generally and the other in cities, and both can probably sustain additional parasitism without much difficulty. But in the case of the other native insects liable to attack by the imported parasite, and already thoroughly well controlled by various agencies, of which parasitism is one, the outcome of the struggle which is likely to ensue is probably going to be different. In the case of such an one it is reasonably safe to predict that one of two things will happen. Either the prevailing abundance of the host will be reduced through the introduction of a new factor into its natural control, or the host will maintain its present relative abundance, and its parasites will suffer directly in the struggle into which they will be forced by the advent of the tachinid.

It is yet too soon to begin to speculate upon what the actual outcome in specific instances will be. An investigation of the parasites of the fall webworm was undertaken in the fall of 1910 on the supposition that Compsilura would find it an acceptable host, but although it is freely attacked when outside of its web in rearing cages in the laboratory, it was not at all commonly attacked in the open, as will be seen by reference to the brief summary of the results of the work in the concluding pages of this bulletin.

There has never been a good opportunity to study the parasites of the Tachinidæ, owing to the fact that some of the species pupate upon or beneath the surface of the soil, and are therefore difficult to find in sufficient quantities to make a comprehensive study possible. So abundant was Compsilura, however, as to make it possible to collect its puparia in considerable abundance and with comparatively little trouble at the base of trees upon which the gipsy-moth caterpillars were common, and accordingly a number was so collected in the late summer of 1910. Not enough attention was given to the work to make the results as definite as is desired, but these were sufficient to indicate that secondary parasitism was undoubtedly of very common occurrence, and that it might be a factor of some consequence in limiting the effectiveness of the parasite. No less than six species of secondaries were reared, including *Monodontomerus æreus*, which was common, Dibrachys, another small chalcidid, a species of Chalcis, a proctotrypid, and a Phygadeuon. It is hoped that circumstances will permit of a more thorough study of this subject in 1911, and should the parasite show an increase proportionate to that which was indicated by its abundance in 1910 over that of 1909, the project should be very easy of accomplishment.

It is fortunate that, under the present circumstances, with the gipsy moth and the brown-tail moth both exceedingly abundant and uncontrolled, there should be at least one parasite which was equally drawn toward both. It is easily possible that the first individuals which are reared upon the brown-tail moth as a host may attack the full-fed caterpillars for a partial second generation the same season, and then, together with the bulk of the brood coming from this host, turn their undivided attention to the gipsy-moth caterpillars. In a similar manner the first individuals to go through their transformations upon the gipsy moth, together with the partial second generation upon the brown-tail moth, may attack the less advanced gipsy-moth caterpillars for a partial third brood before the necessity for an alternate host becomes apparent.

There is thus possible uninterrupted increase for two complete generations at least, and probably for a partial third, but unfortunately the necessity for an alternate host, though delayed until no more than one such host is necessary in order that the seasonal cycle may be rounded out, is not done away with. Unless the parasite hibernates in the brown-tail caterpillars such a host must be found among the native Lepidoptera, and while the number of species available probably runs into the hundreds, they are, with few exceptions, already controlled by their native parasites. Compsilura, if it continues to increase, will have to overcome these parasites in the competitive struggle for possession, and, as already stated, the outcome of this struggle is awaited with interest. Upon it will very largely depend the effectiveness of Compsilura as a parasite of the gipsy moth and the brown-tail moth in America.

TACHINA LARVARUM L.

This rather important parasite of the gipsy moth (fig. 43), and to a more limited extent of the brown-tail moth in Europe, is so similar to the American *Tachina mella* Walk. as to make the separation of the two by structural characters alone difficult at best, and in some instances impossible. It is similarly closely allied to *Tachina japonica* Towns., and the three species or races appear to occupy about the same position in the natural order

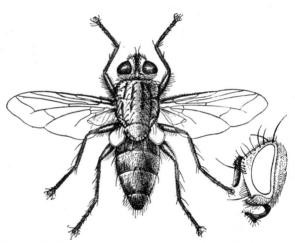

FIG. 43.—*Tachina larvarum:* Adult female and head in profile. Enlarged. (Original.)

of things in the several countries which they inhabit. The European and the American are both quite catholic in their host relations, and while the same can not be said of the Japanese in the present state of our knowledge, it will doubtless be found true when this knowledge shall be more extensive.

From an economic standpoint *Tachina mella* and *Tachina larvarum* are distinct enough specifically, if we are to consider their parasitism from an economic aspect, since the one is habitually and commonly a parasite of the gipsy moth, while the other is not. It would appear that *Tachina mella* attacks the gipsy moth quite as freely in America as *Tachina larvarum* does in Europe, but, as has already been mentioned, the attack is not successful from either the economist's or the parasite's point of view.

This is one of the less striking of several examples of a species which differs from another in biological rather than in structural characteristics. Others are to be found in the European race of Trichogramma, in the Japanese Apanteles parasitic upon *Euproctis conspersa*, which so resembles the brown-tail Apanteles of Europe, or in *Parexorista chelonix*, examples which will be again referred to on subsequent pages.

The large, flattened, and conspicuous eggs characteristic of Tachina and its allies are the most, and in fact the only familiar type of tachinid eggs, and they are deposited before embryological development has taken place in at least a part of the instances which have come under direct observation. The larva issues through an irregular hole in one end, and immediately forces an entrance through the skin of its host. The life cycle is longer than in the case of Compsilura, but just how much longer is not known. Sometimes the larva is carried over into the pupa of its host, but not very often. Very frequently it kills the host after it has prepared for pupation. Nearly always it leaves the host remains before pupating, on its own account, but occasionally puparia within the caterpillar skin or pupal shell are found.

The puparium (Pl. XX, fig. 2), unfortunately, is practically inseparable from that of *Tricholyga grandis* or *Parasetigena segregata* in its structural details, so that it is necessary to rear the fly before the species can be determined.

As a parasite of the gipsy moth, *Tachina larvarum* may and sometimes does take preeminent rank. Caterpillars from Holland have been received from which more puparia were secured than there were hosts, and the same has occurred on at least one other occasion in the instance of a box of caterpillars from Italy. When *Tricholyga grandis* is common, *Tachina larvarum* is rare, or at least has been rare in each instance in which the two species have been specifically determined as they issued from the imported material. It has also been entirely absent from some lots of caterpillars which did not produce Tricholyga.

It was about the first, if not the very first, parasite to be received alive in the course of the parasite-introduction work, and mention will be found in Mr. Kirkland's first report as superintendent of moth work, of its having been reared from Italian material in 1905. It was not secured in sufficient abundance to make colonization possible until 1906, but in that year quite a number of small colonies was planted in various localities in the infested territory. In 1907 it was received, but in not such large numbers, and still smaller numbers were secured and colonized in 1908. In 1909, for the first time, really satisfactory colonies were planted, and one of these colonies was strengthened by the liberation of more individuals in 1910.

So far as it has been possible to determine, no results followed these several attempts at colonization. Of all of the tachinids liberated in

1906, this was colonized the most satisfactorily, or so it is believed (having been confused with *Tricholyga grandis* it is impossible to state definitely which of the two was the more abundantly reared and liberated that year), and it ought to have been recovered by 1910 if it is ever to be recovered as a result of early colonizations. That it has not been recovered as a result of the 1909 colonization work is not at all surprising, because there is every prospect of two or three years elapsing between the liberation and the recovery of any species, and more particularly of those which, like Tachina and many others of the tachinid parasites of both the gipsy moth and the brown-tail moth, are not received from abroad until after the season is so far advanced as to make immediate reproduction upon either of the hosts mentioned impossible.

It is unfortunately true that it would be impossible to distinguish it from *Tachina mella*, should it be reared, since *T. mella* is occasionally reared as a parasite of the gipsy moth or the brown-tail moth, but it is still more unfortunate that no adults of any species which could by any possibility be referred to either were reared in 1910 from the gipsy moth.

In 1910 Messrs. Thompson and Tothill conducted an experiment to determine whether *T. mella* and *T. larvarum* would hybridize. The results were negative, and not of sufficient strength to be at all decisive. If it could be proved that hybridization took place freely, the fact in itself would probably be sufficient to render the European species of no account as an enemy of the gipsy moth in America. Interbreeding with a vastly superior number of another race, the principal and only economically important distinguishing characteristic of which was inability to breed upon a certain host, would undoubtedly result in the sinking of the racial characteristic, and *T. larvarum* as a race would almost immediately cease to exist. This is the more probable in the light of the experiences, yet to be related, which attended the attempted introduction of the brown-tail parasite *Parexorista chelonix*.

On this account, and on no other, *Tachina larvarum* has been tentatively eliminated from the list of promising parasites of the brown-tail moth and the gipsy moth. It may not establish itself here in America, and under the peculiar circumstances, proof to the contrary being lacking, its possible hybridization may make further attempts to import it useless.

TACHINA JAPONICA TOWNS.

Pretty nearly all that has been said of *Tachina larvarum* may be said with equal truth of *Tachina japonica*, in so far as its value in America is concerned. It may possibly be that it is sufficiently distinct as a species to make possible its successful establishment, even presuming that *T. larvarum* should not be established, but the chances are not particularly in favor of such an outcome. It has not been so long nor so satisfactorily colonized, and there is yet a chance that it will be recovered as a result of the colonies which have been planted, or which are likely to be planted in the future. No especial attempt will be made to test its ability to exist as a race apart from *T. mella*, but it is expected that its puparia will be imported in some numbers in 1911 or in 1912, in connection with work involving the importation of other Japanese parasites.

TRICHOLYGA GRANDIS ZETT.

Although generically distinct from Tachina, according to the at present accepted and as is increasingly evident artificial classification of the Tachinidæ, *Tricholyga grandis* is so similar to *Tachina mella* and *T. larvarum* as sometimes to be separated with difficulty from those species. In Europe Tachina and Tricholyga attack the gipsy moth with nearly equal freedom, but relatively a very few Tricholyga have been reared from the brown-tail moth. The fact that Tachina and Tricholyga do not usually occur in the same locality the same year has already been the subject of comment. If may be that a careful review of the rearing records of the two will show that Tricholyga is increasingly important as a parasite in the more southerly localities, but such review has not been made with this point in view.

It was not until 1909 that it was definitely separated from Tachina in the records of the rearing and liberation of the tachinid parasites, and up to 1908 the two species were so inextricably mixed as to make it very difficult to state with any approach to accuracy the relative proportions of the two among the number colonized. There was only a single specimen of Tricholyga among the several Tachina which were preserved for museum specimens from among those imported in 1906 and 1907, and on this account it is probable that Tachina was in the considerable majority.

In habits Tricholyga differs from Tachina in only a single conspicuous respect. It deposits the same sort of eggs, similarly placed; its larvæ appear to have the same feeding habits, and about the same length of life cycle; but unlike Tachina it seems habitually rather than occasionally to pupate within the caterpillar skin or pupal shell of its victim. On this account some of its puparia have been difficult to find in the boxes of imported caterpillars, and it has been found advisable when they are present at all, to keep the dead caterpillars inclosed until such flies as are present have emerged.

Like Tachina, it probably hibernates in the puparium, but neither of the two has ever attempted to hibernate when reared from imported European gipsy-moth caterpillars. The introduction and establishment of Tricholyga will depend upon the existence of an alternate host, and its effectiveness as a parasite upon its ability to make a place for itself in the established American fauna. Several attempts to secure its reproduction in the laboratory on other hosts than the gipsy moth have been measurably successful, and there is good reason to believe that it will find conditions suitable to its continued existence here.

Notwithstanding its similarity to Tachina it appears to be a perfectly good and distinct species, and since it is not known to be represented by any very close ally in America, the objections which have been raised against the probable establishment of Tachina do not apply.

It is unfortunately impossible to say more concerning the likelihood of its becoming established here, since there is much doubt concerning its colonization. If, as is possible, it formed the bulk of the so-called Tachina liberated in 1906 and 1907, it ought to have been recovered before now; if, on the contrary, it was sparingly present among the tachinids reared and liberated during those years, there is no reason to expect its recovery before 1911, and perhaps not until 1912, as the direct result of the large colonies which were liberated in 1909, and which would represent the first satisfactory colonization of the species in America.

In the popular bulletin issued in the spring of 1910, through the office of the State forester of Massachusetts, it was stated that in the fall of 1909 it had already been recovered upon several occasions as a parasite of the gipsy moth, and under such circumstances as to make it possible that it was already established and dispersing rapidly. This statement was in part at least based upon erroneous identification, but at the present date it is expected that 1911, or at the latest 1912, will see its recovery under bona fide circumstances as an established and promising parasite of the gipsy moth.

PARASETIGENA SEGREGATA ROND.

A third species of the group which includes Tachina and Tricholyga, and which deposits similar large, flattened eggs, is to be found in *Parasetigena segregata*, which occurs throughout Europe in very variable abundance as a parasite of the gipsy moth, but not of the brown-tail moth. It is the one species of gipsy-moth parasite which appears to be more common toward the northern limits of the range of this particular host, a fact which may be explained in part by the fact that it is a common parasite of the nun moth (*Liparis monacha* L.) as well. It differs from either Tachina or Tricholyga in that it has but a single generation a year. It hibernates in the puparium, and the flies issue coincidently with, or perhaps if anything a little in advance of, those of Blepharipa in the spring.

Consequently it possesses the material advantage of being independent of an alternate host, and theoretically there is nothing to prevent its rapid increase whenever conditions favor the increase of the gipsy moth. The most that can be said against it is its inability to effect the control of its other and apparently more favored European host, the nun moth, which to a greater extent than the gipsy moth is a pest in the forests of northern and central Europe. Perhaps it may find conditions in America more favorable than in Europe, and thereby be able to do more toward effecting the control of its host here than abroad.

So far as known its larval habits agree very exactly with those of Tachina in all of their essential particulars. It leaves the host caterpillar before pupation, and only upon rare occasions is carried over into the pupa.

The first specimens which were reared in connection with the work of parasite introduction were found mingled with those of Blepharipa, which issued from hibernated puparia in the spring of 1908. There were only a very few of them, but there were enough to make it possible for Mr. Townsend to determine the salient features in its life history and to create a desire to secure more for colonization purposes.

Relatively very few puparia were secured in importations of 1908, and it remained for those of 1909 to produce the number which was necessary to make a satisfactory colony of the species possible. Its puparia being indistinguishable from those of Tachina and Tricholyga, it was necessary to await the emergence of those species before attempting to count upon Parasetigena, but after the others of the Tachina group had ceased to issue, it was found that a very satisfactory number of unhatched and healthy puparia remained. This number was subsequently increased by the importation of several hundred which had been reared from the nun moth, and which subsequently proved to be specifically identical with, or at least indistinguishable from those from the gipsy moth.

For the most part these puparia successfully hibernated, and in excess of 1,000 of the flies were reared in the spring and colonized in one locality where there was every opportunity for them to multiply to the limit of their powers upon the gipsy moth. An attempt to recover the species in the locality later in the season failed, but since it was not expected that it would be recovered so soon the disappointment was not very keen. It would undoubtedly be more encouraging from a practical standpoint if it were positively known that the species was reproducing freely, but the failure to recover it is in no way so significant as would have been the failure in the case of Blepharipa.

Blepharipa, which was colonized in the same locality and under the same circumstances and subsequently recovered, does not leave its host until after the latter has pupated, and since the collection of pupæ is very much less difficult than the collection of caterpillars, its recovery was that much more easy and certain. Although upward of a dozen Blepharipa were secured from collections made in this colony, all of them were from pupæ and none from the caterpillars, which had to be depended upon for Parasetigena.

A determined effort will be made to recover both species in 1911, and the results of the season are anticipated with much interest.

CARCELIA GNAVA MEIG.

This is probably the least understood of the tachinid parasites of the gipsy moth. It appears to be not at all well distributed throughout Europe and has never appeared in sufficient abundance to give it rank as among the important parasites except in the material from southern France. From that region it has been secured in sufficient numbers to make its colonization possible on a scale that is quite as satisfactory as the colonization of Tricholyga, or, for that matter, of Compsilura until after Compsilura was found to be established.

It was received in gipsy-moth caterpillar importations as early as 1906, but in very small numbers in that year, and in still smaller numbers in 1907 and 1908. In 1909 the very large and until then unprecedented importations from the Hyères region produced several thousands of flies, and more were received in 1910, which went to strengthen colonies of the previous year. Curiously enough, in 1910 it was almost the only tachinid parasite secured from this region, on account of which the gross number colonized is in excess of any other species. Like Tachina, Tricholyga, and Compsilura, it is practically certain that an alternate host will be a requisite if it is to complete its seasonal cycle in America. If this disadvantage can be overcome, there is every reason to expect its recovery in 1911 or 1912. That it was not recovered in 1910, in spite of the fact that some 10,000 caterpillars of the gipsy moth were collected in the immediate vicinity of the most satisfactorily liberated colony of the summer before, entirely loses its significance when it is taken into account that neither was Compsilura recovered from these 10,000 possible hosts, and Compsilura was also colonized at the same time and in the same place and under circumstances very much more favorable to its establishment than those which accompanied its original and effective colonization two or three years before. Better than Compsilura has done is expected of none of the tachinids, and neither Tricholyga, Carcelia, nor Parasetigena, nor Zygobothria, which is the next parasite to be considered, has had the opportunity which Compsilura has demanded in each instance in which it has been colonized, to prove itself of value.

ZYGOBOTHRIA NIDICOLA TOWNS.

Pretty much everything which has been said of Carcelia may be said of Zygobothria, not so much because it is similar in its habits as because we have very little first-hand knowledge of its habits. It probably deposits living maggots upon the body of its host or else very thin-shelled eggs containing maggots ready to hatch; but this is not certainly known. It always leaves its host before pupation and forms a free and characteristic puparium with roughened surface and protruding stigmata very unlike that of any of the other tachinid parasites of the same host.

It is not quite so common as a parasite of the gipsy moth as is Carcelia and not so many have been colonized, but the colonies have been very satisfactory notwithstanding, and there is about as much reason to expect the establishment of this species as in the case of any of the others. Like several of the others, it was not colonized until 1909, and its recovery is hardly to be expected until 1911 or 1912, and as in the case of these others its establishment and value as a parasite will very largely depend upon its ability to find a sufficient supply of acceptable hosts.

CROSSOCOSMIA SERICARIÆ CORN.

Many years have passed since Dr. Sasaki published the most interesting and surprising results of his investigations into the life and habits of the so-called "uji" parasite of the silkworm in Japan, and his account of the manner in which this serious enemy of that insect gained access to its host was so extraordinary in the light of that which was known concerning the oviposition of tachinids in general as to cause the truth of his discovery to be questioned by several eminent entomologists.

His work has been most carefully reviewed in connection with the investigations which have been carried on at the laboratory into the life and habits of the allied species, *Blepharipa scutellata*, and it was with much satisfaction that his account of the biology of Crossocosmia was found to apply almost equally well in nearly all of its details to the biology of the European parasite of the gipsy moth. There was one important point of difference, however, in that the first-stage Blepharipa was never found ensconced in the ganglion of its host, while Crossocosmia, according to Dr. Sasaki, habitually chooses this position.

In 1908 quite a number of the puparia of a Japanese parasite of the gipsy moth was received from that country, which, so far as external characteristics were concerned, were indistinguishable from those of Blepharipa from Europe. None of the flies issued the following spring owing to the bad conditions under which the puparia were received, but an examination of the pupæ, which like those of Blepharipa developed adult characters in the fall, was sufficient to convince Mr. Townsend that the species was nothing else than *Crossocosmia sericariæ* itself.

Mr. Townsend's determination of the species, was partially confirmed in the spring of 1910 when several hundred of the flies were reared from puparia received the previous summer. Later the same year, through the kindness of Dr. Kuwana, specimens of the bona fide "uji" parasites, reared from silkworms, were received at the laboratory. No differences whatever were discernible and the confirmation appears complete.

There was an opportunity, during the summer of 1910, to dissect a few of the caterpillars of *dispar* from Japan, and among those so dissected by Messrs. Thompson and Timberlake were found several which contained the young larvæ of Crossocosmia in the ganglia, exactly as described by Dr. Sasaki. Thus it was that his account of the life of the "uji" was confirmed in its every particular in which his remarks were based upon actual observation and not in part upon speculation as to the significance of certain obscure phenomena. To Mr. Townsend, and perhaps more particularly to Mr. Thompson, who has devoted considerable time and performed a vast amount of tedious and in some instances unremunerative dissection work, is the credit due for thus removing all reflection upon the accuracy of Dr. Sasaki's remarkable observations.

In practically every respect, except in the location of the first-stage maggots in the body of their host, the life and habits of Crossocosmia as a parasite of the gipsy moth agree with those of Blepharipa. In Japan it is of about the same relative importance as a parasite as Blepharipa in Europe. Its habits of pupation and the difficulties experienced in providing for its successful hibernation are identical.

Its value as a parasite of the gipsy moth in America depends very largely upon the success which attends the attempts to import and establish the European parasite. Should this be accomplished, as now appears probable, any special efforts to import Crossocosmia might well be deemed unnecessary. It is highly improbable that two species having habits so exactly similar would be any more effective than one.

But it is pretty evident that in one other and very important respect the habits of Blepharipa are different from those of Crossocosmia. It is apparently quite as abundant in Europe as is Crossocosmia in Japan, but even in the most important silk-producing regions it is yet to be recorded as an enemy of the silkworm. It would appear that in their respective host relations the two species possess a difference, and it is probable that it will be found to extend to other hosts than the silkworm when all the hosts of both species are known. In consequence it is not only well to have Crossocosmia to fall back upon in case Blepharipa fails to come up to expectations, but it is well that it be given a trial in order that the relative value of the two species may be determined.

CROSSOCOSMIA FLAVOSCUTELLATA SCHINER (?).

It was with considerable surprise, accompanied with no small degree of doubt as to the accuracy of our records, that the presence of a species of Crossocosmia was recognized among the flies issuing from European puparia in the spring of 1910. At first it was thought that there must have been some Japanese puparia mingled with them, and when reference was made to the notes it was found that something like 15 or 20 larvæ of *Crossocosmia sericariæ* had been received the summer before, and that their disposition was not indicated. Accordingly, for a time it was supposed that the Crossocosmia issuing were from these, but it was not long until more adults had issued than could possibly be accounted for in that manner. There were as many Japanese Crossocosmia puparia producing Crossocosmia as the notes called for, with never a Blepharipa among them, and when after a time it became apparent that the number of European Crossocosmia would run into the hundreds and that they came from a variety of lots of puparia under several numbers and received at different times, it was finally decided that the existence of what has every appearance of being an European race of *C. sericariæ* could no longer be doubted.

Its occurrence in Europe is the more surprising because, like Blepharipa, it has never been recorded from the silkworm in any of the silk-producing districts. In its distribution it also exhibited peculiarities, practically all that issued having come from a lot of puparia received in gipsy-moth caterpillar importations from the vicinity of Charroux, a town in western central France, and one which would hardly be expected to differ particularly in its fauna from other localities from which material was received.

Only a very few specimens of this European Crossocosmia were pinned for the collection, but so far as the closest scrutiny manifests there is not the slightest structural difference between the bona fide "uji" parasites reared from the silkworm—that which is consequently believed to be the same species reared from the gipsy moth in Japan—and the species under present consideration from France, which is seemingly not present, or, if present, not common in other parts of Europe from which parasite material has been received.

The specimens reared, to the number of several hundred, with several hundred of the Japanese Crossocosmia, were colonized together, and under favorable circumstances, as indicated by the recovery of Blepharipa from the immediate vicinity as the result of coincidental colonization. Should the two species be in very truth the same, they will probably hybridize, and enough have been liberated to make one good colony. Should they refuse to intermingle, there is not a sufficient number to make what past experience has indicated as a "satisfactory" colony of either.

UNIMPORTANT TACHINID PARASITES OF THE GIPSY MOTH.

There are not as many unimportant dipterous as there are unimportant hymenopterous parasites of the gipsy moth in Europe, and there are other reasons why they need not be considered at so much length. One of them, *Pales pavida* Meig., which is occasionally present in shipments of gipsy-moth caterpillars, is much more commonly received as a parasite of the brown-tail moth, and *Dexodes nigripes* Fall., which is very rarely associated with the gipsy moth, is a very common parasite of the other host. Both of these species will be discussed later, and something will be said of their life and habits and of what has been done toward securing their establishment in America.

Of the remaining tachinids which have been reared from imported material from Europe, none has been positively associated with the gipsy moth itself. There is always the chance that one or two caterpillars of some other species may have been accidentally included amongst those of the gipsy moth, and while the number of such has always been very small, the chance that a strange parasite should be reared from them rather than from the gipsy-moth caterpillars is large.

To date at least 98 per cent of the tachinid puparia which have been received from Japan as parasitic upon the gipsy moth have been either of Tachina or Crossocosmia. The remaining 1 or 2 per cent have been of various species, among which was one that resembled *Pales pavida* and another has been described as "Compsilura-like." There have been so few of these strange forms as to make impossible a definite statement as to their host relations. It seems rather curious that against the 8 European tachinids, all of which are of at least local importance as parasites of the gipsy moth, Japan should be able to produce only two. It may be that the tachinid fauna of Japan is much less extensive than that of Europe or of America. It may also be that a more thorough survey of the Japanese situation will reveal the presence of species which have not been received hitherto on account of the inadequacy of the methods of collection and shipment. It is rather expected that the latter may be the true explanation and that the apparent scarcity of tachinids in the parasite fauna of the gipsy moth in Japan may not prove to be real.

PARASITES OF THE GIPSY-MOTH PUPÆ.

THE GENUS THERONIA.

The discussion of the pupal parasites of the gipsy moth may well begin with mention of the most generally distributed of all—Theronia. The genus has already been the subject of brief comment in the account of the American parasites, and something was said of the habits of *Theronia fulvescens* in its relation to this host in America, and of its unimportance. The form which by courtesy is thus specifically designated is very imperfectly differentiated from *T. atalantæ* Poda, which prevails throughout Europe in relatively about the same abundance in relation to the gipsy moth. It is readily distinguished from the American form by its habitat and to a less satisfactory extent by color.

In Japan occurs still another, indistinguishable biologically (so far as its biology is known) or morphologically, but differing in color from either the American, from which it is most distinct, or from the European. It has been described as *Theronia japonica* Ashm.

The rôle played by these so-called species in the countries to which they are severally native is nearly identical and at the same time unimportant, when economically considered. The likelihood that either the European or the Japanese would become relatively more effective in America than the American itself seems so very remote as to make unworthy of consideration any serious attempts to introduce and colonize either. Quite a good many of the European have been liberated in America from time to time, but in a purely incidental way. More will probably be received in the future and similarly liberated.

It was in the winter of 1907–8 that the late Mr. Douglas Clemons, of the laboratory, found a large number of the females of *T. fulvescens* congregated beneath old burlap bands in a tract of woodland in which the gipsy moth was actively being fought. Some of these females were dissected some days later and found to be without fully developed eggs, and on the basis of these inadequately conducted dissections it is supposed that, as in Monodontomerus, the males die in the fall, leaving the females to hibernate. It would, in other words, mean that the species is single-brooded.

The subject ought to have been still further investigated, but the unimportance of the species from an economic standpoint has robbed it of interest other than that which has attached to the remarkable and suggestive vagaries which it has exhibited in its host relations.

If it is, in truth, single-brooded, like its host, it ought to multiply much more rapidly than it has done, in view of the superlative opportunities which the past 10 years have afforded.

THE GENUS PIMPLA.

The several forms of the genus Pimpla which have been reared from gipsy-moth pupæ received from Europe and Japan are not, like the forms of Theronia, confusing and indefinitely separable, but good and distinct species. There are 3 European, and a like number of Japanese, making together, with the 2 American, a total of 8 of the genus known to attack this host. Notwithstanding their variety, all the species acting together in any one locality have never effected the degree of parasitism resulting from the attack by Theronia in the same locality. Being collectively of so little importance it is unnecessary to say more concerning their relative importance individually.

Quite a little has been learned at first hand concerning the two European species most frequently encountered, *Pimpla instigator* Fab. and *Pimpla examinator* Fab. Both have been received in considerable numbers in shipments of brown-tail moth pupæ, and have been liberated to the number of several hundred each in 1906, 1907, and 1909. Neither has since been recovered from the field.

Both have been carried through all of their transformations in the laboratory upon the gipsy moth, the brown-tail moth, or the white-marked tussock moth, and in the case of *P. instigator* upon all three above-mentioned hosts. The early stages of the larvæ have not been seen. In nearly every respect, so far as observed, they resemble each other in habit and biology and also *P. (Hoplectis) conquisitor* Say. and *P. pedalis* Cress., their American congenors. The one point of difference between them is the tendency of *Pimpla instigator* to hibernate within the pupa of the brown-tail moth. A very few have been reared each spring since 1908 from cocoon masses received the summer before. The proportion thus hibernated is very small.

Pimpla instigator, like the American *P. (Hoplectis) conquisitor*, may become hyperparasitic on occasion. On August 7, 1907, five female specimens of *P. instigator* were confined with several tussock-moth cocoons which contained the cocoons of *Pimpla (Epiurus) inquisitoriella* Dalla Torre, from some of which adults were emerging, and all of which had been spun for several days. Oviposition was immediately attempted. It was certainly successful, for on August 29, at least two weeks after the Epiurus had ceased to issue, a greatly dwarfed male *P. instigator* appeared and it was followed by another similarly small male on September 3. There is not the slightest doubt that the European parasite attacked the native and that its larvæ fed to maturity. At the same time it is not likely that it would have done so had the cocoons of the native not been associated with the proper host of the other.

The third species, *Pimpla brassicariæ* Poda, is much less commonly reared from either the gipsy moth or the brown-tail moth than the other two. Apparently its habits are identical.

Hardly enough have been received of the three Japanese species to indicate their relative abundance. The most striking of them, *Pimpla pluto*, appears to be the only one of the trio which has been described, and to the others Mr. Viereck has given the names *P. disparis* and *P. porthetriæ*. It is possible that they are just a trifle more common in connection with the gipsy moth in Japan than are the corresponding species in either Europe or America. At the same time Theronia has outnumbered all three together in the Japanese material studied at the laboratory.

Hardly anything is known about them. Not enough have been received to make colonization possible, and only upon one occasion to permit of laboratory reproduction with fertilized females, and upon this occasion there was no time to devote to their further study.

Presumably, except for minor differences, all of the Japanese Pimpla will be found to conform very exactly in biology and habit to the American and European. All will probably be found to attack a very large variety of hosts, and all will defer their attack until their host has entered the prepupal or pupal state. The females of all will probably be ready to oviposit for a new generation almost immediately following their emergence, and the length of life cycle, dependent upon temperature, will be about three or four weeks. There will necessarily be more than one generation each year unless the hibernating individuals should live long enough to deposit eggs for another hibernating generation, as might easily be possible in the case of *Pimpla instigator*, and conceivably possible in the case of each of the others.

Pimpla conquisitor and *Pimpla pedalis* are among the most generally effective of the pupal parasites of the medium-sized cocoon-spinning Lepidoptera in the Northeastern States. The first named is perhaps the most common and effective of all the parasites of the tent caterpillar and about as effective as any other one as a parasite of the tussock moth. It does not vary much in relative abundance from one year to the next, and appears to play a part which is rather to be compared to that taken by the birds than to that taken by most of the parasites. It is, like Theronia, so impartial in its attentions to all of the different species of its hosts as scarcely to be affected by an unusual abundance or unusual scarcity of any one among them in particular.

The same is very likely to be true of the European and Japanese species. The part played by each in the localities where it is native is probably similar to that taken by *P. conquisitor* or *P. pedalis* in America. On this account it is not considered as probable that either *P. examinator* or *P. instigator* will ever become established in America as a result of the not very satisfactory colonies which have been liberated. They will, of necessity, enter into direct conflict with the American species for a share in the business of being parasites upon a certain section of the insect community, including a large number of species of which the gipsy moth is but one. Competition may result in cut rates and more and cheaper parasitism for a time, but eventually, if the newcomers ever secure a foothold at all, they will either drive the natives out of the business or else share and share alike with them in accordance with an amicable and natural agreement.

In consequence, no assistance is expected from the various foreign species of the genus Pimpla as parasites of the gipsy moth or of the brown-tail moth. They are merely liberated when received, under the best conditions which can be afforded looking for their establishment, and if they are ever recovered from the field, the most that is expected of them is that the circumstances surrounding such recovery will exemplify the truth of the above remarks.

ICHNEUMON DISPARIS PODA.

One of the most distinctive of the gipsy-moth parasites, and one of the first, if not the very first, described as attacking that host, *Ichneumon disparis* is at the same time one of the less common, if dependence is to be placed upon the rearing records at the laboratory. It may be that it is never common, or it may be that it is eastern and southern in its distribution in Europe, rather than central and western; some few incidents in connection with its importation have indicated that perhaps its scarcity in European imported material was due to such material having been collected outside of its natural range. In any event not more than two score of individuals have been reared in the course of the five years since the work was begun.

Very little is known of its life and habits, other than that it probably attacks the pupæ or perhaps the prepupæ, and never the active caterpillars. It is thought possible that it hibernates as an adult, and if this is true, it might conceivably be a parasite of importance could enough be secured to make possible a sufficiently strong colony. To date there never has been a single mated pair available for liberation at any one time.

THE GENUS CHALCIS.

The first few boxes of parasite material which were received in 1905 produced among other things quite a large number of Chalcis, a part of which issued from the pupæ of the gipsy moth and a part from dipterous puparia, supposed at that time to be those of tachinid parasites of the same host. All of them appeared to be of one species, *Chalcis flavipes*, and on the supposition that those which appeared to issue from the gipsy-moth pupæ might actually have come from tachinids which were inside, all were destroyed.

In 1906 and 1907 very few Chalcis were received from any source and there was no opportunity to determine the true host relations of the European species. In 1908 a considerable shipment of gipsy-moth pupæ from Italy arrived in good condition for the first time since 1905, and another shipment from Japan, also in good condition, reached the laboratory almost coincidently. Both were soon found to contain Chalcis in some numbers, and, as it soon developed, in considerable variety.

It is not necessary to go into any details as to the steps through which it was finally decided that no less than six species of Chalcis were present in these two shipments, of which two were easily separable by conspicuous structural and color characters. The others were more or less confusing to one who had only a few doubtfully identified specimens in the collection, and little knowledge of what were the characteristics of a species in the genus.

With the assistance of biological and geographical characters, the separation was finally effected, and the 6 have been since definitely identified by Mr. Crawford, as below. To the list are added 2 more, 1 of which is Japanese and the other American, making a total of 8 in all that have been definitely associated with the one host.

Chalcis flavipes Panz. Primary parasite of the gipsy moth in Europe.

Chalcis obscurata Walk. Primary parasite of the gipsy moth in Japan.

Chalcis callipus Kirby. Primary parasite of the gipsy moth in Japan, according to rearing note attached to a specimen forwarded to the laboratory through the kindness of Mr. Kuwana.

Chalcis fiskei Crawf. Parasite of the tachinids *Crossocosmia sericariæ* and *Tachina japonica* in Japan, and thereby a secondary parasite of the gipsy moth.

Chalcis compsiluræ Crawf. Parasite of *Compsilura concinnata* in America, and therefore a secondary parasite of the gipsy moth. (*Chalcis ovata* Say has never been reared as a parasite of the gipsy moth, although it is not improbable that it will be found to attack it when the moth shall extend its range southward into territory where the Chalcis is more common than it appears to be in eastern Massachusetts.)

Chalcis minuta L. Parasite upon sarcophagids associated with the gipsy moth in Europe. Since the status of the sarcophagids themselves remains to be determined, it is impossible to state that of the Chalcis. It is believed that the sarcophagids are scavengers, and neutral, in which case the Chalcis would also be neutral.

Chalcis fonscolombei Duf. Also a parasite of sarcophagids associated with the gipsy moth in Europe.

Chalcis paraplesia Crawf. Parasite upon sarcophagids associated with the gipsy moth in Japan.

It is thus seen that the genus Chalcis is a little of everything in its relations to the gipsy moth. Of the 8 species, 3 are enemies, 2 are friends, and 3 are undertaker's assistants. To round out the series, one may expect to find a species attacking tachinids in Europe, 1 attacking the gipsy-moth pupæ as a primary parasite, and another attacking sarcophagid puparia in America.

So far as known tachinids are never attacked by the species which prey upon the sarcophagids, although this statement presupposes a discriminating instinct which has rarely been encountered among the parasites of the Diptera generally. For the most part, and in fact with no other exception, so far as the experiences of the laboratory have gone, the parasites which will attack the one will attack the other family also. There are several records, including that already mentioned which was made in 1905, of the rearing of Chalcis from tachinid puparia, but these have either been made before a distinction was made between the puparia of the two families, or else there have been a large number of mixed tachinid puparia involved, and in such instances it is always possible and usually the case that a few sarcophagids are present.

FIG. 44.—*Chalcis flavipes:* Adult. Enlarged. (From Howard.)

As parasites of the gipsy-moth pupæ, *Chalcis flavipes* and *C. obscurata* are closely allied, and exceedingly similar in every respect. *Chalcis flavipes* (figs. 44, 45) appears to be rather restricted in its range in Europe and has never been received from any localities outside of the watershed of the Mediterranean, if an exception is made of the portion of southern France which drains into the Atlantic. The Japanese *C. obscurata* (fig. 46) has been present in every shipment of pupæ from Japan, but the exact localities from which these shipments came is not known.

Both are, or appear to be, invariably solitary, notwithstanding that there is an ample food supply in one pupa for several individuals. Invariably there is an abundance of unconsumed matter in the host pupa, and on this account the parasite has rarely been successfully reared from any of the imported pupæ except the small males, in which this matter is in such small amount as partially to dry before receipt at the laboratory. In the large female pupæ the decomposing contents of the pupal shell form a semiliquid mass, which is shaken about while the material is in transit, and completely overwhelms the larva or pupa of the parasite. The parasite is able to withstand this condition to a remarkable extent, but not to the extent frequently brought about by the unnatural conditions incident to transshipment.

FIG. 45.—*Chalcis flavipes:* Female. Hind femur and tibia, showing markings. Greatly enlarged. (From Crawford.)

Partly on this account, but still more owing to the difficulties which have stood in the way of securing an adequate supply of gipsy-moth pupæ in good condition from localities where Chalcis occurs, it has not yet been possible to colonize either the European or the Japanese species satisfactorily, nor, so far as known, successfully. Only a few hundred have been received, all told, since their status as primary parasites was first established in 1908. Had they all been of one species, received at one time, and colonized in the same place, there would be some reason to expect that the colonization would be followed by establishment. There were two species, however, they were not all colonized in one place, and colonization has extended over three years. The best and largest colony was liberated in 1909 and strengthened by the addition of the small number received from abroad in 1910.

FIG. 46. — *Chalcis obscurata:* Female. Hind femur and tibia, showing markings. Greatly enlarged. (From Crawford.)

A single specimen was reared from a lot of gipsy-moth pupæ collected in the immediate vicinity of the colony shortly after it was founded in 1909, but none issued from similar collections made in 1910.

Both *Chalcis flavipes* and *C. obscurata* have been carried through all of their transformations in the laboratory on American pupæ. The females are able to oviposit very shortly after emergence, and will do so with considerable freedom in confinement, making possible the artificial multiplication of either species were it possible to secure a supply of host pupæ. In the act of oviposition the female firmly grasps the active host-pupa with her powerful hind legs and resists all of its efforts to dislodge her. The egg has not been observed nor

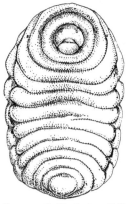

FIG. 47.—*Chalcis flavipes:* Full-grown larva from gipsy-moth pupa. Much enlarged. (Original.)

FIG. 48.—*Chalcis flavipes:* Pupa, side view. Much enlarged. (Original.)

FIG. 49.—*Chalcis flavipes:* Pupa, ventral view. Much enlarged. (Original.)

the early-stage larvæ. The full-fed larva is quite characteristic in appearance, and well represented in the accompanying illustration (fig. 47). The pupa (figs. 48, 49) is almost invariably located in the

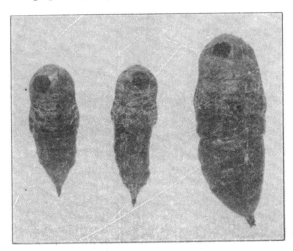

FIG. 50.—Gipsy-moth pupæ, showing exit holes of *Chalcis flavipes*. Enlarged. (Original.)

anterior portion of the host pupa, and the exit hole (fig. 50) of the adult is characteristic, being smaller than that of Pimpla or Theronia, and rarely at the extreme end, as is the case with the ichneumonid parasites. The pupal exuvium is also characteristic and, curiously enough, yellow in the case of the Japanese, and black in that of the European species. Between three and four weeks of ordinary summer weather are necessary for the complete life cycle.

The adults are very long lived, and a few of both species were kept in confinement from early in August, 1910, until December of the same year. During this time they were offered numerous sorts of pupæ, but after it was no longer possible to secure those of the gipsy moth there was no further reproduction. It has always been supposed that it was the adults which hibernated, and the longevity of the individuals mentioned above lends strength to this supposition. If

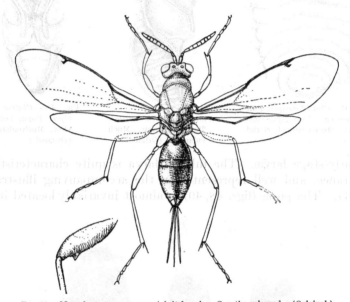

FIG. 51.—*Monodontomerus æreus:* Adult female. Greatly enlarged. (Original.)

correct, there need be only a single generation annually, and the species would therefore be independent of any other host. Neither is known to attack the brown-tail moth, but both have been reared through their transformations upon pupæ of the white-marked tussock moth.

Both of the species of Chalcis are of considerable importance as parasites of the gipsy-moth pupæ in their respective habits, and so far there has nothing occurred to destroy confidence in their ability to become of importance here provided a sufficiently large number may be secured to enable them to become established. It is confidently expected that they will disperse at a very rapid rate, and on this account it will be necessary that the colonies be large and strong, so that extinction through too great scarcity during the first or second season following colonization will not result. Renewed efforts to make this possible will be made this coming season, and at the same

Fig. 1.—View of Laboratory Interior, Showing Cages in Use for Rearing Parasites from Hibernating Webs of the Brown-Tail Moth in 1910-1911. (Original.)

Fig. 2.—"Sifting" Gipsy-Moth Eggs for Examination as to Percentage of Parasitism. (Original.)

time special efforts will be made to recover the species from the field as a result of earlier colonization.

MONODONTOMERUS ÆREUS WALK.

Few among the parasites have been the cause of a larger variety of mingled feelings than this, and the history of its introduction into Massachusetts is in many respects unique and apart from similar histories of the other parasites.

The females (fig. 51) have the curious habit of hibernating in the winter webs of the brown-tail moth, and the species is rather a parasite of the brown-tail moth pupæ than of the gipsy-moth pupæ, although it is sometimes common in the latter connection as well. It was first received at the laboratory in the winter of 1905 in shipments of brown-tail-moth hibernating nests and was reared from these nests in the spring. It has been recorded as a parasite of the gipsy moth, and a colony was planted by Mr. Titus early in the spring of 1906. The records of this colony have apparently been lost and it will never be known exactly how many individuals were included in it.

Some 1,700 issued from the imported nests that first spring, but not all of them were liberated. Dr. W. H. Ashmead,[1] to whom the specimens were sent for determination, stated it as his opinion that it was a secondary parasite rather than a primary, since few or none of the group to which it belonged were definitely known to be primary parasites upon lepidopterous hosts. Accordingly the work of colonization was stopped almost as soon as it was begun, and for a period of more than two years Monodontomerus was treated as a secondary parasite, and destroyed whenever found. During this period, many thousands issued from importations of brown-tail-moth cocoons, and much doubt was felt as to its actually being a secondary, on account of the numbers alone, since it enormously outnumbered all other hymenopterous parasites (whether primary or secondary) reared. For reasons which would be obvious to anyone who has ever had any experience in handling the cocoons of the brown-tail moth, no serious effort was made to determine its host relations by the dissection of the brown-tail moth pupæ. A few pupæ were sought out from which it had issued, and no trace of any other host was found, but such was the state of our technical knowledge at that time as to render questionable such evidence. We were not sufficiently familiar with the appearance of pupæ from which Monodontomerus as a secondary parasite had issued, and dared not give any more than negative weight to the fact that no remains of any other primary host than Monodontomerus could be found. Moreover, against this negative evidence indicative of primary parasitism, was much that

---

[1] Now deceased.

was positive, indicative of secondary parasitism, because every little while the Monodontomerus would issue from tachinid puparia which had been sorted out from the cocoons and pupæ. Still more frequently it was reared from puparia of sarcophagids.

In the summer of 1908 the shipment of gipsy-moth pupæ from Italy, which served the purpose of establishing the status of the European species of Chalcis in their relation to the gipsy-moth, served also in establishing the status of Monodontomerus as a primary parasite of this host. A large number of the pupæ which were examined was found filled with the larvæ (fig. 53) or pupæ (figs. 54, 55) of the parasite, and even when the larvæ were still immature and feeding there was absolutely no trace of any other parasite present in the majority of instances. There was such trace in a few, and it was found that the former presence of Pimpla, Theronia, or any tachinid was very easy of determination, no matter how completely it might have been destroyed.

It was felt that a mistake had been made in not liberating the very large number of Monodontomerus which had been secured through the earlier shipments, and it was resolved to colonize them as fast as they were secured in the future. Hardly anything was less expected than that the species should even then be established.

Each winter since that of 1906–7 (Pl. XXI) large numbers of the winter nests of the brown-tail moth had been collected in a vain endeavor to secure evidences of the establishment of *Pteromalus egregius*, but without results. In the winter of 1908–9 this work was undertaken anew, and almost the first lot which was brought into the laboratory was shortly productive of a number of Monodontomerus, exactly as lots collected in the open in Europe had been productive of the species each season since their importation had been begun. The circumstance, surprising and unexpected, was also gratifying, coming as it did so soon after the investigations which had served to demonstrate the primary parasitism of the species. The surprise and gratification was increased materially when it was discovered, through the collection of a large quantity of the winter webs, that the parasite was distributed over a considerable territory indicated by the area I on the accompanying map (Pl. XXII), and though the actual number recovered was small, the rapid rate of dispersion was sufficient to indicate a very rapid rate of increase. It was estimated, in fact, that at least a 25-fold per year increase and a 10-mile per year dispersion had followed the colonization three years before.

In 1909 an examination of the pupæ of the gipsy moth in the field revealed the presence of what was actually a small, but under the circumstances a gratifyingly large, number which contained the larvæ or pupæ of the parasite, and the results of the winter scouting work were awaited with confidence and interest. They were quite as satisfactory as could be expected. The collections of nests from areas II and III on the map (Pl. XXII) produced the parasite in abundance, and in area I, throughout which it was found the winter before, it was very much more abundant, as will be seen by reference to the tabulated summary of the results of the work for the winter in Table X.

TABLE X.—*Monodontomerus æreus as distributed over its area of dispersion.*[1]

Section.	Year.	Number of brown-tail nests collected.	Number of Monodontomerus recovered.	Monodontomerus per 1,000 brown-tail nests.	Section.	Year.	Number of brown-tail nests collected.	Number of Monodontomerus recovered.	Monodontomerus per 1,000 brown-tail nests.
1	1908	5,574	39	6	4	1910	1,698	234	137
	1909	2,200	708	376	5	1910	701	215	305
	1910	1,508	495	328	6	1910	2,836	521	183
2	1908	947	0	0	7	1910	1,050	260	246
	1909	1,107	124	112	8	1910	555	86	151
	1910	700	182	260	9	1910	825	538	652
3	1909	770	34	49	10	1910	500	1	2
	1910	260	13	50	11	1910	1,600	95	59

[1] Refer to the map (Pl. XXII) for the area included in each section.

TABLE REPRESENTING THE RAPID MULTIPLICATION OF MONODONTOMERUS IN THE FIELD.

	Brown-tail nests collected.	Monodontomerus recovered.	Monodontomerus per 1,000 brown-tail nests.
Average of sections 2 and 3 in 1909	1,927	168	87
Average of sections 2 and 3 in 1910	960	190	199

In 1910 a fairly satisfactory number of the parasites was reared from collections of brown-tail moth cocoons made in the field, but when the gipsy-moth pupæ were examined in the field as in 1909, scarcely if any more were found to be parasitized. This was anything but encouraging, because it had been expected that parasitism would amount to at least 1 per cent, if the rate of increase which had prevailed up to 1909 had continued. It appeared that the Monodontomerus was either inclined to pass over the gipsy-moth pupæ in favor of other hosts, or else that its rate of increase had received a sudden check before it was sufficiently abundant to become of aid in the control of the moth. As before, that winter's work was anticipated with interest since its results would be more directly comparable with those of the year before than was that summer's work.

The collections of winter webs were first made in the territory included within the range of the parasite in the winter of 1909–10 (areas I, II, and III), and the fact soon became manifest that instead of increasing it had actually decreased in abundance throughout that territory in the course of the year. It was inexplicable, in view of the unlimited opportunities for increase, and it was, to say the least, discouraging.

Following these preliminary collections, which were intended for no other purpose than to indicate the rate of increase, collections from towns and cities to the westward of its known distribution the previous winter and to the northward in southern New Hampshire and southernmost Maine were made. It was rather confidently expected that it would be found in Maine just over the New Hampshire line, and also that this would mark the limits of its distribution in that direction.

How far removed the expectations were from the reality is well indicated by the accompanying map[1] (Pl. XXIII), and still more by a study of the table. It will be seen that instead of stopping at the Maine State line, Monodontomerus has extended its range for a full hundred miles to the northeastward, and that to the north and west it has pretty nearly reached the limits of the present known distribution of the brown-tail moth itself. But what is more surprising, it is actually much more abundant in a large part of this new territory than it was in Massachusetts a year before.

It will also be observed that the distribution has been much more rapid toward the north and east than toward the west and south, which is true also of that of the gipsy moth and the brown-tail moth. Whether this will prove to be the rule with others of the parasites remains to be seen. It is not indicated in the instance of any other as yet.

Monodontomerus appears to pass through but a single generation annually. The females are sometimes, perhaps habitually, fertilized before they actually issue from the pupal shell of the host. The males invariably die before the winter, or at least out of many thousands of individuals which have been secured in the winter from brown-tail-moth nests at home and abroad, only females have been present. Dissection of a considerable number of hibernating females has failed to result in the finding of even partially developed eggs. Neither has it been found possible to keep females alive in the spring until eggs should develop, although some have remained in a state of activity in confinement for several months.

Beginning in 1906, and each year thereafter until 1909, numerous attempts were made to secure reproduction in confinement. Dipterous larvæ and puparia as well as pupæ of the gipsy moth and the brown-tail moth were supplied as hosts, and females from hibernating nests as well as those from gipsy-moth and brown-tail moth pupæ and other sources were used in these experiments. Failure resulted in every instance, due, apparently, to the impossibility of keeping the parents alive until eggs should be developed.

---

[1] The maps and tables have been prepared by Mr. H. E. Smith, to whom the work of caring for the nests as they have been received at the laboratory has largely been intrusted.

The egg which is figured (fig. 52) was dissected from a female which was imported in 1909 with cocoon masses of the brown-tail moth and which was evidently hibernated. She was given no opportunity to oviposit. In 1910 several females were collected in the open in June, and these, upon being supplied with fresh pupæ of the brown-tail moth, immediately oviposited.

The very characteristic larvæ (fig. 53) feed externally upon the pupæ of tachinids within the puparium, but internally within the pupæ of Lepidoptera. The pupæ (figs. 54, 55) are also characteristic, and the appearance of that of the female is indicated by the accompanying illustrations. The exit hole (fig. 56) left in the gipsy-moth pupæ is invariably smaller than that left by Chalcis, and larger than that of Diglochis. It may be located anywhere, in which respect it differs from any of the larger of the pupal parasites.

FIG. 52.—*Monodontomerus æreus*: Egg. Greatly enlarged. (Original.)

As a secondary parasite, Monodontomerus has been reared from tachinid puparia upon numerous occasions both from those which have been received from abroad and from those collected in America. It was rather expected of it that its attack would be confined to those

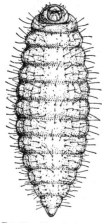

FIG. 53.—*Monodontomerus æreus*: Larva. Greatly enlarged. (Original.)

FIG. 54.—*Monodontomerus æreus*: Pupa, side view. Greatly enlarged. (Original.)

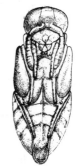

FIG. 55.—*Monodontomerus æreus*: Pupa, ventral view. Greatly enlarged. (Original.)

which were immediately associated with one or another of its chosen hosts, but as usual it did the unexpected, and it has been reared from Compsilura puparia which were collected at the base of trees upon which the caterpillars of the gipsy moth had been common. It has also been reared from tachinids parasitic upon the tussock moth (and from the tussock moth as a primary parasite), from the tent caterpillar, in which it was apparently parasitic upon Pimpla, and from the cocoons of *Apanteles lacteicolor* Vier., the imported brown-tail moth parasite. Like another anomalous species, *Pteromalus egregius*, it betrays a distinct partiality for anything that savors of the brown-tail moth. It is thereby led to seek out the molting webs of the brown-tail caterpillars in the spring and consequently comes into contact with the Apanteles cocoons.

THE SARCOPHAGIDS.

There has been considerable controversy in the past concerning the habits of the Sarcophagidæ, and a wide difference of opinion as to whether they were to be considered as truly parasitic, or whether they were merely scavengers, attacking and feeding upon insects which had died through some other cause. In the case of those species which are reared from grasshoppers there seems to be no further question that they are to be classed as true parasites or at least that they are as truly parasitic as many of the more degraded among the hymenopterous parasites. This seems not to have been proved of any of the species which are found within the pupæ of the larger Lepidoptera.

Fig. 56.—Gipsy-moth pupa, showing exit hole left by *Monodontomerus æreus*. Enlarged. (Original.)

If judgment were to be based upon the occurrence of sarcophagids in the shipments of gipsy-moth pupæ from abroad, it would certainly be judged that the sarcophagids were parasitic. Their puparia (Pl. XX, fig. 3), have frequently outnumbered the tachinid puparia, and even the tachinid puparia and hymenopterous parasites together. Unfortunately, there is nothing known of the circumstances under which this material was collected in any instance, and for all that is known to the contrary, the sarcophagids actually entered gipsy-moth pupæ which had been attacked and killed by another parasite, Chalcis for example, and by feeding, first upon the unconsumed contents of the pupal shell, and later upon the body of the true parasite, which might be destroyed either through accident or design on the part of the intruder, would become, in effect, secondary parasites.

If judgment were to be based upon the results of a quite elaborate series of investigations into the relations between the native sarcophagids and the gipsy moth in America, it would unavoidably be to the effect that these sarcophagids were scavengers and nothing more. We are confronted with conflicting evidence, presented by a much greater abundance of sarcophagids associated with the gipsy moth in Europe than is similarly associated with it in America, which is suggestive of two things: Either the sarcophagids are associated with the gipsy moth because they are parasitic upon it or because of the presence of its parasites, which is quite as reasonable an explanation. It will require much careful work in Europe before it will be possible to settle this point at all definitely. Meanwhile it does not seem to be advisable to attempt the introduction of the European sarcophagids until we know whether they are an aid in the control of the moth or a possible hindrance to the work of the parasites.

The special investigations which were conducted for the purpose of determining the exact status of the sarcophagids in America in relation to the gipsy moth were conducted by Mr. T. L. Patterson, and have been made the subject of a special report.[1]

Another series of investigations, conducted by Mr. P. H. Timberlake, upon the parasites of the pine "tussock moth" in northern Wisconsin, resulted in the accumulation of evidence which pointed quite convincingly to the parasitic character of certain sarcophagids which he encountered in abundance associated with this insect. Unfortunately it is not wholly convincing. If it could be accepted at its full face value it would mean that in these flies we have a group of dipterous parasites wholly distinct from the tachinids, and working in a wholly different manner. The tachinids are caterpillar parasites, and never, so far as has been recorded, attack the caterpillar after it has spun for pupation. The sarcophagids, like Pimpla, Theronia, etc., are pupal parasites and will be grouped together, and at the same time apart from the hymenopterous pupal parasites, even as the tachinids as a group stand beside but apart from the hymenopterous parasites of the caterpillars.

THE PREDACEOUS BEETLES.

It is very probable that further studies into the subject of natural predatory enemies of the gipsy moth will result in the addition of a considerable number of names to the list of predaceous beetles which attack it in one stage or another of its existence and with more or less freedom. The egg masses received from abroad have very frequently been infested with small dermestids, and in the forests in the vicinity of Kief, Russia, in September, 1910, large numbers of the larvæ of a species not yet determined were found feeding, to all appearances, upon the eggs of the moth as well as upon the covering of felted hair.

That these larvæ do actually eat the eggs was demonstrated by Mr. Burgess during his association with the moth work as conducted by the State board of agriculture in 1899 and later his observations were confirmed by a series of simple experiments conducted at the laboratory.

In the spring of 1908 a large number of cocoons of the tussock moth with egg masses attached was collected in East Cambridge, Mass., and from them in June a number of dermestid beetles issued, deter-

---

[1] U. S. Department of Agriculture, Bureau of Entomology, Technical Series 19, Part III, March 22, 1911.

mined by Mr. E. A. Schwarz as *Anthrenus varius* Fab. and *Trogoderma tarsale* Melsh. The Trogoderma was the more common of the two. Later, in the fall, another collection of old cocoons was made for the purpose of determining the status of these beetles. It was found that both of them fed, as larvæ, upon the eggs of the tussock moth, and when they were confined in vials with eggs of the gipsy moth they fed not only upon the hairy covering of the egg masses, but also upon the eggs themselves. Larvæ apparently of one of these species have several times been received at the laboratory associated with egg masses of the gipsy moth, which were in each instance collected upon the sides of buildings or in other situations different from those under which egg masses are most frequently encountered.

As soon as the gipsy-moth caterpillars hatch, if, as frequently happens, the egg mass is situated in some particularly well-sheltered spot, the young caterpillars are liable to attack by small carabid beetles, several species of which have been found under burlap bands in the spring apparently feeding upon the gipsy-moth caterpillars in this stage. Several of these species were made the subject of casual study in the summer of 1910, the results of which will be published later.

The elaterid genus Corymbites, though not generally recognized as predaceous, is undoubtedly more or less addicted to a diet of living insects. An adult of one species was once found feeding upon the cocoons of *Apanteles fulvipes;* and the larva of another, upon one occasion, at least, upon the pupæ of the gipsy moth. There are many species in the New England States. Some of them are nocturnal, and it is not at all beyond the limits of probability that they may be found listed among the predatory enemies of the gipsy moth and the brown-tail moth when these lists shall have been finally completed.

Among the coccinellids the large *Anatis 15-punctata* Oliv. has more than once been observed, as a larva, attacking the small caterpillars of the gipsy moth, and it is not at all unlikely that the species is actually of as much consequence as some of the minor parasites in assisting in the control of the pest.

The lampyrids, too, include amongst their numbers many species which are either occasionally or habitually predatory. One such which abounds in eastern Massachusetts in the spring flying about in the tops of the trees and crawling over the foliage was encountered in the spring of 1910 in the act of destroying a small gipsy-moth caterpillar. Probably one beetle would not destroy many caterpillars in the course of its life, but there are such swarms of the beetles as to make an average of even one caterpillar count materially in the end. Some of the lampyrids are nocturnal, as in fact are a great many of the proved or probably predatory Coleoptera, and their association with the gipsy moth is not likely to be established unless special effort toward that end is undertaken. Such studies require time and patience, but are none the less necessary if we are ever to know all that is to be known about the subject.

None of the beetles mentioned is likely to attack the later-stage caterpillars, but among the larger Carabidæ is to be found a variety of species which are not only able, but more than willing to destroy the full-fed caterpillars and pupæ whenever opportunity offers. There are many such in Europe which do not occur in America, and altogether a considerable number of different species has been received from abroad and tested as to ability to assist in the control of the gipsy moth in this country.

Three characteristics in addition to ability and willingness to attack the gipsy moth are necessary if the introduction of a beetle is to be seriously undertaken as an economic experiment. It must breed at the proper season of the year, so that its larvæ may receive the advantage of the practically unlimited food supply which the present superabundance of the gipsy moth gives; it must be able to withstand the rigors of the New England climate, and not only the adult beetles but their young must be arboreal in habit. An abundance of species both native and foreign will feed freely upon the gipsy moth in confinement, but of these only a few will seek out the caterpillars or pupæ in the situations in which they are to be found in America. The adults of a portion of this number do habitually climb into the trees in search of their prey, but not all such are similarly arboreal during their larval stages. Of those which are arboreal, or which appear to be arboreal, during all of their active life, a part appear to breed at the wrong season of the year and another part do not extend their range into a sufficiently high latitude to make them effective as enemies of the gipsy moth. There is not a single species native to America which meets all of the delicate requirements of the situation, but such a species has been found abroad in *Calosoma sycophanta* L. (See Pl. I, frontispiece, adult eggs, larvæ, and pupa.) This, of all of the numerous species of predaceous beetles which have been investigated at the laboratory, bids fair to be of real assistance in the fight which is being waged.

Like all the larger carabids inhabiting the temperate regions, this species is terrestrial during a considerable portion of its life cycle, but both adults and young, which are equally voracious, climb freely into the trees in search of their prey. The eggs are deposited in the earth, and the young larvæ upon emerging are possessed of a remarkable vitality and sufficient strength and cunning to enable them to seek out and successfully to attack, when found, the largest and most active of the gipsy-moth caterpillars. They also attack the pupæ with even greater freedom, and once ensconsed within such a mass of pupæ as is frequently encountered in partially protected situations upon a badly infested tree, will rapidly complete their growth without leaving the spot. The full-fed larvæ seek the earth and, burrowing well below the surface, construct a vaulted pupal cell, within which the final transformation takes place during the late summer or fall. The adult beetles remain quiescent and as a rule do not issue until late in the succeeding spring.

The breeding season coincides almost exactly with the caterpillar season. The hibernated beetles begin egg deposition just a little before the caterpillars are large enough to be easily found and attacked by their young; the height of their activities in this direction is at a time when their young are best provided for, and they cease oviposition very shortly after the gipsy moths themselves begin to deposit eggs for a new brood. Very shortly thereafter, with summer still at its height, the adult beetles, both male and female, burrow deep into the soil and become dormant, awaiting the arrival of another spring.

The data as given above concerning the life and habits of *Calosoma sycophanta* have been accumulated by Mr. A. F. Burgess, who has had full charge of that part of the laboratory work which had to do with the predatory beetles since the late summer of 1907. Up to that time the pressure of other work was so great as to render impossible any systematic studies along that or similar lines. The first of the adult beetles of this species, together with a smaller quantity of another, *Calosoma inquisitor* L., were imported and in part liberated in the spring of 1906. A few were confined within the large out-of-door cages of the type already figured and briefly described (see Pl. XIV), and reproduction was secured in the instance of *Calosoma sycophanta*. Neither Mr. Titus nor Mr. Mosher was able to give this phase of the work the attention which it really deserved, and while their observations were sufficient to cover most of the salient points in the life history of the predator, there was still an abundance of opportunity for further studies.

In 1907 early and not very systematic surveys of the several field colonies established by Mr. Titus the year before failed to result in the recovery of the beetle. Accordingly, when similar experience with others among the introduced insects had indicated that larger colonies were likely to be required, it was determined to liberate all the adult Calosomas in one locality as they were received from abroad, and thus secure its establishment, if this were possible, before attempting further artificial dispersion. This was done, and several hundred had been received and thus liberated by the time Mr. Burgess was ready to take full charge of the work.

Although it was quite late in the season, Mr. Burgess, with the assistance of Mr. C. W. Collins, who has remained associated with him ever since, succeeded in securing the eggs and in carrying to maturity several larvæ of the species in close confinement in jars of earth, and effectually demonstrated the superiority of this method over that involving the use of the large out-of-door cages. The following spring the work was undertaken upon a considerably larger scale, and along still more specially developed lines. From the hibernated parent stock, and from newly imported beetles, he reared large numbers of larvæ, a part of which were allowed to complete their transformations in confinement, while others were colonized directly in the open when about half grown. These larval colonies promised to be successful, and accordingly the work of rearing the larvæ and distributing them throughout the gipsy-moth-infested area in eastern Massachusetts, with an occasional incursion into other parts of the infested area, was continued throughout 1909 and 1910.

Meanwhile, beginning in the late summer of 1907 and continuing uninterruptedly until the close of the season of 1910, Calosoma has been steadily gaining in the confidence of those who have watched its progress. Its larvæ were first recovered from the field at just about the time when Mr. Burgess first took over the beetle work, and its ability to complete its seasonal cycle in America unassisted was thus indicated. The large colony was also proved to be unnecessary.

Its progress in the field was slow at first, even in the instance of the large adult colony founded in 1907 before it was known to have become established. In 1908 its larvæ were found in abundance in the center of this colony, but not to any great distance away from the point where the beetles had first been liberated. In 1909 the spread was more rapid, but at the same time restricted in comparison with that which became evident in 1910. As will be seen by reference to the accompanying map (Pl. XXIV) which has been prepared by Mr. Burgess from the results of the scouting work of three years, its apparent or discernible dispersion has been at a rapidly increasing rate each year in the instance of colonies which, like these, chanced to be so happily located as to allow for unrestricted and uninterrupted increase from the start.

At the present time there is every prospect that a continued rapid increase for a few years more will result in an abundance of the beetles sufficient to render very efficient aid in the fight against the moth. It is not expected that they will be of very much assistance in localities in which the moth is reduced to such numbers as to make control through parasites such as Compsilura and others of its character possible, but it is expected that whenever the moth breaks out of bounds, and increases to such abundance as to afford the beetles and their larvæ an unlimited food supply, first migration and later rapid multiplication of the beetle will result. In this respect the rôle played by Calosoma is similar to that which is rather confidently expected of Blepharipa.

THE EGG PARASITES OF THE BROWN-TAIL MOTH.

THE GENUS TRICHOGRAMMA.

The parasites belonging to the genus Trichogramma, of which several have been reared from eggs of the brown-tail moth, are the most minute of any which have been handled at the laboratory, and are among the smallest of insects. The egg of the brown-tail moth is in form of a flattened spheroid, approximately as large in its greatest diameter as the printed period which ends this sentence. Normally two or three individuals of the parasite pass through all of their transformations from egg to adult upon the substance of a single host egg, and in exceptional instances as many as 10 perfect adults are known to have issued from one egg. This is the more remarkable when it is remembered that the female Trichogramma is sexually mature at the time of issuance, and ready to deposit a large number of eggs for a new generation.

FIG. 57.—*Trichogramma* sp. in act of oviposition in an egg of the brown-tail moth. Greatly enlarged. (Original.)

The mother parasite exhibits little discretion in the selection of host eggs for attack (fig. 57), and if any dependence is to be placed upon observations which have been made in the laboratory, she is quite as likely to select eggs which contain caterpillars nearly ready to hatch as those which are freshly deposited. The feeding habits of her young are such as to permit a considerable latitude in this respect, but there is a certain limit, and after the embryological development has passed beyond a certain point in the host egg, the attack by the parasite is unsuccessful. It is much better that the host egg be dead than that it contain a living embryo in the later stages of its development.

The life cycle, from egg to adult, varies very considerably in length in accordance with the prevailing temperature. In the summer it may be completed in as short a period as nine days, while in the fall three weeks or more may be required. If the temperature falls below certain limits the young parasites will hibernate or attempt to hibernate, and thereafter their development may be delayed for several weeks, or even months, even though they are exposed to continuous high temperature during this period.

After about one-third of the time requisite for the completion of the life cycle has elapsed, the eggs begin to turn dark, and finally become shining, lustrous black (fig. 58). This change is brought about by the preparation of the larvæ for pupation.

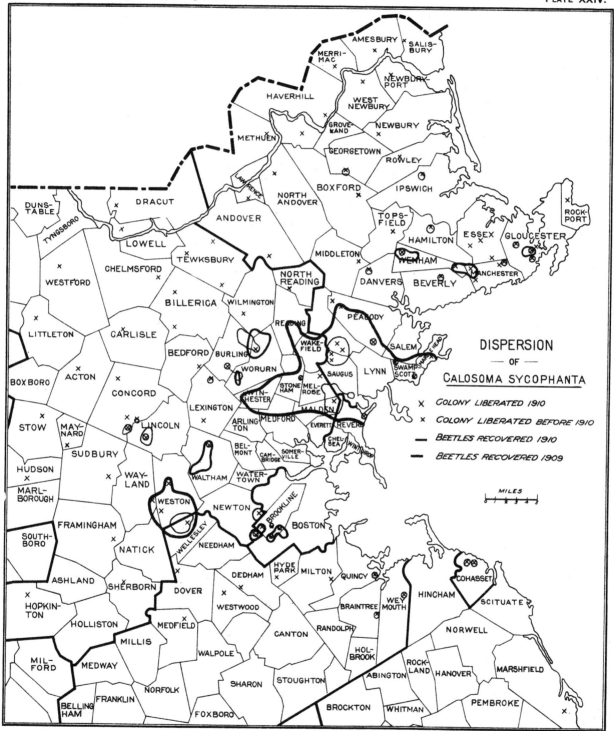

Three races or species have been reared from the eggs of the brown-tail moth, two of them being European and the third American. The American, according to Mr. A. A. Girault, to whom the series of mounted individuals was submitted for determination, is the common and widely distributed *Trichogramma pretiosa* Riley. One of the European, which is here referred to as the *pretiosa*-like form, is or appears to be structurally identical with the American *pretiosa*. It differs in that the progeny of parthenogenetic or unfertilized females is either of both sexes, or else exclusively female, while the progeny of unfertilized females of the American species has always been exclusively male in the very considerable number of reproduction experiments with such females which have been carried on at the laboratory.

The other European species may at once be distinguished from either of its congeners by its dark color, as well as by other characters of taxonomic value. Like the American race of *T. pretiosa* which was studied at the laboratory, it produced males exclusively as the result of parthenogenetic reproduction.

FIG. 58.—Eggs of the brown-tail moth, a portion of which has been parasitized by *Trichogramma* sp. (Original.)

It seems to the writer that in the two morphologically identical but biologically distinct races of Trichogramma (*T. pretiosa*, American or European) we have what is nothing less than two species, quite as distinct as are the species of bacteria, for example, which are founded upon cultural characters. If the manner in which a bacterium reacts when cultivated upon a certain medium prepared after a fixed formula may be considered as sufficient to separate it specifically from an otherwise indistinguishable form which reacts in a different manner under identical circumstances, why may not the same distinctions be made to apply to insects? It may not appeal to the taxonomist and student in comparative insect morphology, but it certainly will appeal to the economic entomologist, who has, or ought to have, a greater interest in the biological than in the anatomical characteristics of the subjects of his investigations. The case of Trichogramma is by no means unique. That of *Tachina mella*, which is practically indistinguishable from *T. larvarum* but which reacts differently in its association with the gipsy moth, is another. Another is to be found in the American and European races of *Parexorista cheloniæ*. There are also others, which need not be mentioned here, but which will receive attention, it is hoped and intended, at some future time.

These statements concerning the behavior of the several forms of species of the genus Trichogramma are based upon the results of something like 275 separate but similar experiments in their reproduction in confinement in the laboratory. It is of course possible, since it was especially desired to continue the experiments as long as possible with individuals of known parentage, that the results are misleading. Possibly had American Trichogramma been collected in the open from a variety of sources, a race might have been found which was arrhenotokous, even as the similar search might have resulted in the discovery of a thelyotokous race in Europe. As it is, the American stock was once renewed. In 1907 a series of experiments was conducted with parent stock reared from brown-tail moth eggs collected in Maine, and in 1908 a similar but more extensive series with parent stock from eggs of the brown-tail moth collected in Massachusetts. In each instance the results were the same.

The longest series of experiments with the arrhenotokous European race was with the progeny of individuals reared from one lot of European eggs from the Province of Carniola, Austria. Similar experiments with one other lot of females upon another shipment of eggs from the same Austrian province and perhaps from the same locality resulted similarly, but the series was not nearly so long. In the first-mentioned series 13 generations were reared in the laboratory, all but the first three being parthenogenetic. Males were secured at one time, and for a limited number of generations, but soon disappeared, even from the progeny of mated females. The results of these experiments will be published in detail later.

Importations of egg masses of the brown-tail moth which had been collected in the open in Europe were first attempted in the summer of 1906, and from almost the first of those which were received at the laboratory a few examples of the *pretiosa*-like European form were reared. Mr. Titus attempted to secure reproduction in the laboratory that first season, but as he had no supply of host eggs in which embryonic development was not considerably advanced, his attempts met with failure.

In 1908 a larger number of egg masses of the brown-tail moth was imported from a great variety of European localities, and as before, the *pretiosa*-like Trichogramma was quickly secured. The failure of the previous season and its cause had early been taken into account, and some time before a large quantity of fresh eggs of the brown-tail moth had been collected and stored at a temperature sufficiently low to prevent embryological development. When supplied with a quantity of these eggs the imported Trichogramma oviposited with the greatest freedom, and in the course of a few generations had increased enormously, so that many thousands were liberated later in the fall. It was conclusively demonstrated that even though the host eggs were dead, abundant reproduction could be easily obtained under laboratory conditions.

A large number of parasitized eggs, containing the brood in various stages of development, were placed in cold storage and kept until the following June and July, when, upon being removed, a few of the parasites completed their transformations. With these as parents large numbers were reared in the laboratory upon the fresh eggs of the brown-tail moth, at that time abundantly available, and the cold-storage experiment was repeated during the winter of 1908–9 with much better results than before. An abundant supply of parent females was available in the summer of 1909, and a great many thousands of the parasite were reared and liberated under the most favorable conditions which could possibly be desired or devised. Many thousands were known to have issued from parasitized eggs contained in small receptacles attached to the branches of the trees upon which the brown-tail moths were even then depositing eggs in abundance.

No false hopes were felt as to the probable success of this venture. It has been amply demonstrated in the laboratory that the females were unable to penetrate the egg mass for the purpose of oviposition, and the location in the mass of the few eggs parasitized by the American race of *pretiosa* indicated sufficiently well the inability of that species to do better in the open than either it or the European would do in confinement.

Accordingly no disappointment was felt, when it was found that the degree of parasitism effected by the European species in the immediate vicinity of the colony sites was hardly, if any, greater than that ordinarily effected by the native species. It is hardly a physical possibility for Trichogramma to effect more than a small percentage of parasitism in the egg mass of the brown-tail moth, and the value of the genus as represented by the three species or forms which have been studied at the laboratory is slight.

At the same time, it is not felt that the labor which has been expended in an attempt to give Trichogramma a fair test has been altogether lost. There are numerous other hosts upon which it is a very efficient parasite, and it is easily conceivable that at some future time it will be found possible to utilize it in some manner which the circumstances themselves will suggest.

As a possible example may be mentioned the tortricid *Archips rosaceana* Harris, which at times becomes a pest in greenhouses devoted to the growing of roses. In Volume II, No. 6, of the Journal of Economic Entomology, Prof. E. D. Sanderson describes such an outbreak in a large rose house in New Hampshire under the heading of "Parasites." Prof. Sanderson says:

The outbreak observed by us furnished a case of the most complete parasitism we have ever seen. When first observed, in late July, from one-third to one-half of the eggs were parasitized by a species of Trichogramma. Two weeks later it was difficult to find an egg mass in which over 95 per cent of the eggs did not contain the black pupæ of the parasite and in most cases 99 to 100 per cent were affected. So effective were the parasites that the control of the outbreak was undoubtedly due to them much more than to any remedial measures.

At about the time when the American parasite was reaching a state of efficiency, a large number of eggs of the brown-tail moth containing the brood of one of the European species was sent to Prof. Sanderson for liberation in the rose house. They were received too late for service, but had they been sent at an earlier date it might easily have been claimed, and with perfect confidence, that the final results were the direct outcome of the colonization experiment.

In such circumstances as these it would (or at least it seems from this distance as though it would) easily be possible and practicable to collect masses of the parasitized eggs and by keeping them in cold storage have ready at hand within the following twelvemonth a supply of the parasites which would be available should the natural stock perish through lack of food, and the destructive increase of the host follow. Parasitized eggs could be sent from one greenhouse to another, and stock could be kept in cold storage in one city to be drawn upon by a florist in any other part of the country when need arose.

Another possible use for the parasite is as an enemy of Heliothis, which is causing serious injury to tobacco in Sumatra. Dr. L. P. De Bussy, biologist of the tobacco growers' experiment station at Deli, has already undertaken its introduction there, and will attempt to handle it after somewhat the same manner as that above described.

TELENOMUS PHALÆNARUM NEES.

A small number of this species was reared from imported eggs of the brown-tail moth from several European localities in 1906, and an attempt was made to secure reproduction in the laboratory. Oviposition was secured, as in the instance of similar attempts with Trichogramma, but it did not result successfully, and apparently for the same reason.

In 1907 a somewhat larger number was reared, and an abundant supply of suitable host eggs having been provided, this number was soon increased several fold, and one large, and several smaller colonies of the parasite were liberated under very satisfactory conditions late in the summer. It was found that the reproduction could be secured upon host eggs which had been killed through exposure to cold, and the experiment was made of hibernating the brood in cold storage, but without success.

In 1908 the quantity of eggs of the brown-tail moth imported was smaller than during the previous year, and only a very small proportion of them proved to be attacked by the Telenomus. Not nearly enough for a satisfactory colony were reared, and again it was attempted to hold the brood over winter in cold storage, and again the attempts failed.

If judgment is based upon the percentage of parasitism by this species in the lots of egg masses of the brown-tail moth which have been received from abroad, it is an unimportant parasite in Europe. Partly on this account, and more, perhaps, because it was colonized so satisfactorily in 1907, no further attempts to secure its introduction into America have been made. Neither has a serious attempt to recover it from the field in the vicinity of the 1907 colony site been made, and it may have become established from this colony.

The plans for field work in 1911 include the collection of a large number of eggs of the brown-tail moth from the general vicinity of the larger colonies of 1907 and, if arrangements can be perfected, for a study of the extent to which the eggs of the brown-tail moth are attacked by parasites in Europe. As in the case of every other class of parasite material received at the laboratory, nothing is known of the circumstances under which those egg masses which were received from 1906 to 1908 were collected. It may easily be that they were collected too soon following their deposition to permit of their having been parasitized to anything like the extent which would have come about had they been allowed to remain in the open for a few days longer, and in at least one instance the receipt of the masses with a dead female moth accompanying each was sufficient to more than justify such doubts.

PARASITES WHICH HIBERNATE WITHIN THE WEBS OF THE BROWN-TAIL MOTH.

Partly because it has been practicable to import the gipsy moth and the brown-tail moth in the hibernating state in better condition than it has been possible to import their active summer stages, but equally because there has been ample time and opportunity to study them during the winter months when only a limited amount of field work could be done, it has been possible to learn more of the parasites which hibernate within the gipsy-moth eggs and the nests of the brown-tail moth than of those parasites which are only associated with the same hosts during a more or less limited time in the summer. The winter nests of the brown-tail moth have from the beginning been the subject of an increasingly intensive study, and as a result more is known of the parasites which hibernate within them than of any other group of brown-tail moth or the gipsy-moth parasites without excepting even the parasites of the gipsy-moth eggs.

Very large numbers of these nests, amounting in the aggregate to more than 300,000, have been imported each winter from that of 1905–6 to that of 1909–10, inclusive, Now that all of the primary parasites known to hibernate within them are believed to be thoroughly well-established in America, their importation has been discontinued.

These parasites, including two which are secondary, number eight species in all, and will be first considered as a group. In the subsequent pages each will be taken up separately, and the story of its importation and progress in America will be told. The species are as follows:

Monodontomerus æreus Walk. Adult females hibernate within the nests, but do not attack the caterpillars.

Pteromalus egregius Först. Females enter the nest in the fall and oviposit upon the caterpillars after they have become dormant. Their eggs are deposited, and the larvæ feed externally (fig. 59), becoming full fed before cold weather puts a stop to their activity. Transformations are completed in the spring, and adults of the new generation leave the nests about two or three weeks following resumption of activity on the part of the caterpillars.

FIG. 59.—Larvæ of *Pteromalus egregius* feeding on hibernating caterpillars of the brown-tail moth. Much enlarged. (Original.)

Apanteles lacteicolor Vier. Attack is presumably made upon the very small active caterpillars in the fall before they enter the nests for the winter. The parasitized caterpillars hibernate and resume activity in the spring. About the time when, had they remained healthy, they would have molted for the first time, they die, and the parasite larva soon issues and spins a white cocoon within the molting web, which may or may not be upon the winter nests. There is no second generation upon the caterpillars of the brown-tail moth the same season.

Meteorus versicolor Wesm. Habits essentially the same as those of Apanteles until after the caterpillars have resumed activity in the spring. The parasitized individuals usually live to molt once, and are overcome and destroyed away from the molting web or nests. The cocoons, which are characteristic of the genus, swing from the end of long threads. The adults issuing from them immediately attack the larger caterpillars of the brown-tail moth for a second generation.

Zygobothria nidicola Towns. Hibernating habits similar to those of Apanteles and Meteorus. The affected caterpillars become full grown and spin for pupation before being overcome by their parasite. Sometimes they pupate. The parasite adult issues at about the time when the moth would have issued had the caterpillar completed its transformations. There is but one generation annually, and no alternate host is necessary.

Compsilura concinnata Meig. Hibernating larvæ are occasionally found, but apparently do not complete their transformations in the spring.

Mesochorus pallipes Brischke. Occasionally reared as a parasite of *Apanteles lacteicolor*. The Apanteles larva reaches full maturity and spins its cocoon, but is overcome before pupating. The Mesochorus adult issues from the cocoon a very few days later than would the Apanteles had it remained alive.

Entedon albitarsis Ashm. An internal parasite within the larvæ of *Pteromalus egregius*.

The appearance of the hibernating larvæ of the Pteromalus is indi-

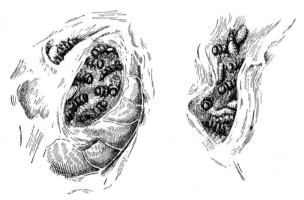

FIG. 60.—Portion of brown-tail moth nests, torn open, showing caterpillars attacked by larvæ of *Pteromalus egregius*. Enlarged. (Original.)

cated fairly well in the accompanying illustration (fig. 60), which represents a "pocket" of parasitized caterpillars torn open. Very little of interest is associated with the life and feeding habits of these larvæ. The female pierces the host caterpillar with her ovipositor preliminary to the deposition of her egg externally, and the caterpillar thus stung is frequently rendered quiescent, and may even die before the hatching of the parasite larvæ.

The hibernating larva of Apanteles is so small as to be very difficult of detection until after it has resumed activity in the spring and increased in size. Its

FIG. 61.—*Apanteles lacteicolor:* Immature larva from hibernating caterpillar of the brown-tail moth. Much enlarged. (Original.)

exact appearance during the hibernating stage can not be described, because nearly every specimen found has been injured more or less in the removal. The accompanying illustration (fig. 61) is from a sketch made by Mr. Timberlake of a half-grown larva from life. None of the preserved specimens shows the curious projection beneath the anal bladderlike appendage which latter is characteristic of the early stage larvæ of the subfamily to which Apanteles belongs. The head and mouthparts are strikingly dissimilar from those of the first-stage Meteorus, and are so little differentiated as to be indescribable.

The hibernating stage of Meteorus is in remarkable contrast to that of Apanteles. The accompanying drawing (fig. 62) is from a balsam mount, and represents an individual which has resumed activity and grown very slightly larger and plumper than is character-

FIG. 62.—*Meteorus versicolor:* Immature larva from hibernating caterpillar of the brown-tail moth. Much enlarged. (Original.)

istic of its hibernating condition. These larvæ are curiously anomalous, in that though they are actually first-stage, the head alone is considerably larger than the original egg as deposited by the mother.

FIG. 63.—*Zygobothria nidicola:* First-stage larvæ in situ in walls of crop of hibernating brown-tail moth caterpillar. Greatly enlarged. (Original.)

An interesting series of dissections made by Mr. Timberlake in the spring of 1910 served to explain this apparent anomaly. The eggs are very small when first deposited and almost globular. Apparently with the beginning of embryological development they begin to grow and by the time the inclosed embryo begins to assume the characteristics of the larva they have reached a diameter at least four times greater than that of the newly deposited egg. The enormous chitinized head, with strong, curved mandibles, is in strange contrast to the undifferentiated cephalic segment of Apanteles and is apparently closely analogous to the large-headed, heavily mandibled larvæ of the Platygasters, as described by Ganin, Marchal, and others. There are many points of resemblance between the two forms, and it would seem, without going into the matter at all deeply, as though the type of embryological and early larval development characteristic of Meteorus were essentially the same as that of the Platygasters and many ichneumonid genera, while that of Apanteles would have to be considered as of an essentially different type.

In both Apanteles and Meteorus the later larval stages are much more conventionalized and more like the familiar type.

The position assumed by the Apanteles larva is not very definitely known. The Meteorus larva usually lies superior to the alimentary canal, its axis parallel to the axis of the body of the host caterpillar, and its head in the ultimate or penultimate body segment, and pointed toward the rear.

The larvæ of Zygobothria are similarly assigned to a definite position, and in otherwise healthy caterpillars have invariably been found embedded in the walls of the crop, as indicated by figure 63. In appearance they are typical of the tachinid first-stage larvæ generally, and with no extraordinary points of difference from most others of the group to which they belong. Those of Compsilura (fig. 64) may be found in similar positions, but they are easily distinguishable from Zygobothria by the presence of the three chitinous anal hooks or spines, as indicated in the accompanying figure.

Nothing is known of the hibernating stage of Mesochorus. It does not seem probable that it should resemble the planidium of Perilampus, which, like Mesochorus, is a secondary parasite which gains access to its host before the latter has left the body of the caterpillar which harbors both primary and secondary. It is presumed that it will be representative of a highly specialized type of development which fits it for the peculiar rôle which it plays, but that this development will have been along wholly different lines from that which has taken place in the case of Perilampus. The whole genus, apparently, possesses habits similar to those of *Mesochorus pallipes*. A very beautiful and, according to Mr. Viereck, an undescribed species has been reared from the cocoons of *Apanteles fiskei*, parasitic upon a species of Parorgyia, under circumstances which indicate positively that attack was made while the primary host was still alive. The same may be said of another undetermined species which has similarly been reared from *Apanteles hyphantriæ*.

Fig. 64.—*Compsilura concinnata:* First-stage larva. Greatly enlarged. (Original.)

Mesochorus pallipes is not an uncommon parasite of *Apanteles lacteicolor* Vier., having been reared from only a few among the many localities from which its host has been secured in numbers, but the average proportion of parasitized individuals has been only about 2 per cent.

The interrelations of these several parasites thus closely associated with one stage of the same host, and consequently with each other, are interesting and peculiar. Pteromalus, of course, cares little whether the host caterpillar selected for attack is parasitized by one or more of the endoparasites which hibernate as first-stage larvæ. The female will undoubtedly attack parasitized as freely as it will attack unparasitized caterpillars and its larvæ develop as satisfactorily upon the one as upon the other, and at the expense of the other internal parasites as well as of the primary host. But the matter does not stop here. The adults issue from the hibernating nests at just about the time when the Apanteles are issuing from the young caterpillars and spinning their cocoons in the molting webs, which are very frequently in the outer interstices of the very same nests from which the Pteromalus are also issuing. The females of the latter are ready to oviposit almost immediately following their eclosion, and will oviposit with the greatest freedom in the cocoons of Apanteles or of Meteorus whenever they chance to encounter them. Thus it comes about that the Pteromalus, after passing one generation as a primary parasite of the brown-tail moth, immediately passes another as a secondary upon the same host. Undoubtedly it would thrive equally as well upon Mesochorus as upon Apanteles. Proof of this could unquestionably be secured through the careful dissection of the very large number of cocoons from which it had issued in the laboratory, some of which, it is certain, must have contained Mesochorus as well, but proof is really unnecessary. By doing so, it becomes tertiary upon the same host as that upon which it is habitually and regularly a primary and secondary parasite.

Entedon, were it to follow Pteromalus through its varied adventures, would in like manner (as it probably does) become successively secondary, tertiary, and quaternary.

Monodontomerus, commonly a primary parasite upon the pupa of the brown-tail moth or gipsy moth and only present as a regular guest in the winter nests, is none the less pretty intimately connected with them in other ways. It directly attacks the cocoons of the Apanteles, acting in all respects like a secondary parasite, and thereby comes into direct conflict with Pteromalus, one of the other of which must develop at the expense of its competitor. It also will become tertiary whenever it chances to attack a cocoon containing Mesochorus as a secondary parasite on Apanteles. It is also a parasite of tachinid puparia, and especially of tachinid puparia which it encounters associated with the gipsy moth, or the brown-tail moth, and thereby becomes a parasite of Zygobothria and in consequence a secondary parasite of the brown-tail moth.

Should Apanteles and Meteorus, or Apanteles and Zygobothria chance to become located in the same host, the Apanteles, because of its more rapid development in the spring, would certainly be the winner.

When Meteorus and Zygobothria enter into competition for possession of the same host individual, Meteorus is invariably the winner and is in no way affected by the presence of the other parasite. In fact, Zygobothria is twice apt to be the victim of Meteorus, which goes through two generations before the tachinid has entered its second stage.

Table XI has been prepared for the purpose of showing these interrelations graphically. It is to be understood, of course, that not in every instance have the exact relations thus set forth been actually observed; but it is perfectly safe to say that they are not only within the bounds of probability, but that they actually occur in nature. The only point concerning which doubt is felt is in the hyperparasitism of Entedon upon Pteromalus, when Pteromalus itself is hyperparasitic upon Apanteles or Meteorus.

TABLE XI.—*Possible interrelations between parasites hibernating in brown-tail caterpillars.*

Primary parasites.	Secondary super- or hyper-parasites.	Tertiary super- or hyper-parasites.	Quaternary super- or hyper-parasites.	Quinquinary super- or hyper-parasites.
Pteromalus egregius Apanteles lacteicolor.	Entedon albitarsis.[1] Pteromalus egregius.[1][2] Mesochorus pallipes.[2] Monodontomerus æreus.[2]	Entedon albitarsis.[1] Pteromalus egregius.[2]	Entedon albitarsis.[1]	
Meteorus versicolor.	Apanteles lacteicolor.[2]	Pteromalus egregius.[2] Mesochorus pallipes.[1] Monodontomerus æreus.[1]	Entedon albitarsis.[1] Pteromalus egregius.[2]	Entedon albitarsis.[1]
	Pteromalus egregius.[1][2]	Entedon albitarsis.[1]		
Zygobothria nidicola.	Apanteles lacteicolor.[2]	Pteromalus egregius.[1][2] Mesochorus pallipes.[1] Monodontomerus æreus.[1]	Entedon albitarsis.[1] Pteromalus egregius.[2]	Do.[1]
	Meteorus versicolor.[2] Monodontomerus æreus.[1] Pteromalus egregius.[2]	Pteromalus egregius.[1][2] Entedon albitarsis.[1]	Entedon albitarsis.[1]	
Monodontomerus æreus.				

[1] Hyperparasitic relations.   [2] Superparasitic relations.

PEDICULOIDES VENTRICOSUS NEWP.

During the winter of 1908–9 trouble was experienced in the work of breeding Pteromalus, the exact nature of which was not immediately apparent. There were numbers of the reproduction experiments in which the proportionate number of progeny to parents used was much below that which had hitherto been secured as the result of similar work in the previous spring. An examination of the nests of the brown-tail moth which had been used in these experiments disclosed the presence of vast numbers of the adults and young of a mite, determined by Mr. Nathan Banks as *Pediculoides ventricosus* Newp. The gravid females were attached to the caterpillars of the brown-tail moth, or to the larvæ or pupæ of the parasite, indiscriminately, and in some of the reproduction cages practically every host and parasite had been attacked.

It was not known where these mites came from, but it was presumed that they were brought in from the field upon nests of the brown-tail moth. By the time that they had been discovered they were in practically everything in the laboratory. Even tachinid puparia were not immune to attack, and there were numerous instances in which the wandering young had forced their way through tight cotton plugs, which would ordinarily have prevented the passage of bacteria.

Much time and trouble was necessary before the laboratory was finally cleared of the pest; but it was finally accomplished by the rigid separation of every rearing cage containing life which had been present before the invasion became apparent from those which were begun afterwards. The general cleaning up and policy of segregation proved effective, and by spring the last of the mites appeared to have died; nor has a single specimen been observed since.

As parasites of the brown-tail moth the mites were singularly effective. If it were possible to bring about a general infestation of the nests in the early fall, it would doubtless result in the destruction of a very large proportion of the hibernating caterpillars; but unfortunately this seems to be not at all practicable. It is not even certain that the parasite was actually brought into the laboratory in nests of the brown-tail moth, though this would seem to be the most likely explanation of its presence.

The fact that its presence has never once been detected in any of the many thousands of similar nests which have been brought in at other times indicates rather conclusively that it is not actually an enemy of any consequence in the field.

PTEROMALUS EGREGIUS FÖRST.

It was quite late in the spring of 1905 before the senior author was able to organize a corps of European collectors, and as a consequence only a very small quantity of parasite material was imported during the summer of that year; but during the fall and winter following, well within a year after the work was first authorized by the Massachusetts Legislature, importation was begun in earnest. More than 100,000 hibernating nests of the brown-tail moth were received from abroad that winter, and since scarcely anything was surely known of the parasites which were likely to be reared from them, the early discovery of the hibernating brood of *Pteromalus egregius* (fig. 60, p. 263) was hailed with satisfaction. The circumstance has already been the subject of comment in an earlier section.

In the spring of 1906 some 40 large tube cages (Pl. X, fig. 1), each capable of accommodating several thousand nests, were constructed after the model of a cage which had been successfully used for a somewhat similar purpose in California. Hardly had the nests been placed in these newly constructed cages before the caterpillars began issuing in extraordinary numbers, and with them many thousands of adult parasites, representing a great variety of species. *Monodontomerus æreus* was about the first to issue, and with it was a quantity of *Habrobracon brevicornis*. A little later *Pteromalus egregius* (fig. 65) appeared in an abundance which exceeded that of all the other parasites taken together, and it was followed shortly afterwards by swarms of its own little parasite, determined by Dr. Ashmead as *Entedon albitarsis*.

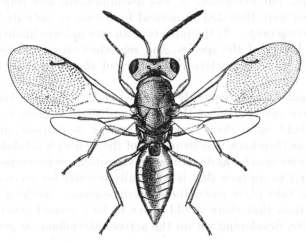

FIG. 65.—*Pteromalus egregius:* Adult female. Greatly enlarged. (Original.)

Mr. Titus at once recognized Entedon as hyperparasitic and proceeded as assiduously to destroy it as he was assiduous in saving the Pteromalus. Of the myriad of other parasites issuing, the vast majority were represented by so few individuals as to render it very improbable that any among them were enemies of the caterpillars of the brown-tail moth. Nearly all of the more common species, aside from Pteromalus and Entedon, were representative of genera or groups of genera well known to be parasitic upon Cynipidæ, of which large numbers issued from the galls on oak leaves that had been used by the caterpillars in the construction of their nests. There remained as possible parasites of the caterpillars of the brown-tail moth only *Habrobracon brevicornis*, *Pteromalus egregius*, and *Monodontomerus æreus*.

It looked for a time as though the Habrobracon might be parasitic upon the hibernating caterpillars, and quite a large number of them was liberated in the spring of 1906, but it was later discovered that, like Monodontomerus, they merely sought the hibernating nests for the protection which was thus afforded during the winter. One colony of Monodontomerus was also established early in the spring of 1906, but almost immediately thereafter the action was regretted on account of the doubts which Dr. Ashmead expressed concerning the host relations of this species. He was certain that it was a parasite of Diptera, and that it could not be a primary parasite of the gipsy moth or of the brown-tail moth. As events have since abundantly proved, he was right and wrong at one and the same time.

The separation of the parasites from the exceedingly large number of caterpillars which issued coincidently, and the subsequent separation of Pteromalus from the remaining species, was a task of huge proportions, but eventually it was accomplished, and some 40,000 Pteromalus were liberated in several localities, as indicated on the accompanying map. At the same time an attempt was made to determine the habits of the species, and reproduction experiments were conducted, using the active caterpillars of the brown-tail moth as hosts.

The females were frequently observed to take peculiar interest in these active caterpillars of the brown-tail moth. They would frequently alight upon their backs and appear to oviposit, and since nothing was then known or suspected of the well-nigh total depravity of this species in so far as its habits of oviposition are concerned, it was only natural to suppose that it was really possible for successful oviposition to take place under these circumstances. Nothing less was expected than that there would prove to be a second generation of the parasite, developing within the active caterpillars, or perhaps in the pupæ.

Attempts to discover some trace of this generation were futile, but failure could not altogether be attributed to the fact that such a generation did not exist. As it happened, every one of the several colonies of the parasite was situated within a territory to the northward of Boston over which the brown-tail moth was exceedingly abundant. Late in the spring the host of caterpillars was suddenly destroyed by an epidemic of a fungous disease which was so complete and overwhelming as to leave very few survivors. Even now, four years later, the brown-tail moth has not reached its former abundance over a considerable portion of the territory affected, notwithstanding that there has been steady and fairly rapid annual increase throughout this period. It looked, in fact, as though the parasites had suffered to an even greater extent than their hosts (since they were not so thoroughly well established), and failure to recover Pteromalus from the field during the summer, or even during the winter following, was thought to be the result of the epidemic of disease.

Another importation of the hibernating nests consisting, like the first, of about 100,000 from various localities in Europe, was received the next winter and handled in the same manner as was the other, but affairs at the laboratory did not run as smoothly as they might in the spring of 1907 at about the time when the Pteromalus were issuing. Mr. Titus was absent on account of sickness which eventually forced him to resign from his position at the laboratory, and neither Mr. Crawford, who first took his place, nor the present incumbent, who finally assumed charge the latter part of May, was sufficiently familiar with the work to carry it on to as good advantage as Mr. Titus would have done had he retained his health. Partly on this account and partly on account of weather conditions which were very unfavorable to the issuance of the parasites, only about 40,000 of the Pteromalus were reared and liberated. As before, they were colonized in various localities within the infested area as soon after their emergence as was practicable, and as before attempts to secure laboratory reproduction were made.

All of these attempts to secure the reproduction of the parasite in 1906 or in 1907 failed, since only active caterpillars of the brown-tail moth or gipsy moth were used. All sorts of theories to explain this were formulated, but that which seemed the most reasonable at the time, namely, that the parasite did not actually reproduce upon active caterpillars or pupæ, but only upon inactive caterpillars after the construction of their nests in the fall, could not be given an actual test, since inactive caterpillars were not available. An attempt to carry the living Pteromalus adults through the summer did not succeed, and with the death of the individuals in confinement, and the almost immediate disappearance of those which were liberated in the field, the investigations were necessarily brought to a close.

Meanwhile, as will be detailed later on, a variety of other parasites was found to be present as minute larvæ which hibernated within the still living caterpillars, and for the purpose of securing these as well as an additional supply of the Pteromalus, further extensive importations of the nests of brown-tail moths were made during the winter of 1907-8. A radical modification in the policy of the laboratory was inaugurated at the same time, and instead of discontinuing its activities during the winter months, the experiment was made of keeping it open for the purpose of conducting a series of winter investigations, and the study of the hibernating caterpillars of the brown-tail moth and of their parasites was selected as the subject for the first winter's work.

The first lot of nests arrived from abroad in December, and instead of awaiting the coming of spring they were immediately brought into a warmed room in the hope that the parasites might thereby be forced into activity. The experiment was successful. The first of the

Pteromalus began to issue coincidently with the beginning of the new year, and they were at once supplied with a quantity of nests of the brown-tail moth collected in the open and containing living caterpillars. In most cases the females almost instantly entered these nests and oviposited upon the still dormant caterpillars (fig. 66) with the result that in three and four weeks large numbers of a second generation began to issue. This successful outcome to what was considered to be, until that time, an experiment of rather doubtful utility, was very encouraging, since it was at once evident that any desired number of Pteromalus might easily be reared in captivity. Accordingly the work of rearing it on a large scale was begun, with the result that by the end of March American nests which contained the progeny of some 100,000 individuals were available for colonization.

Meanwhile large numbers of nests of the brown-tail moth—several thousand, in fact—had been collected in the neighborhood of the colonies which had been planted in 1906 and 1907 and no Pteromalus issued from them. It was evident that the colonization experiments of the spring of 1907 were no more successful than those of the spring before, and it was no longer possible to consider the bad results as due to the unusual mortality of the brown-tail moth in the vicinity of the colonies. It was necessary to seek some other explanation for this apparent failure to establish the one parasite which had been imported and colonized in wholly satisfactory numbers, and it was thought that this might be found in the circumstances under which the parasites were reared and liberated.

In 1906 and 1907 the adults had been liberated in the field some two or three weeks sooner than they would normally have issued as adults on account of their development having been hastened by the storing of the nests of the brown-tail moth at an artificially high temperature during the time that they were in transit from Europe. This, it was believed, might be responsible for the fact that the species had failed to establish itself and it was planned to do things very differently in the spring of 1908.

In accordance with these plans the nests containing the brood (as well as quantities of healthy caterpillars) were placed in large tube cages, which were fitted with a "tanglefooted" shield within, intended to prevent the emergence of the caterpillars without hindering the egress of the winged parasites, and four colonies, each of which was estimated to consist at the very least of 50,000 of the parasite larvæ, were located in four widely separated localities in eastern Massachusetts. The cages were simply taken into the field and left, so that the parasites were free from the moment of their emergence.

Considerable trouble was experienced at first on account of the "tanglefooted" shields failing to do all that was expected of them in the matter of preventing the escape of the caterpillars, but aside from that, the experiment promised to be highly successful. Instead of losing track of the parasites immediately following their liberation, they were found to be present in abundance in and about these cages throughout May and June, and even in July Mr. Mosher (who conducted this work) observed a few alive and apparently waiting until the next generation of hibernating caterpillars would be open to their attack.

Not all of the Pteromalus brood was liberated in this manner, but a part of the artificially infested nests was placed in cold storage at a constant temperature of approximately 30° F. and kept during the summer and until the formation of the brown-tail moth nests in the fall. Then a part of them was removed as a check on the condition of the remainder, and when it was certain that many, if not most of the Pteromalus had survived, a considerable number of them was allowed to issue in the open in a locality where they would find an abundance of fresh nests of the brown-tail moth ready at hand. Others of the stored Pteromalus were held for the purposes of winter reproduction, in case the further colonization of the parasite seemed worth attempting.

At first it appeared that the colonies of 1908, both spring and fall, were successful. In the vicinity of each of them (but particularly of that which was planted in the fall) the larvæ of the parasite were found in the nests of the brown-tail moth, and for the first time it was known to have lived over summer out of doors. Extensive rearing work was organized in the laboratory, with the intention of securing at least 1,000,000 for colonization in 1909, and certain technical investigations into the life of the parasite, which were begun in the spring of 1908, were continued.

The results of these biological investigations soon became startling in their nature. Gradually, as they were continued, and the results of one experiment after another became apparent, a tale of insect duplicity was unfolded the like of which has never been quite equaled in any similar investigation. It is not possible to give the story in anything like complete detail, but a brief summary ought to be presented, if for no other purpose than to illustrate the degradation to which a parasite may sink.

It was found that the instinct of the female Pteromalus was first to seek the immediate vicinity of the feeding caterpillars, or of the nests or molting webs which they had deserted, and second to oviposit upon nearly anything which she encountered, providing it resembled in the slightest degree a dormant caterpillar of the brown-tail moth inclosed in its hibernating web (fig. 66). Attempted oviposition upon active caterpillars was only one of innumerable illustrations of the lengths to which she would go in satisfying her crude and unreasoning instincts. She would oviposit with as much apparent freedom upon a dead and decomposing caterpillar, or the fragment of skin torn from such a one, as upon a living caterpillar. Quite an extensive series of experiments was to have been made to determine the lengths to which she would go, but these experiments were discontinued after a time because there hardly seemed to be any limit beyond the purely physical. No Pteromalus was ever induced to oviposit in any tachinid puparium, nor in any other insect protected by a hard shell, but almost any small, inactive, and soft-bodied insect, especially if it were inclosed in a thin silken web or cocoon, and provided it was in the near vicinity of caterpillars of the brown-tail moth or of their webs, would be attacked and usually

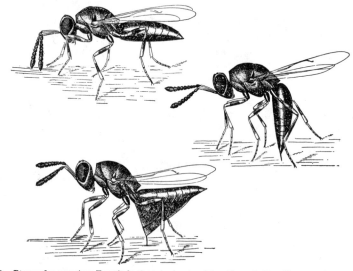

FIG. 66.—*Pteromalus egregius:* Female in the act of oviposition through the silken envelope containing hibernating caterpillars of the brown-tail moth. Greatly enlarged. (Original.)

without hesitation. The cocoons of small Hymenoptera, such as Apanteles and Limnerium, were especially attractive, and would be attacked whether associated with the brown-tail moth or not.

The results of this indiscriminate oviposition were very varied. Eggs deposited upon dead caterpillars of the brown-tail moth invariably perished except in one instance, in which the caterpillars were freshly killed and "pasteurized." Upon this occasion a small proportion of the larvæ lived, and at least one went through to maturity. Upon active caterpillars of the brown-tail moth and even upon inactive caterpillars removed from the silken envelopes with which they surround themselves within their nests, oviposition was never successful if the caterpillar moved to any extent afterwards. Even the hibernating caterpillars, removed from their nests, will move about a very little as a rule, and the least motion was sufficient to dislodge the egg or young larva of the parasite.

Oviposition upon any other host was equally unsuccessful, provided that the host was free to move about to any extent, but whenever it was confined within the limits of a cocoon, and was not too large, it usually fell a victim to the parasite. Especially was this true of the hibernating larvæ of hymenopterous parasites within their cocoons, and from these the largest and finest Pteromalus were reared.

If the parasite or hymenopterous larva was very small, as in the instance of the larva of Apanteles, it was very likely to be killed by the Pteromalus in the process of oviposition and, as a common result, her progeny would perish also.

Evidence to indicate that the female parasite possesses discriminative powers which enable her intelligently to select suitable hosts for her young is wholly lacking, and in consequence, when several individuals are given access to a single nest of the brown-tail moth, the chances are that all of them will concentrate their attack upon the few caterpillars which chance to be most readily accessible, to the exclusion of all others. The outcome is one of the manifold phases of superparasitism. The larvæ hatching from the superabundance of eggs are unable to reach their full development. They may complete their transformations but the adults produced are small, weak, and in extreme instances wholly unfit for further reproduction.

In the work of rearing the parasite for colonization purposes, no matter how many parent Pteromalus were used, the number of caterpillars which were parasitized by them would be a small percentage of those in the nests exposed to their attack, and invariably when more than a few females were used as parents the nests had to be torn open, so as to expose a large number of caterpillars equally. Otherwise the progeny would be so small as to be practically worthless for further reproduction, colonization, or anything else. This in itself was sufficient to render Pteromalus of very much less value from an economic standpoint, and the extraordinary avidity with which it attacked the cocoons of other hymenopterous parasites was anything but a point in its favor. Most especially was this true when the life and habits of *Apanteles lacteicolor* Vier. were taken into consideration. It soon became evident that Pteromalus was peculiarly fitted to act as its most dangerous enemy, and since, between the two, Apanteles was much the more promising parasite, it was decided to abandon all further effort toward the introduction of Pteromalus, and the work of rearing was discontinued.

Some 250,000 larvæ and pupæ were on hand at the time when this decision was reached, and these were placed in cold storage. It was considered probable that the species was already introduced, if it were possible to do so, as the result of the elaborate colonization work already described, and that any harm which might result was probably already done, so it was determined to use these larvæ and pupæ for the purpose of giving the parasite one more opportunity to retrieve a lost reputation. The brood lived through the summer in cold storage without much loss, and in the fall one tremendous colony of some 200,000 individuals was established in the midst of a tract of small oak, well infested with nests of the brown-tail moth. The adults issued at a time when there was nothing to prevent their entering these nests and ovipositing immediately, and there were enough of them to destroy all of the caterpillars of the brown-tail moth within a considerable radius. There were many larvæ to be found in the nests that winter, but, as was the case in the laboratory, only a few of the more exposed caterpillars were attacked.

A rather elaborate series of nest collections was made within a radius of a mile of the center of the colony, but the data obtained were of little consequence. From only a part of the many lots of nests did any of the parasite issue, and its probable rate of dispersion was not definitely indicated. One lot of nests collected a little over a mile away produced a few individuals, and this was the only instance in which it could be shown to have traveled so far.

At the same time large collections were made in the vicinity of the 1908 colonies, from which, it will be remembered, some few parasites had been recovered the winter before. In no instance was it again recovered, and there was everything to indicate that it had failed to establish itself.

No attempt whatever was made to rear it for colonization in 1910, and until the beginning of the winter of that year it was considered that the story of Pteromalus in America was complete. It is the unexpected which usually happens in the gipsy-moth parasite laboratory, however, and even as the rough manuscript for the last few pages was being prepared, Chapter II of the history of *Pteromalus egregius* in America was about to begin.

Every winter since that of 1906–7, to and including the present, an increasingly large number of the hibernating nests of the brown-tail moth have been collected from various localities throughout eastern Massachusetts and confined in tube cages in the laboratory. In the first two winters this was done for the express purpose of recovering Pteromalus and, as has been already stated, without result. In the winter of 1908–9, it was found that Monodontomerus was to be recovered in this manner over a considerable territory and under conditions which were both interesting and instructive. Accordingly, beginning with that winter, the collections have been made general throughout the territory in which it was thought likely that Monodontomerus would occur, and with less reference to the localities in which Pteromalus had been colonized. Several thousands were thus collected in 1909–10 (as may be seen by reference to Table X) and a much larger series of collections was planned for the winter of 1910–11.

On the face of the results of this work during the two previous winters, nothing was much less likely than that Pteromalus should be recovered from any of these collections of nests. When a few specimens of a pteromalid which looked very much like it did issue early in December, they were accorded a rather cool reception, and made to identify themselves by reproducing upon hibernating caterpillars of the brown-tail moth in confinement. Before such identification was complete, it was rendered unnecessary through the issuance of considerable numbers of what could no longer be questioned as the true *Pteromalus egregius* from no less than 10 lots of nests collected in different towns scattered all the way from Milford, Mass., down near the Rhode Island line, to Dover and Portsmouth, N. H., just across the Piscataqua River from Maine. At the time of writing they are still emerging from the collected nests, and the extent of their dispersion is not yet known, but Mr. H. E. Smith, who is attending to the rearing cages, has prepared a map (Plate XXV) showing the location of the original colonies as well as the towns from which recovery has been made the present winter.

Sufficient data have already been accumulated to make certain the astounding fact, that as a result of the colonization work conducted between 1906 and 1908, the parasite is now thoroughly established over a territory which undoubtedly includes portions of four States, and during the period of its dispersion it spread itself out so thin as to make its recovery impossible except in the immediate vicinity of the colony sites, and for a short period immediately following colonization. Until this time Monodontomerus has held the record for rapid dissemination, but this record is now eclipsed.

It is impossible to determine whether the first of the colonies were after all successful, or whether they actually died, as was supposed, and success finally resulted from the very much larger colonies in 1908. If the early colonies lived, it means that no less than four years elapsed before any evidence to that effect was forthcoming. This fact, in its relation to circumstances attending the colonization of another parasite, *Apanteles fulvipes*, which seems not to have succeeded in establishing itself any more than Pteromalus appeared to have established itself as a result of those early colonizations, will sustain some hope for the ultimate recovery of this parasite until 1912 or 1913.

If, on the other hand, the establishment of Pteromalus resulted from the very much larger and in every way satisfactory colonizations of 1908, it may mean, in its reference to *Apanteles fulvipes*, that very much larger colonies will be necessary before we can hope to see that species established in America. To colonize it under more satisfactory conditions than those which prevailed in 1909 would be well-nigh impossible except at a very heavy expenditure, because the favorable conditions in 1909 were primarily due to the unusual coincidental circumstance of an early season in Japan, and a late season in Massachusetts. Such coincidences can not be depended upon, and without them, a tenfold expenditure over that of 1909 would be insufficient to secure equally favorable conditions for the establishment of the species, and a proportionately larger expenditure to better them.

The story of Pteromalus has been given at length because of the bearing which it has upon the question of what constitutes a satisfactory colony of any species of parasite. Except in a few instances, of which Calosoma and Anastatus are conspicuous, we frankly do not know the answer, and it is only through the study of such phenomena as those which have accompanied the recovery of Pteromalus that we are able to judge the probable character of the answer in the instance of those parasites which for some obscure reason or another have failed to make good their establishment in America.

APANTELES LACTEICOLOR VIER.

The story has already been told of how, during the winter of 1905-6, some 100,000 hibernating nests of the brown-tail moth were imported, placed in large tube cages in the laboratory at North Saugus, and how some 60,000 Pteromalus and countless thousands of caterpillars of the brown-tail moth issued into the attached tubes, and were sorted with difficulty. There is not a single published record outside of those emanating from the laboratory, so far as was then known, or is known now, which suggested the possibility of this particular sort of caterpillar harboring other parasites than those which issued as adults from its nests. Mr. Titus recognized that this might well be possible, however, and rather with the purpose of determining the fact than with the expectation of securing such parasites in any quantity for liberation, he caused some of the caterpillars to be fed in confinement and under observation. His foresight was well rewarded when, in the course of time, a number of cocoons of an Apanteles (fig. 67) was found in these cages, and the fact that at least one parasite hibernates within the living caterpillars was demonstrated.

The following spring he laid his plans for the wholesale rearing of this parasite and whatever other parasites might chance to be present. A considerable number of wood and wire-screen cages (Pl. XXVI, fig. 1), modifications of the familiar Riley type, was procured, and as the caterpillars issued from the cages containing the second large importation of hibernating nests in the spring of 1907 they were placed in these cages and fed. As a result of his enforced absence from the laboratory at a critical period these cages lacked the proper attention, and things went wrong with many of them. A few, however, were measurably successful, and eventually about 1,000 of the Apanteles were reared and colonized. Meteorus was discovered to have similar hibernating habits, and Zygobothria was also reared under circumstances which were sufficient to indicate its hibernating habits to the satisfaction of the junior author of this bulletin, but not to that of the senior. The Apanteles, in accordance with what was then the policy of the laboratory with regard to parasite colonization, were liberated in no less than three widely separated localities. None of the colonies, so far as known, was successful.

As anyone who was unfortunate enough to be associated with the laboratory during the spring and summer of 1907 will undoubtedly be willing to testify, the discomfort caused by handling quantities of caterpillars and cocoons of the brown-tail moth was literally dreadful. The poisonous spines upon the young caterpillars are neither so abundant nor so virulent as those upon the older caterpillars, but they are bad enough, and the task of feeding the inmates of the numerous cages which contained some thousands was a task of no little magnitude and one involving much physical discomfort.

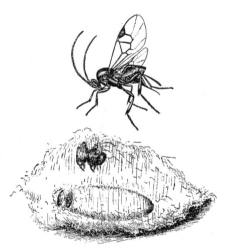

FIG. 67.—*Apanteles lacteicolor:* Adult female and cocoon. Much enlarged. (Original.)

The instant the door of one of these cages was opened, if the day was warm and its occupants active, a variable, but usually a large number would crawl outside, and to attempt to brush them back was but to afford opportunity for more to escape. Consequently thousands did escape and had to be brushed up and destroyed after each day's feeding. To keep the cages clear of débris was well-nigh out of the question, and every time that some attempt was made to clean them out more thousands of caterpillars escaped and had to be destroyed.

When the Apanteles and the Meteorus cocoons were discovered to be present in variable abundance in several of the cages trouble began in earnest, because they were for the most part firmly attached to the sides, or cunningly concealed in the midst of an accumulation of unconsumed food, so that much time was required to find and remove them. During this operation the caterpillars, stirred into unusual activity, were crawling over everything in the immediate vicinity, but more particularly over the outside of the cage and the person of the operator.

If a sufficient number of these caterpillar parasites were to be reared to make possible satisfactory colonies another year, it was obviously exceedingly desirable to devise some other means of feeding the caterpillars than that afforded by the closed cage, and accordingly, in the winter of 1907-8, when the first active caterpillars began to emerge from the nests which had been kept in the warmed part of the laboratory for the purpose of securing Pteromalus, all sorts of experiments were made in the hope of discovering some method whereby the disadvantages above recounted might, at least in part, be obviated. The feeding tray illustrated herewith was the result of these experiments, and as soon as it was found to be practicable, enough to accommodate several thousand caterpillars were constructed, and one wing of the laboratory "annex," illustrated in Plate XXVI, figure 2, and Plate XXVII, was fitted for their accommodation.

In all respects these trays were a success. There was occasionally some trouble caused by the caterpillars finding or constructing a "bridge," by which they passed from the interior of the tray directly to the frame above the concealed band of "tanglefoot," but when sufficient care was used in feeding and in searching for bridges before they were completed this was almost completely done away with.

It was manifestly impossible to feed more than a very small part of the caterpillars from the many thousands of nests which had been imported during this winter, and accordingly the caterpillars from a few nests in each lot were fed in small trays in the laboratory during the late winter and early spring, and the extent to which they were parasitized by Apanteles was thus determined. The most highly parasitized nests were saved, and the larger part of those less highly parasitized were destroyed forthwith, since it was no longer desired to save the Pteromalus which might be reared from them.

A good many Apanteles were reared in the course of this work, and since they issued long before the resumption of insect activities out of doors they were used in a series of reproduction experiments upon active caterpillars of the brown-tail moth feeding upon lettuce indoors. It was found to be easy to secure reproduction when caterpillars which had not molted since leaving the nest were used as hosts, but if they had molted once successful reproduction was secured with great difficulty or not at all. The adult Apanteles were very far from being as strong and hardy as the adult Pteromalus and could not be kept alive to deposit more than a small part of their eggs. There were other reasons, too, why reproduction upon caterpillars in confine-

Bul. 91, Bureau of Entomology, U. S. Dept. of Agriculture.   PLATE XXVI.

FIG. 1.—RILEY REARING CAGES AS USED AT GIPSY-MOTH PARASITE LABORATORY. (ORIGINAL.)

FIG. 2.—INTERIOR OF ONE OF THE LABORATORY STRUCTURES, SHOWING TRAYS USED IN REARING APANTELES LACTEICOLOR IN THE SPRING OF 1909. (ORIGINAL.)

VIEW OF THE LABORATORY INTERIOR, SHOWING CAGES IN USE FOR REARING PARASITES FROM HIBERNATING WEBS OF THE BROWN-TAIL MOTH IN THE SPRING OF 1908. (ORIGINAL.)

ment could not be looked upon as a feasible method for obtaining the parasites for liberation, and all ideas of laboratory reproduction work on a large scale were regretfully abandoned before the spring was far advanced.

A program for the colonization of Apanteles during the year 1908 was definitely formulated as the direct result of this experimentation, by which it was hoped to afford the parasite the best possible opportunity for speedy establishment. In accordance with this plan the nests which had been found to contain the more highly parasitized caterpillars were divided into three lots. The larger of these was placed in the same form of tube cage which was used for the Pteromalus-rearing work in 1906 and 1907; the next larger was placed in cold storage, and the nests remaining were brought into the laboratory toward the end of March and the caterpillars forced into premature activity. There was not very much room available for this indoors, so that the number of nests thus treated was decidedly limited, but from the caterpillars issuing from them no less than 2,000 Apanteles cocoons were secured during the latter part of April, and the adult Apanteles, to the number of about 1,300, which issued from them were liberated in the one colony in the field just as the caterpillars of the brown-tail moth were issuing from their nests and beginning to feed out of doors.

It was known that under natural conditions the parasite never issued as an adult at this season of the year, but it was reasonably certain that it would immediately reproduce upon the small caterpillars which in a week or two more would be so large as to make reproduction impossible. It was hoped in this manner to give the individuals liberated in this colony a certain advantage over those liberated later by allowing them superior opportunities for immediate reproduction and incidentally an opportunity for one more generation during the year than would be possible in colonies established at a later date.

To a certain extent these expectations were realized. It was positively ascertained that the parasite did take advantage of the opportunity offered and that it did actually pass one generation upon the newly active caterpillars of its chosen host. The experiment is not known to be a practical success, however, because all subsequent attempts to recover Apanteles from nests of the brown-tail moth collected in the vicinity of this colony have failed.

As soon as the caterpillars in the nests which had been placed in the large cages became active they were transferred to the larger trays which had been provided especially for them (Pl. XXVI, fig. 2), and fed first with lettuce and later with fresh foliage collected in the field. They did remarkably well at first, and about May 20 the cocoons of Apanteles began to appear in the trays in large numbers. The collection of these cocoons and their removal to small cages for the rearing of the Apanteles was simplicity itself, compared with the similar process the year before. A large sheet of paper, thickly perforated with small holes,[1] or what was equally suitable, a strip of ordinary mosquito netting, would be spread over the pile of débris in each tray and fresh food placed on it. In the course of 24 hours the great majority of the caterpillars would have crawled upon this paper or netting, and could be removed instantly, and with scarcely any disturbance, to a fresh tray. The sorting over of the contents of the tray in which they had been feeding, for the cocoons of their parasites, could then be conducted without the annoyance of their presence, and with a minimum of discomfort. This is, of course, the same method used in feeding the silkworm of commerce.

In all some 15,000 cocoons were secured in this manner, but only about 10,000 of the adults were reared and liberated. Some 100 of the cocoons produced the secondary parasite *Mesochorus pallipes*, and a considerably larger number a small pteromalid, which was vaguely familiar in appearance, but which was not at that time recognized as identical with *Pteromalus egregius*, concerning which so much has already been written.

The Apanteles were carefully separated from their enemies and three colonies were established in the field. Two of these were rather small, but one of them was made very large, and to comprise more than two-thirds of the total number reared. It was no longer a question that the small colony was sometimes a mistake, and that it was invariably safer to liberate large colonies and to establish the species first of all, and to bring about dispersion later, if artificial dispersion should appear to be necessary.

It is interesting to note, in this connection, that neither in 1909 nor 1910 was it possible to find any trace of the Apanteles in the neighborhood of either of the two smaller colonies mentioned above, while from the larger it was recovered in 1909, and by 1910 had spread to a distance of several miles at least.

The third lot of nests, which was placed in cold storage before the caterpillars became active in the spring, was left there until early in July, when it was removed. A part of the caterpillars immediately became active, but it was at once evident that many of them had died as a result of the unnatural conditions. The weather was exceedingly hot immediately following and suitable food for the young caterpillars could not be obtained. In consequence, a great many of them died from one cause or another, or from a combination of several; the larger part of the caterpillars died soon after having become active, and it seemed as though those containing the Apanteles suffered much greater proportionate mortality than the others; in any event, only about 250 of the cocoons were secured, when it was

---

[1] This paper was originally imported from France for use in a similar manner in rearing silkworms.

hoped to secure 10 times that number at the very least. The few that were reared were placed in the field in accordance with the program mapped out the winter before, just about the time when the new generation of brown-tail caterpillars was beginning to construct winter nests, and when there was no possible excuse for failure on the part of the Apanteles to reproduce to the full extent of its powers. In so far as the puny colony thus planted could possibly be expected to succeed, this one was a success. Quite a number of cocoons was found the next spring in the molting webs of the caterpillars from the near-by nests, and it was evident that if the Apanteles had been reared as successfully at this season of the year as it had been hoped would be the case, no better plan for the rearing and colonization of the species could be devised.

There was no possibility of judging the success or failure of the Apanteles colonies of 1908 until the following spring at the earliest, and whether they succeeded or failed it was obviously desirable to continue the work at least one year more. Accordingly, more nests were imported in the winter of 1908-9, and from among them those which were the most highly parasitized were selected for rearing the Apanteles. No attempts to establish colonies out of season were made this time.

Partly as a result of experience gained the year before, partly because more caterpillars were fed, and partly because several among the lots of nests received this year were very heavily parasitized, the number of Apanteles reared and colonized was about 23,000, or twice the number of the year before. They were distributed in three colonies, one of which was near the site of the only successful late spring planting of 1908. There was no apparent necessity for this, but the accuracy of the theory of the large colony and establishment at any cost was becoming more and more evident, and it was resolved to let no opportunity slip by which a possible advantage might be lost. The two remaining outlying colonies were each as large as the successful colony of 1908, and Apanteles was recovered in the spring of 1910 in the vicinity of both.

By the spring of 1909 the pteromalid, which had commonly been reared from the Apanteles cocoons, was identified beyond question as *Pteromalus egregius*. It was found that the females persistently haunted the trays in which the caterpillars were feeding, and that they were very free in ovipositing in the cocoons of Apanteles whenever they encountered them. It was discovered, furthermore, that if the weather was hot and humidity low, the Apanteles larva or pupa in the cocoon attacked would die and dry up before the Pteromalus was full-fed, so that nothing would emerge. A few of the unhatched cocoons, of which there were more than 25 per cent in the summer of 1908, were saved and examined after these facts were known, and in several of them what were almost certainly the eggs and very young larvæ of Pteromalus were found. It thus became evident that at least a portion of this unfortunate mortality was due to hyperparasitism by Pteromalus, a considerable number of which had been free in the compartment where the caterpillars of the brown-tail moth had been feeding. In 1909 pains were taken to prevent a recurrence of these circumstances, and as a result only a very few of the cocoons were lost through attack by Pteromalus. That Pteromalus was to be considered as an aggressive enemy of Apanteles could no longer be doubted, and when it was remembered that the adults naturally emerged from the nests of the brown-tail moth in the open at almost the precise time (Pl. XXVIII, fig. 1) when the Apanteles larvæ were emerging and spinning their cocoons, more often than otherwise in the outer interstices of these same nests, and that, furthermore, the Pteromalus was prone to linger in the vicinity of these nests in preference to any other place, its true duplicity was at last realized.

A much smaller number of over-wintering nests of the brown-tail moth was imported during the winter of 1909–10 than during any other since the beginning of the work, more for the purpose of securing Zygobothria, if possible, than for the rearing of additional Apanteles. The large trays were used as before (Pl. XXVI, fig. 2) for the rearing of a number of the caterpillars in the spring, and 10,000 or more Apanteles were reared and liberated in one colony at some distance from any of the others.

Until late in the summer of 1910 considerable doubt was felt as to the ability of this Apanteles to pass through the summer months successfully in large numbers. That it was able to live from June to August or September at all was rather more than was expected when it was first liberated. When, in 1908 and 1909, it proved its ability to do that much it remained to be determined whether it was going to be dependent upon an alternate host during that period or not, and if dependent whether a sufficient abundance of such hosts would be found in America to support as many of the parasites as would needs be carried through the summer, if it were to become an aggressive enemy of the brown-tail moth when this insect is in abundance.

It was with much satisfaction, therefore, that *Apanteles lacteicolor* Vier. was recovered as a parasite of Datana and Hyphantria late in the summer of 1910. Both hosts are common at that season of the year in Massachusetts, and both are parasitized to a considerable extent by tachinids. It is certain that the Apanteles will develop at the expense of these parasites as well as that of their hosts, and the chances are good that it will replace them to a certain extent, without bringing about a serious reduction in the prevailing abundance of these hosts; in short, that it will find a permanent place for itself in the American fauna.

Fig. 1.—Cocoons of Apanteles lacteicolor in Molting Web of the Brown-Tail Moth. (Original.)

Fig. 2.—View of Laboratory Yard, Showing Various Temporary Structures, Rearing Cages, etc. *A, A,* Out-of-Door Insectary Used for Rearing Predaceous Beetles. (Original.)

Fig. 1.—Outbreak of Amaryllis Caterpillar in Meeting Without the Index of Tall Moth (Original).

Fig. 2.—View of Laboratory Yard, Showing Various Temporary Structures, Rearing Cages, etc., Out-of-Doors Insectary Used for Rearing Ips and other Beetles. (Original.)

The diversity in the host relations of the parasite thus indicated is also encouraging. If it is capable of attacking arctiid as well as notodontid caterpillars with as much apparent freedom as it does liparids, there ought always to be plenty of available hosts to carry the species over the two or three months which must elapse after its emergence from the hibernating brown-tail moth before it can attack the young caterpillars of the same species for a second generation. It was hoped for a time that it would succeed in passing one generation upon the gipsy moth, but although it has been forced to oviposit in gipsy-moth caterpillars, and its larvæ have upon a single occasion attained their full development upon this host, there is no indication that it ever attacks it voluntarily in the field.

It is interesting and perhaps significant that in its relations with Datana it affects the host caterpillars exactly as in its relations with the brown-tail moth. The caterpillars died before the emergence of the parasite larvæ, and were left as nothing more than mere skins containing a small quantity of a clear liquid. In this respect, *Apanteles lacteicolor* Vier. differs materially from *A. solitarius*, or from many other among its congeners, which leave the host in a living condition but so seriously affected as to be unable to feed again.

Just as the proof of this bulletin is being read (June 12, 1911) word is received from the laboratory at Melrose Highlands that 4,000 cocoons of this parasite have been secured from brown-tail moth webs taken in the field in Malden and other towns.

APANTELES CONSPERSÆ FISKE.

In the summer of 1910 several boxes of the cocoons of an Apanteles parasitic upon the Japanese brown-tail moth, *Euproctis conspersæ* Butl., were received at the laboratory through the kindness of Prof. S. I. Kuwana. All of them had hatched at the time of receipt, and the circumstance would hardly be worthy of mention were it not for the fact that the adults which were dead in the boxes proved upon examination by Mr. Viereck to be identical in all structural characteristics with *Apanteles lacteicolor* Vier. It would appear that here was still another example of that phenomenon which has several times been mentioned without having been particularly designated, but which is, in effect, the existence of what has been termed "physiological" or "biological" species.

It is not so difficult to conceive as to find proof of the existence of two species which are so nearly alike structurally as to be indistinguishable by any taxonomic characters commonly recognized, but which are, at the same time, different. This difference may be exemplified by the sex of the parthenogenetically produced offspring, as in the instance of the European and American races of *Trichogramma pretiosa*. It may lie in the instincts of the female, which lead her to select certain hosts in preference to certain others, as in the instance of the American and European races of *Parexorista cheloniæ*. Again, it may be in the ability of the young larvæ to complete their development upon a certain host, as in the case of *Tachina mella* and *Tachina larvarum*. Or again, it may be that the difference lies as between *Apanteles lacteicolor* Vier. and *A. conspersæ* Fiske in the methods of attacking the host.

As has already been recounted, no less than 45,000 adults of *Apanteles lacteicolor* Vier. have been reared at the laboratory and liberated in the field. In addition a very large number has been reared under close observation during the winter or spring, and there has been a large number of more or less successful reproduction experiments conducted, in most instances with great care. In all this time there has not been a single exception to the rule, that the larva of *Apanteles lacteicolor* Vier. is solitary, and kills its host before issuing from its body. Nothing whatever, either in the field, or in the many experiments in reproduction, or in the occurrence of the parasite in shipments of larger caterpillars from Europe, has indicated in any way that it may ever attack the large caterpillars successfully, or that it is ever anything else than solitary.

Had *Apanteles conspersæ* Fiske been received as a parasite of the Japanese brown-tail moth without other data than the mere rearing record it would undoubtedly have been considered as identical with *Apanteles lacteicolor* Vier., but it is impossible so to consider it in view of the fact that it is not solitary but gregarious; that it attacks, not the small but the large caterpillars, and, if appearances of the material from Mr. Kuwana were not deceiving, that the host is left alive instead of being killed before the emergence of the parasite larva. These differences are, or ought to be, sufficient to make of it another species.

It is not at all improbable that if it were given the opportunity it would attack the caterpillars of the European brown-tail moth, and it is hoped that enough can be collected in Japan and forwarded to America to make the experiment possible.

METEORUS VERSICOLOR WESM.

A very few specimens of this parasite were imported in 1906 with caterpillars of the brown-tail moth and the gipsy moth from several European localities. In 1907, as already stated in the account of *Apanteles lacteicolor* Vier., a few specimens of Meteorus (fig. 68) were reared from caterpillars imported in hibernating nests the winter before. There were very few, less than 100 all told, and not enough to colonize with any likelihood of success. It was therefore decided to use them in a series of reproduction experiments, on the chance that a much larger number might be reared for colonization.

There were plenty of caterpillars of the brown-tail moth available and the smallest that could be found were confined in a cage with the first of the parasites that were reared. Oviposition was not observed, and the parent adults did not live very long, but the caterpillars did very well for about 10 days, after which the cocoons of Meteorus began to be found in the cages in most gratifying numbers. It seemed as though success was assured, and other similar experiments were immediately begun in the hope that some method would be found for prolonging the life of the adult parasites in confinement and securing more abundant reproduction.

The days of rejoicing over this, the first successful reproduction experiment with any of the parasites imported in 1907, were very few. In about a week the adults began to issue from the cocoons, and all proved to be males. It looked like a curious coincidence at first, but when one after another of the various lots of cocoons hatched and out of the total of 156 every single individual was of the one sex, it was evident that something serious was the matter. Where the trouble lay was not ascertained at that time, nor has it been determined as the result of other experiments similarly conducted in later years. In all, 244 adult Meteorus have been reared in confinement, and among them there have been just 5 females, not one of which was secured until the late summer of 1908. Breeding Meteorus on a large scale for colonization purposes under circumstances like these can not be considered as an economically profitable venture.

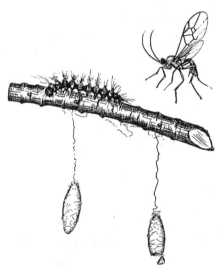

FIG. 68.—*Meteorus versicolor:* Adult female and cocoons. Much enlarged. (Original.)

The numbers of Meteorus reared from the caterpillars imported in the hibernating nests were increased by the addition of some few more secured from importations of full-fed and pupating caterpillars later in the season, and a small colony was planted in 1907, but it was so small as to make its success more than doubtful, and it was determined to rear enough for at least one good colony in the spring of 1908.

A description has already been given of the methods which were perfected during the winter of 1907-8 for the rearing of *Apanteles lacteicolor* Vier. in large numbers from the caterpillars of the brown-tail moth imported in hibernating nests, and these methods applied equally well to Meteorus. It was not nearly so common as the Apanteles, and only about 1,000 adults were secured for colonization. These were all liberated in one colony at a convenient place from the laboratory, and in order that they might have an opportunity for immediate reproduction a very large number of retarded caterpillars of the brown-tail moth from nests which had been placed in cold storage during the winter were liberated upon trees in the immediate vicinity.

These caterpillars seemed to be not at all injured as a result of their abnormal experience, but immediately began to feed voraciously and to grow apace. That they were injured soon became evident, but it could not be determined whether such injury was due to the enforced lengthening of their period of hibernation, to the hot weather which then prevailed, or, possibly, to the fact that the foliage was much more advanced than that upon which caterpillars newly emerged from hibernation usually fed. They began to die at an alarming rate inside of two weeks, and when it was time to make a collection for the purpose of determining whether the Meteorus had found them or not hardly more than 300 could be found out of the thousands which had been liberated. These were removed to a tray in the laboratory, and from June 23 to July 15 no less than 76 Meteorus cocoons were removed. From these 43 adults, of which 16 were females, were reared.

These were the first females of the second generation which had been secured at the laboratory, and a part of them was used in a reproduction experiment similar to those which had resulted in the production of males the previous year. Curiously enough, the adults of the third generation reared from these parents, under circumstances identical with those which had been used in earlier reproduction experiments, consisted of both sexes, there being 5 females out of a total of 40. These were the first females of the species ever reared from adults in confinement.

In the spring of 1909 the caterpillars from a few nests which had been collected the winter before in the vicinity of this first satisfactory field colony were fed in the laboratory, and from them a few cocoons of Meteorus were secured. It was certain that the species had completed the cycle of the seasons in the open, but it was also rather evident that it was not very common. If this were due to widespread dispersion, as might easily be the case, it might possibly result in the species spreading out so thin as to be lost, and it was resolved to place the Meteorus reared in 1909 in the same general vicinity, on the theory that by spreading over the same territory the colony might be materially strengthened throughout. This was done, and about 2,000 individuals were liberated during that spring and summer, the most of which came from hibernating caterpillars, but a part of which was imported as parasites of the full-fed and pupating caterpillars. The experiment of colonizing large numbers of retarded caterpillars in the vicinity was repeated, and with similar results to those secured in the previous season.

In 1910 a larger number of the cocoons was found, but at the same time a very few were secured from the caterpillars which had been collected in the vicinity of the colony. It did not look as though much was to be expected from the parasite at first, but when, toward the end of June, collections of full-fed caterpillars were made from various localities for the purpose of determining the status of the tachinid parasites, the results were much more encouraging. Cocoons of the second generation of Meteorus were soon found in some numbers and to a distance of a mile or more from the original colony center. Within a rather limited area near the colony center they could almost be said to be abundant, so abundant that 50 were collected in the course of about two hours' work. They are far from being conspicuous objects, being wholly disassociated from the caterpillar which served as host, and on this account the number collected was considered to indicate a very satisfactory abundance.

Its rate of dispersion, so far as indicated by the results of the summer work upon the caterpillars of the brown-tail moth, was too slow to be satisfactory, but in the early fall a single specimen, definitely determined by Mr. Viereck as of this species, was secured from a lot of caterpillars of the white-marked tussock moth collected in the city of Lynn, some 7 miles from the colony site. This would indicate a rapidity of dispersion in excess of that of Compsilura, and one which is distinctly satisfactory.

Another specimen was reared in the fall of 1910 from a caterpillar of the fall webworm collected in the open, and this was also considered as satisfactory evidence of its ability to exist here. At the present time there seems to be every reason to expect that it will be found in 1911 over a more considerable territory and in a much greater abundance than in 1910.

ZYGOBOTHRIA NIDICOLA TOWNS.

The few caterpillars which Mr. Titus saved from among those emerging from the hibernating nests in the spring of 1906 all died before pupation, and no other parasite than Apanteles and a single specimen of the Apanteles parasite, *Mesochorus pallipes*, was reared from them. In 1907 trouble was again experienced in carrying the caterpillars from imported nests through to maturity, but among the thousands which were fed in the cages at the North Saugus laboratory, as described in the account of the introduction of Apanteles, a few did reach the point of pupation, and from them a very few Zygobothria adults (fig. 69) were reared. There was no ground for doubting that the tachinids actually issued from the imported caterpillars of the brown-tail moth and that they had actually been present as hibernating larvæ within the caterpillars when they were received from Europe, but at the same time the circumstance seemed so improbable as to be refused immediate credence. Confirmation of the records was accordingly sought in 1908, and preparations were made to carry large numbers of the caterpillars from imported nests through to maturity in the large trays, already mentioned in the discussion of Apanteles.

For a time everything went well, and the caterpillars passed through three of the spring stages and assumed the colors characteristic of the last with scarcely any mortality. Then, for some reason, they ceased to feed freely, and began to die, and even those which did feed ceased to grow. Eventually practically all of them died, but of the few which survived to pupate, a very few contained the parasite, and although only about half a dozen of the adult Zygobothria were reared, they were sufficient to prove beyond question the validity of the earlier conclusions. The death of the caterpillars from imported nests in 1906 was supposed to be due to the epidemic of fungous disease which affected those in confinement quite as generally as those in the open, and in 1907 death was presumed to be the result of the unsanitary conditions which resulted from the use of the closed cages. In casting about for a cause in 1908, the drying of the food in the open trays before the caterpillars fed upon it was deemed to be sufficient, and consequently, in 1909, it was determined to use extraordinary precautions and to rear a large number of the tachinids if it were possible.

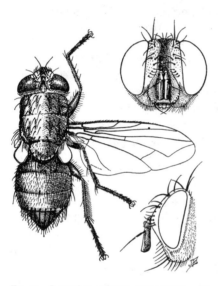

FIG. 69.—*Zygobothria nidicola:* Adult female, with front view of head above and side view below. Much enlarged. (Original.)

In the early spring of 1909 a considerable number of the imported caterpillars was dissected before they began to feed, and in some lots a high percentage was found to contain the hibernating larvæ of the Zygobothria (fig. 63, p. 264). These lots were to be given especial care, and little doubt was felt as to the success of the outcome, because no particular difficulty had been experienced in feeding small numbers of the caterpillars from native nests through all of their spring stages.

As was the case in 1907, the caterpillars passed through the first three spring stages with scarcely any mortality, and, as before, trouble was finally encountered. In the first place a considerable proportion of the trays was infected with the fungous disease, which had been accidentally brought in from the field, and these had to be destroyed summarily. There were still a number of the trays unaffected, however, and these were given the very best care which previous success with native caterpillars and failure with imported caterpillars suggested. In spite of all the results were exactly as before, and, as before, only an insignificant number of the Zygobothria completed their transformations. It was all the more surprising because there were several of the smaller and choicer lots which were kept in a cool, airy place, side by side with trays of native caterpillars, fed upon the same food and given identically the same attention, and yet every single individual of the one lot died, while nearly every individual of the others went through to maturity.

It began to look as though there was something wrong which was outside of the power of anyone at the laboratory to remedy, and it was resolved to test the matter thoroughly in 1910.

The caterpillar-dissection work which was begun in the spring of 1909 was carried on quite extensively in the winter of 1909–10, and among the several lots of hibernating nests imported that winter those which came from Italy and France were found to contain a very large percentage of caterpillars bearing the larvæ of Zygobothria (fig. 63, p. 264). These caterpillars, as soon as they emerged from these nests in the spring, were separated into two lots. A part of them was fed in trays, as before, and another part was immediately placed in the open, upon small oak trees which had previously been cleared of native nests of the brown-tail moth with this end in view.

The caterpillars, as usual, did remarkably well in both cases, and as usual the three spring stages were passed in the normal manner. At the end of that time those which had been fed in trays began to die, and those in the open to disappear. Mr. Timberlake, who was assiduously trying to follow the development of the Zygobothria maggots throughout their later stages, found it increasingly difficult to find the caterpillars in very large numbers in the field where they had been colonized, and finally of the thousands originally present only about 150 could be found. These had reached their last stage by this time, and they were collected and brought into the laboratory. Within a few days all but a very small number had died, and as there was a good chance that a few native caterpillars were present, there was nothing to indicate that all of the survivors were not native instead of imported.

At the same time that the imported caterpillars were liberated, as above described, a number of experiments in the similar colonization of native caterpillars was begun, and in every instance in which they were not overtaken with some well-defined calamity—fire in one instance, starvation in others—they went through to maturity in large numbers and in a perfectly normal manner.

It is no longer to be doubted that in the case of·tne imported caterpillars some element other than any which is operative during the feeding period of the caterpillars in the spring is to be held responsible for their wholesale demise. The uniform ill success which has invaribly attended the attempts to feed the brown-tail caterpillars from imported nests through to maturity can no longer be considered as either coincidental or the result of inexperience in this sort of work. Something else is responsible, and in looking about for parallel instances the results which have attended all attempts to feed caterpillars of the brown-tail moth, no matter from what source, out of season, are possibly to be considered as comparable.

Hundreds of experiments involving the feeding of native and imported caterpillars upon lettuce during the late winter and early spring have invariably resulted in carrying the caterpillars through their first three spring stages and in their death before pupation. This may be due to the character of the food.

A smaller number of experiments in feeding caterpillars of the brown-tail moth, which had been retarded in their emergence from the winter nests, have always resulted in a manner not altogether incomparable. Many thousands of these caterpillars have been kept in cold storage for about one month after they would normally have issued and then placed upon their favored food plants in the open. Upon several occasions when this has been done the caterpillars have fed very freely at first, grown rapidly, and appeared to be perfectly healthy. Then they would begin to die, almost exactly as the imported caterpillars would begin to die in the fourth spring stage, and it does not appear that any of them have ever completed their transformations. This may be due to the weather conditions and unsuitable food. It is believed that it is indirectly due in this instance, and in the instances of the caterpillars from imported nests, to the fact that both the one and the other have been subjected to abnormal conditions during hibernation. The imported nests are always exposed for a considerable period during the winter to an unduly high temperature. The caterpillars are almost upon the point of becoming active—sometimes they are beginning to become active—when the nests are received at the laboratory. As soon as possible after their receipt they are placed under out-of-door temperature again, with the result that the caterpillars become inactive and remain so until the time when they would normally have issued from the nests had they not been exposed to undue warmth during the winter.

It makes little difference whether the nests are exposed to one temperature or another during the winter so long as the caterpillars are not actually stirred into activity; the date of final emergence in the spring remains practically unchanged. Roughly speaking, if brown-tail nests are exposed to a constant high temperature beginning at any time during October the caterpillars will die without becoming active; during November they will die if kept too warm, but become active in a little over a month if kept warm and humid; in December they will sometimes become active by the 1st of January if they are kept fairly humid, and during January they will nearly always become active in a little less than a month, no matter what the conditions of humidity; after the 1st of February activity is resumed in something like two weeks; after the 1st of March in about one week, and later in a few days. If kept at a high temperature for three weeks in December or two weeks in January and then placed under natural conditions for the rest of the winter, their emergence will not be appreciably hastened in the spring, but if the attempts to rear Zygobothria from imported caterpillars which have been handled in much this manner are to be properly interpreted, subjection to such abnormal conditions results in a subtle disarrangement of the vital processes, and the insect is metabolistically unbalanced.

It is hardly necessary (to return to the story of Zygobothria) to state that these successions of almost total failures were not only puzzling, but decidedly exasperating. In 1910, for example, we estimated the number of apparently healthy Zygobothria larvæ on hand in apparently equally healthy caterpillars to be something like 40,000, of which something like 10,000 or 15,000 were in the caterpillars which were feeding and growing in a perfectly natural manner in the open. Long before it was time for these caterpillars to pupate we had given up all hope of more than an insignificant number of these parasites going through to maturity, and, as a matter of fact, there is no record of a single one among them going through. Every resource had been exhausted the winter before in attempting to secure a shipment of nests of the brown-tail moth in good condition from some locality where there was a likelihood of Zygobothria occurring in abundance as a parasite, and the failure was even more complete than usual. There remained only the alternative of importing large numbers of full-fed and pupating caterpillars of the brown-tail moth, collected in the same localities, and the prospect that this would be successfully accomplished was far from brilliant. The senior author was in Europe at the time when these conclusions were formed and was putting forth his utmost endeavors to bring about this very thing, but June passed, and with the advent of July it became certain that no shipments of any consequence would be received.

It was known that the parasite could be secured in this manner because small numbers had been reared from the imported quantities of full-fed and pupating caterpillars which were received at the laboratory in 1906 and several hundred from similar shipments in 1907. This latter year no accurate records had been made of the number of each species of tachinids emerging from the importations of brown-tail moth material, but it was known that somewhere between 300 and 500 individuals had been reared, the most of which were colonized at North Saugus. This was the only lot of adult flies of any consequence which had been reared and liberated, and since special efforts which had been made to recover this and other species liberated at the same time and place had failed in both 1908 and 1909, it was not considered to be at all likely that the attempted colonization was successful.

The situation, in so far as Zygobothria was concerned, could hardly have appeared worse than it was at the beginning of July, 1910. No one species of anything like equal importance had been quite so difficult to secure in adequate numbers and, moreover, there was no immediate prospect of finding a way to overcome the difficulties attending its importation. Consequently no similar circumstance, except perhaps the recovery of the gipsy-moth parasite, *Apanteles fulvipes*, could have caused a livelier satisfaction than was felt when several bona fide specimens of Zygobothria were reared from a lot of cocoons of the brown-tail moth which had been collected in the field some time before. The first specimen to issue was a male and it was followed by several more of the same sex. The males are markedly different from the females in appearance and not quite so distinctive, and we did not feel absolutely sure of their identity at first, but when after a few days a female was secured in the same manner from American cocoons there was no possible doubt that the species was not only established in America as firmly as three generations from a small beginning would permit, but dispersing with considerable rapidity, since of the seven specimens reared none was from less than 1 mile of the original colony site and one was from at least 3 miles distant. It is certain that the species must have spread over at least 30 square miles since its colonization three years ago, and when the millions of brown-tail moth caterpillars which are present in that territory are compared with the few thousands which produced the seven Zygobothria reared in 1910, it is equally certain that its increase has been at the same time enormous.

It bids fair, judging from this, to do exceedingly well in America. Unlike *Compsilura concinnata*, *Pales pavida*, and other tachinids, which rank of some importance as parasites of the brown-tail moth and gipsy moth in the Old World, it is wholly independent of any host other than the brown-tail moth, and its rate of multiplication, being unquestionably more rapid than that of the brown-tail moth, ought not to be checked until it has become a factor in the control of its own particular host.

It may be added as a postscript that a few days after writing the above a few hundred caterpillars of the brown-tail moth collected in the field from hibernating nests were dissected in the laboratory. In them were found several of the characteristic first-stage Zygobothria larvæ (fig. 63, p. 264) embedded in the walls of the gullet. The evidence presented by this small number of dissections is less satisfactory than though the number were larger, but if it is to be accepted the rate of increase of Zygobothria in 1910 is considerably better than was expected.

PARASITES ATTACKING THE LARGER CATERPILLARS OF THE BROWN-TAIL MOTH.

HYMENOPTEROUS PARASITES.

In Europe after the caterpillars of the brown-tail moth resume activity in the spring they become subject to attack by a variety of tachinid parasites, but so far as has been determined by rearing work with imported material the only hymenopterous parasite of any consequence is Meteorus, which passes the winter as a first-stage larva in the hibernating caterpillars.

In fact, only a single other parasite has ever been reared from imported caterpillars which may not have come from some other accidentally included host, and this is the *Limnerium disparis*, which has already received attention as a minor parasite of the gipsy moth. It would certainly seem as though there were likely to be others attacking the caterpillars of the brown-tail moth in Europe in spite of the fact that none has been secured, and this supposition is upheld by the published results of a study in the parasites of the brown-tail moth which was made a few years ago by a Russian entomologist, Mr. T. W. Emelyanoff. He mentions a number of parasites which have not been reared at the laboratory from imported material, and among them one, *Apanteles vitripennis* Hal., which is so common, according to his account, that the "cocoons are sometimes accumulated together in great numbers." Any suspicions that the Apanteles thus observed by him is identical with *A. lacteicolor* Vier, as reared at the laboratory and which is the only representative of the genus that has been reared from caterpillars collected in Russia or elsewhere, is at once dispelled by his detailed account of the early life and habits of the species which he had under observation and which differ in all essential particulars from the life and habits of *A. lacteicolor*. The caterpillars are attacked soon after they leave the nests. Instead of dying in their molting webs they crawl down the trunks of the trees, and the cocoons of the parasite are found in splits and holes in the bark, rarely higher than from 1 to 1½ yards from the ground. The host caterpillar is left alive and remains for some time clinging to the cocoons of its parasite, something which has never been observed in the case of *A. lacteicolor*.

The plans for the coming season, if they materialize, call for a thorough study of the Russian parasitic fauna of the brown-tail moth, and it is sincerely hoped that the observations of Mr. Emelyanoff may be confirmed.

TACHINID PARASITES.

Several of the tachinids which attack the brown-tail moth have already been mentioned in the course of the discussion of the gipsy-moth parasites. Among them *Compsilura concinnata* is the only species which is of real importance in connection with both hosts.

Tachina larvarum is not uncommonly encountered as a brown-tail moth parasite, but never so commonly as it frequently is in its other connection. *Tricholyga grandis* has also been reared in small numbers from cocoon masses of the brown-tail moth.

The tachinid parasites of the brown-tail moth, which are either unknown as parasites of the gipsy moth or which are rarely encountered in that connection, include a considerable variety of species, several of which appear to be of little or no real importance. As will be seen, they include amongst their number species which represent the extreme of diversity in habit.

DEXODES NIGRIPES FALL.

Another example of the artificiality of the present accepted scheme of classification of the tachinid flies is to be found in the separated positions therein occupied by the two exceedingly similar species *Compsilura concinnata* and *Dexodes nigripes*. So similar are these two that if a few hairs and bristles were to be rubbed from the head of one it would be practically impossible to distinguish it from the other, even though everything in connection with the early stages and life of each was known. The one point of difference of any consequence from an economist's standpoint is the more restricted host relationship of Dexodes, which, though equally common with Compsilura as a parasite of the brown-tail moth in Europe, is exceedingly rare as a parasite of the gipsy-moth caterpillars. In every other respect, except host relationship, the habits of the two are identical, and so far as known their earlier stages are absolutely indistinguishable.

Dexodes was first received and liberated as a parasite of the brown-tail moth in 1906, and it was the first of the tachinid parasites to be carried through all of its transformations in the laboratory upon American hosts. This was accomplished by Mr. Titus in one of the large out-of-door cages in 1906, and again with somewhat more success in one of the smaller indoor cages in 1907. As with Compsilura, only two weeks are required for the larval development, a week or ten days for the pupal stage, and three or four days for the female to reach her full sexual maturity. As is also true with Compsilura the larvæ are deposited by the female beneath the skin of the host caterpillar.

Several small colonies were planted by Mr. Titus in 1906, followed by several more small ones and one larger one in 1907. None was liberated in 1908, but in 1909 one very large and satisfactory colony was put out. In 1910 only a single specimen of the parasite was received from abroad and this, curiously enough, in a shipment of gipsy-moth caterpillars.

It was confidently expected that in 1910 at least a few specimens would be recovered from the field as a result of the earlier colonization work, but these expectations were not realized. Of all of the tachinid parasites of the brown-tail moth, not excepting *Compsilura concinnata*, it was the one most satisfactorily colonized in 1906 and 1907, and on this account it was expected to find it established in the field.

It is considered as one of the most likely of the as yet unrecovered parasites to be recovered from the field in 1911 or 1912.

PAREXORISTA CHELONIÆ ROND.

No brown-tail moth material was received from abroad during the summer of 1905, and consequently nothing was known of the hibernating tachinids which attack this host until the spring of 1907, when they began to issue from the puparia of the previous summer's importations. All that were reared that spring were of the one species, which has since been determined as *Parexorista cheloniæ* Rond., and to date no other species hibernating as a puparium and with but one annual generation has been reared from this host. Nothing to compare with the difficulties which attended the hibernation of the principal gipsy-moth parasite having similar habits was encountered in the case of Parexorista. Its puparia (Pl. XX, fig. 4) were carefully covered with earth the first winter and the second; but it was then found that this precaution was unnecessary and that the percentage of emergence was quite as large when the puparia were kept dry as when they were damp. The difference appears to be associated with the state in which the pupæ themselves hibernate. Those of Blepharipa and Crossocosmia develop adult characters in the fall, and it is in reality the unissued adults which hibernate. Those of Parexorista do not develop adult characteristics until spring, and besides in Parexorista the space between the pupa or nymph and the shell of the puparium is dry and does not, as in Blepharipa, contain a small quantity of colorless liquid.

Accordingly the percentage of emergence of *Parexorista cheloniæ* has aggregated nearly or quite 90 per cent each year as against the relatively small percentage of Blepharipa which has been carried through its transformations.

A very few of the flies were liberated in 1907, but there were too few puparia of the species received the summer before to make anything like a satisfactory colony of the species possible. In 1908 in excess of 2,000 of the flies issued from the previous season's importations, and of these about 1,500 were liberated in one colony under circumstances which were the most favorable that could be imagined. The remaining ones were used by Mr. Townsend in a successful series of experiments which have already been summarized in an account of his first year's work, published as Part VI of Technical Series 12 of the Bureau of Entomology. Other equally satisfactory colonies were established later, but of these nothing more need be said at this time.

The large colony liberated by Mr. Townsend in the spring of 1908 consisted of flies of both sexes, very many of which had mated before they were given their freedom. This circumstance, which was not considered as particularly of interest at the time, has acquired significance more recently, as will be shown.

Later in the spring and early in the summer caterpillars collected from the immediate colony site were found to contain the larvæ of the parasite, and a calculation involving the number of caterpillars within a limited area immediately surrounding the point of liberation, the number of flies liberated, and the percentage of parasitism prevailing in this area indicated a very satisfactory rate of increase. It will be remembered that the flies were in part ready or nearly ready to oviposit when they were given their freedom, so that dispersion did not have to be taken into consideration to the extent which is necessary when a long period elapses between the time of liberation and the time of recovery.

In 1909 similar collections of caterpillars and cocoons were made in the same and in nearby localities, and the number of Parexorista which was secured from them was gratifyingly large. These collections had not been made with the view of determining the rate of dispersion, but it was apparent that the increase had been accompanied by a rate of dispersion that was, at the very least, satisfactory and which, for all evidence to the contrary, might be phenomenal.

Accordingly in 1910 a series of collections was planned, some of which were to be made in exactly the same localities as those from which the flies were recovered the year before and which were designed to be indicative of the prevailing rate of increase, while others at varying distances and in different directions from the colony center were designed to show the rate of dispersion. No doubt whatever was felt concerning the recovery of the parasite, which was considered to be as firmly established as the Calosoma or *Compsilura concinnata*. The results afforded another example of the obtrusiveness of the unexpected. Not a single Parexorista puparium was secured from any of the material included in this series of collections.

This was, all things considered, the most serious setback of any which the parasite work has experienced since its inception. It was never doubted from the first that some among the parasites would be unable to exist in America, and no species was really credited with having demonstrated its ability to do so until it had lived over at least one complete year out of doors. Parexorista had done this and more, having gone through two complete generations, unless, what was not at all likely, its puparia had all been killed some time during the fall or winter.

Without indulging in unnecessary speculation as to the reason for its disappearance, the following facts are presented for consideration:

There is in America a tachinid known as *Parexorista cheloniæ*, which is morphologically identical with the European race so far as may be determined through a painstaking comparison of the two. It is a common parasite of the tent caterpillars *Malacosoma americana* Fab. and *M. disstria* Hübn. The adult flies issue at the same time in the spring as do those of the European parasite of the brown-tail moth. The same type of egg is deposited; the larvæ are indistinguishable in any of their stages or habits during their several stages; the third-stage larvæ issue at the same time and form puparia which are apparently the exact copies of the European, and the hibernating habits are the same. The one and only difference is that the American *Parexorista cheloniæ* does not attack the caterpillars of the brown-tail moth, while the European *Parexorista cheloniæ* is perhaps the most important of the tachinid parasites of this host.

Mr. W. R. Thompson, whose excellent and painstaking work makes possible the above comparison between the two races, went a step further in his investigations. He found by actual experiment that in confinement, at least, the European males would unite with the American females with as much freedom as with those of their own species. Granted that similar intermingling of the races takes place in the open, and the reason for the nonrecovery of *Parexorista cheloniæ* as a parasite of the brown-tail moth in the summer of 1910 is no longer a mystery.

It was stated a few paragraphs back that the flies which were colonized in the spring of 1908 were largely mated at the time of liberation. Their progeny, which issued in the spring of 1909, would therefore be of the pure-blooded European stock. Issuing at the same time were a vastly larger number of the American race, because as it happened there was an incipient outbreak of *Malacosoma disstria* in that very locality, which was quite heavily parasitized by Parexorista. There were easily 50 or 100 of the American flies to one of the European race present in that general vicinity in the spring of 1909. The chances that the pure-blooded European females were fertilized by American males were therefore a good 50 or 100 to 1 at the most conservative estimate.

Being of the European race, their instincts led them to attack the caterpillars of the brown-tail moth, and the attack was successful, as witnessed by the number of puparia which were secured from the collected caterpillars and pupæ in the summer of 1909, but these puparia, instead of representing the pure-blooded European stock, as was then supposed, represented the half-breed stock resulting from the promiscuous mating of their mothers. Evidently the females issuing from them in the spring of 1910 lost the cunning which is characteristic of the European race, which makes possible the deposition of the soft-shelled eggs amongst the bristling poisonous spines of the host without injury.

Mr. Thompson, in his experiments with the American female which had been fertilized by an European male, found that she was neither anxious to oviposit upon the caterpillars of the brown-tail moth nor able to do this successfully. A proportionately large number of the eggs deposited upon this host were either pierced by the poison spines or else the young larvæ came in contact with these and died before entering. A few larvæ did succeed in gaining entrance, and one or two passed through their transformations, but when the natural disinclination to attack the caterpillars of the brown-tail moth was associated with a heavy mortality following occasional attack the percentage of parasitism is reduced to the minimum.

In consequence of these observations in field and laboratory, the name of *Parexorista cheloniæ* has been erased from the list of promising European parasites of the brown-tail moth and placed at the head of the list of the imported parasites which are proved unfit.

It is a pity, too, as has incidentally been stated, because it is about the most common of any of the tachinid parasites in Europe, and, moreover, is one which is entirely independent of any alternate host.

PALES PAVIDA MEIG.

There is a very considerable group of tachinid parasites of the brown-tail moth which appears to be more commonly encountered in material from southern European localities than from those in the north. One of these, *Zygobothria nidicola*, has already been the subject of lengthy discussion. The fact that though apparently southern in its distribution in Europe, it has manifested a strong tendency to become thoroughly acclimatized here, has lent encouragement to the attempts which have been made, and which will be renewed, looking toward the establishment here of others having a somewhat similar distribution.

Pales pavida (fig. 70) is perhaps as promising as any among these, although it is possible that it appears so on account of a somewhat larger knowledge which we possess concerning its life and habits. It was first imported in not very large numbers in 1906. In 1907 about as many were secured and colonized as of the successfully introduced Zygobothria, and more were colonized in 1909. The fact that it has not been recovered is by no means to be taken as positive asssurance that it is not established, and it is well within the bounds of possibility that it will be recovered in 1911 or 1912.

It is one of the species which deposits its eggs upon the leaves to be eaten by its host (fig. 36, p. 214) and was the first species having this habit to be carried through all of its transformations in the laboratory. In 1908 Mr. Townsend succeeded in carrying some of the flies through the period allotted for the incubation of their eggs, but he did not succeed in securing oviposition. In 1909 Mr. Thompson had better fortune, and not only secured eggs in abundance,

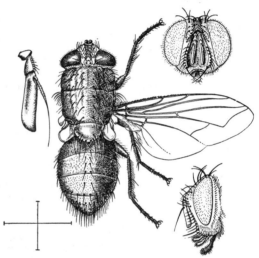

FIG. 70.—*Pales pavida:* Adult female, with front view of head above and side view below, and antenna at left. Much enlarged. (Original.)

but fed these eggs to a variety of caterpillars and secured either the puparium or the fly in nearly every instance. He also secured much interesting data upon the early stages, and upon the life and habits of the early stages, a story of which is left for him to tell. The accompanying illustrations of the eggs and larvæ were prepared under his direction. That of the egg (fig. 37, *b*, p. 214) is of interest in comparison with that of the egg of *Blepharipa scutellata* (fig. 37, *a*, p. 214), as showing the difference in the characteristic microscopic markings. That of the larva will give a good idea of the integumental "funnel" (figs. 71, 72), formed by the ingrowing epidermis, as differing from the tracheal "funnel" characteristic of the larva of Blepharipa, as figured on pages 215 and 216.

Not very much that is definite can be said of the seasonal history of Pales. It undoubtedly will require another host than the caterpillar of the brown-tail moth in order that it may complete its seasonal cycle, but that it will find such a host is pretty certain. It would rather appear, from what has been observed, that it will attempt to hibernate as an adult. Whether or not it will be able successfully to do this in New England remains to be proved.

FIG. 71.—*Pales pavida:* Second-stage larva *in situ* in basal portion of integumental "funnel." Much enlarged. (Original.)

It has occasionally been reared as a parasite of the gipsy moth, and if successfully introduced into America it ought to be of some assistance in this rôle also. Unfortunately, as a parasite of the caterpillar of the brown-tail moth, it does not issue until about the time when the moth would have issued had the individual remained healthy. It requires some little time for the females to develop their eggs, and it is not at all likely that, like Compsilura, it will be found to pass one generation upon the caterpillars of the brown-tail moth, and the next upon the gipsy-moth caterpillars.

ZENILLIA LIBATRIX PANZ.

This parasite, like Pales, deposits its eggs upon foliage to be eaten by its host, but, unlike Pales, it has not been reared through its stages in the laboratory. Like Pales, it is southern in its distribution, and in relative importance they are about equal, judging from the numbers of each which have been reared at the laboratory.

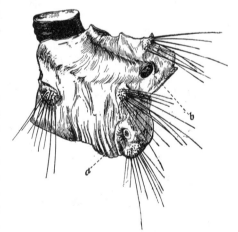

FIG. 72.—*Pales pavida:* Integumental "funnel," showing orifice in skin of host caterpillar. Much enlarged. (Original.)

It was colonized in small numbers in 1906, in larger numbers in 1907, and in very small numbers subsequently. The circumstances attending its colonization are as satisfactory, so far as known, as those attending the colonization of Pales and Zygobothria, and it is hoped that it may be recovered in the course of 1911 or 1912. It is also hoped that a large number will be imported in 1911.

MASICERA SYLVATICA FALL.

This tachinid appears not to be uncommon as a parasite of the brown-tail moth in Italy, but has not been received from other countries in more than the most insignificant numbers.

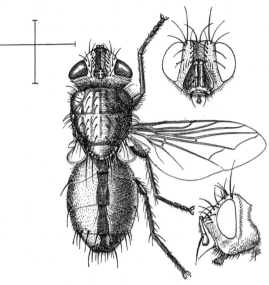

FIG. 73.—*Eudoromyia magnicornis:* Adult female, with front and side views of head at right. Much enlarged. (Original.)

Not enough have been received to make anything like colonization possible, and it is one of the species which it is hoped to receive in 1911.

EUDOROMYIA MAGNICORNIS ZETT.

This (see fig. 73) is the most distinctive of the tachinid flies parasitic upon the brown-tail moth, and the only one among the parasites of either the gipsy moth or the brown-tail moth which has the habit of depositing its active larvæ upon the food-plant of its host. This habit was first discovered by Mr. Townsend, who gives an account of the manner of the discovery in Technical Series VI, part 12, of this bureau, from which the accompanying figure (fig. 74) was taken.

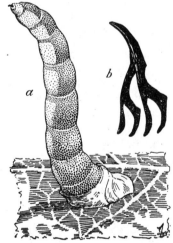

FIG. 74.—*Eudoromyia magnicornis: a,* First-stage maggot attached to leaf, awaiting approach of a caterpillar; *b,* mouth-hook of maggot. *a,* Greatly enlarged; *b,* highly magnified. (From Townsend.)

It is another of the group of tachinid parasites which appear to be southern rather than northern in distribution, on account of which it has been found impossible to secure a sufficient number to make adequate colonies practicable. It was colonized together with *Zygobothria nidicola*, *Pales pavida*, and *Zenillia libatrix*, in about the same numbers in 1906 and 1907, and, like the two last named, it is hoped to recover it in 1911. It is also hoped to import and liberate a much larger number than hitherto during that year.

CYCLOTOPHRYS ANSER TOWNS.

Mr. Townsend described this species as new from specimens reared in 1908 from brown-tail moth material received from the Crimea. It has not been detected in shipments of similar character from any other locality in sufficient numbers to indicate it as being an important parasite, nor have enough been received from the Crimea to make possible its colonization. It is hoped that this may be done in the course of the year 1911.

It is one of the relatively few species of tachinids attacking the larvæ of the brown-tail moth which deposit large, flattened eggs upon the body of the host caterpillars.

BLEPHARIDEA VULGARIS FALL.

This is almost the only tachinid parasite of either the gipsy moth or the brown-tail moth which is of no apparent importance in connection with either host and which at the same time has been reared a sufficient number of times to make its host relationship reasonably certain. The few specimens which have been received have mostly come from various parts of the German Empire. Very little is known of its life and habits, and it is not considered as being of sufficient importance to warrant further investigation.

PARASITES OF THE PUPÆ OF THE BROWN-TAIL MOTH.

By far the most important of the parasites of the pupæ of the brown-tail moth in Europe appears to be Monodontomerus, an account of which has already been given in the discussion of the parasites of the gipsy moth. It is more frequently reared in connection with the brown-tail moth than with the gipsy moth, and some of the shipments of cocoons have produced it in extraordinary numbers.

Theronia, also mentioned as a parasite of the gipsy moth, is about the next in importance, but the European *T. atalantæ* Poda is no more frequently reared than the American *T. fulvescens* Cress.

The same species of Pimpla already mentioned as parasites of the gipsy moth in Europe attack the brown-tail moth as well. Like Theronia and Monodontomerus, they are more frequently encountered in this connection than in the other.

No species of Chalcis has been reared from any European material received to date, and in this respect the parasitism of the pupa of the brown-tail moth differs from that of the gipsy moth. It further differs in that two small gregarious chalcidids, both of them closely allied or identical with American species of the same respective genera, have occasionally been reared from imported cocoon masses. Neither of these is common. One, *Diglochis omnivora* Walk, appears to be specifically indistinguishable from the form which goes under the same name in America, where it has occasionally been reared from the gipsy moth and abundantly from the brown-tail moth. The other is a species of Pteromalus, which, according to Mr. Crawford, is hardly to be distinguished from the tussock-moth parasite, *Pteromalus cuproideus* How.

Enough of the latter species have been reared to make small colonies possible, but these colonies have been so very small as to make its establishment improbable. It is hoped that a larger number will be imported in 1911, but since it appears to be of very slight importance in Europe no great enthusiasm is felt over the prospect.

SUMMARY AND CONCLUSIONS.

The work of introducing into America the parasites and other natural enemies of the gipsy moth and the brown-tail moth has been more arduous than was anticipated when it was begun. It was soon found that the published information concerning these enemies was deficient and unreliable, and that much original research was necessary in order that they might be intelligently handled. Later it developed that the rate of dispersion of the introduced species was so very rapid as to necessitate larger and stronger colonies than had been contemplated.

The policy originally adopted of employing foreign entomologists to collect the eggs, caterpillars, and pupæ of these pests abroad for shipment to the Massachusetts laboratory, where the parasites which they contained might be reared, has resulted in the successful importation and colonization of a considerable number of the parasites which a study of this material, after its receipt at the laboratory, has indicated as being of importance. Numerous others successfully imported have been colonized, but so recently as to render the success of the experiment uncertain. On account of the rapidity of dispersion, which results in the parasites being very rare over a large territory instead of being common over a restricted territory, as long a period as four years may elapse before it is possible to recover them after colonization. It has been found impossible to secure certain of the parasites in adequate numbers for colonization under satisfactory conditions. The proportion of such is very small, it is true, but at the same time it may easily be that ultimate success or failure may depend upon the establishment, not of the most important among the parasites and other natural enemies, but of a group or sequence of species which will work together harmoniously toward the common end. Viewed in this light, the importance of parasites which otherwise might be considered as of minor interest is greatly enhanced.

It is impracticable to determine certain facts in the life and habits of those parasites which have been colonized under conditions believed but not known to be satisfactory. Further detailed knowledge is necessary before we can judge whether the circumstances surrounding colonization were in truth the best that could be devised. Furthermore, so long as original research is confined to the study of material collected by foreign agents, some of whom are technically untrained, it is practically impossible to secure the evidence necessary to refute published statements concerning the importance of certain parasites abroad which the results of first-hand investigations have not served to confirm. It is believed that these statements are largely based upon false premises, but should this belief prove ungrounded it would mean that there are important parasites abroad of which little or nothing is known first-hand.

A determined effort was made in 1910 to better the deficiencies in the foreign service without going to the lengths of adopting a radically changed policy, but the results were not satisfactory. Lack of assurance that a continuation of the work in 1911 along similar lines would bring more favorable results made its continuation inadvisable. It therefore became a question of adopting new and radically different methods in so far as the foreign service was concerned.

In favor of a policy of inactivity was the prospect of an immediate reduction, as opposed to an increase, in expenditures should renewed activity be decided upon. There was the chance that the parasites already introduced and colonized would be sufficient to meet the demands of the situation.

On the other hand, the vast majority of defoliating native insects, which rarely or never become so abundant as to be considered injurious, prove upon investigation to support a parasitic fauna similar in all its essential characteristics to that supported by the gipsy moth in countries where it is similarly a pest at very rare intervals or not at all.

Parasitism appears to be unique among the many factors of control, in that no other agency similarly increases in efficiency in direct proportion as the efficiency of other agencies, such as climatic conditions, miscellaneous predators, etc., diminishes. In short, the apparent importance of parasitism as a factor in the natural control of defoliating insects has been decidedly enhanced as a result of these more or less technical and intensive studies. It can be said with the utmost assurance that if a sufficient number and variety of parasites and other natural enemies of the gipsy moth which act in a manner comparable to the true facultative parasites, as above described, can be introduced into America, the automatic control of the gipsy moth will be permanently effected.

During the past four years the "wilt" disease has been increasing somewhat in efficiency, but notwithstanding that it is and has been prevalent in every locality in which the gipsy moth has been allowed to increase unchecked, the gipsy moth still continues to be a menace to the life and health of valuable trees which have been protected during this time at a considerable cost. In parts of Russia, where parasitic control is obviously inefficient, control through disease is not sufficient to keep the gipsy moth from increasing until defoliation of large areas results. Similarly, in America, the destruction of very large numbers of caterpillars of several sorts of the larger Lepidoptera has been observed, but in no instance until after the caterpillars involved had increased to such numbers as to become a pest.

It may be that the parasites already introduced and established, or likely to become established, will prove to be sufficient for the purposes intended. Only events themselves can be depended upon to answer this question, and from five to six years must pass before the answer is known. During this period the gipsy moth will continue to disperse and multiply, and large expenditures will be necessary to prevent much more rapid dispersion and multiplication than has prevailed in the past.

Expenditures amounting to a very small percentage of the total will suffice to carry on the parasite work. If the parasites already introduced are sufficient to meet the needs of the situation, the expenditure projected will have been needless and unnecessary. If the parasites already introduced are not sufficient, it may be that this deficiency can be made up in time to avoid much if any delay in the day of final triumph.

THE PRESENT STATUS OF THE INTRODUCED PARASITES.

PARASITES OF THE GIPSY MOTH.

EGG PARASITES.

ANASTATUS BIFASCIATUS Fonsc.

Received first in 1908. Colonized unsuccessfully in 1908 and successfully in 1909. First recovered in immediate vicinity of colony in 1909. Increased notably in 1910, but indicated dispersion is only about 250 feet per year. Artificial dispersion necessary. Apparently well established.

SCHEDIUS KUVANÆ How.

Received first in 1907, dead, and in 1909, living. Successfully colonized in 1909. Recovered in immediate vicinity of colony site in 1909. Doubtfully recovered in 1910. Establishment very doubtful on account of climatic conditions.

HYMENOPTEROUS PARASITES OF CATERPILLARS.

APANTELES FULVIPES Hal.

Received first in 1905, dead, and in 1908, living. Colonized unsatisfactorily in 1908 and under exceptionally favorable conditions in 1909. Two generations recovered in immediate vicinity of colony site in 1909. Not recovered in 1910 except from recent colony. Establishment doubtful on account of lack of proper alternate hosts.

TACHINID PARASITES.

COMPSILURA CONCINNATA Meig.

First received in 1906 and colonized same year. Colony strengthened in 1907. Recovered doubtfully in 1907 from immediate vicinity of a colony site. Certainly recovered and found to be generally distributed over considerable territory in 1909. Marked increase in 1910. Apparently established.

CARCELIA GNAVA Meig.

Doubtfully colonized in 1906. Satisfactorily colonized in 1909. Not recovered from field. Establishment hoped for.

ZYGOBOTHRIA GILVA Hartig.

Doubtfully colonized in 1906. Satisfactorily colonized in 1909. Not recovered from field. Establishment hoped for.

TACHINA LARVARUM L.

First received in 1905 and colonized in 1906. Much more satisfactorily in 1909. Not recovered. Establishment doubtful on account of hybridization with similar American species.

TACHINA JAPONICA Towns.

First received and poorly colonized in 1908. A better colony put out in 1910. Recovery doubtful on same account as above.

TRICHOLYGA GRANDIS Zett.

Doubtfully received and colonized in 1906. Satisfactorily colonized in 1909. Recovered from immediate vicinity of colony site in 1909. Not recovered in 1910, but establishment hoped for.

PARASETIGENA SEGREGATA Rond.

First received in 1907 and colonized in 1910. Not recovered. Establishment hoped for and expected.

BLEPHARIPA SCUTELLATA R. D.

First received in 1905. Colonized under very unsatisfactory conditions in 1907. Satisfactory colonization for first time in 1909. Recovered from immediate vicinity of colony site in 1910. Establishment confidently expected.

CROSSOCOSMIA spp.

First received in 1908 and colonized in 1910 under fairly satisfactory conditions. Not recovered. Establishment rather doubtful on account of unsatisfactory colony.

PARASITES OF THE PUPA.

MONODONTOMERUS ÆREUS Walk.

First received in 1906. Colonized in 1906. Recovered, generally distributed over considerable area, in winter of 1908-9. Firmly established and dispersing at a very rapid rate.

PIMPLA spp. (See Parasites of the brown-tail moth.)

CHALCIS OBSCURATA Walk.

First received in 1908. Colonized in 1908 and 1909, but not satisfactorily. Establishment doubtful on account of small size of colony.

CHALCIS FLAVIPES Panz.

First received in 1905. Colonized in 1908 and 1909 but in unsatisfactory numbers. Recovered from immediate vicinity of colony site in 1909. Not recovered in 1910. Establishment doubtful on account of small colony.

PREDACEOUS BEETLES.

CAOSOMA SYCOPHANTA L.

First received in 1906. Colonized same year. Recovered from immediate vicinity of colony site in 1907. Found generally distributed over limited area in 1909. Firmly established and increasing and dispersing rapidly.

PARASITES OF THE BROWN-TAIL MOTH.

PARASITES OF THE EGG.

TRICHOGRAMMA spp.

First received in 1906. Colonized in 1907; more satisfactorily in 1909. Recovered from immediate vicinity of colony site in 1909. Not recovered in 1910. Establishment probable, but the species of no importance as a parasite.

TELENOMUS PHALÆNARUM Nees.

First received in 1906 and colonized satisfactorily in 1907. No attempts toward recovery since made. Establishment hoped for. Not an important parasite.

PARASITES ATTACKING HIBERNATING CATERPILLARS.

PTEROMALUS EGREGIUS Först.

Received first in 1906. Colonized in large numbers that year and in 1907 but much more satisfactorily in 1908. Recovered from immediate vicinity of colony site in 1909. Not recovered in 1910. Found to be generally distributed over very extended territory in 1911.[1] Apparently well established.

APANTELES LACTEICOLOR Vier.

Received first in 1906. Colonized in 1907. Satisfactorily colonized in 1908. Recovered in immediate vicinity of colony site in 1909. Generally distributed over considerable area in 1910. Apparently firmly established.

METEORUS VERSICOLOR Wesm.

First received in 1906. Colonized satisfactorily in 1908. Recovered in immediate vicinity of colony site in 1909. Generally distributed over limited area in 1910. Apparently firmly established.

ZYGOBOTHRIA NIDICOLA Towns.

First received in 1906. Colonized unsatisfactorily only in 1906-7, but notwithstanding was recovered in 1910 over a considerable territory. Apparently firmly established.

TACHINID PARASITES OF LARGER CATERPILLARS.

COMPSILURA CONCINNATA Meig. (See Gipsy-moth Parasites.)
TACHINA LARVARUM L. (See Gipsy-moth Parasites.)
DEXODES NIGRIPES Fall.

First received in 1906 and colonized satisfactorily in 1906 and 1907. Still more satisfactorily colonized in 1909. Not recovered. Establishment hoped for.

EUDOROMYIA MAGNICORNIS Zett.

First received in 1906. Colonized in about the same numbers as *Zygobothria nidicola* in 1906 and 1907. Not recovered. Establishment doubtful on account of small size of colonies.

---

[1] Winter of 1910-11.

PALES PAVIDA Meig.
   Status same as that of Eudoromyia.
PAREXORISTA CHELONIÆ Rond.
   Received first in 1906. Colonized very unsatisfactorily in 1907 and satisfactorily in 1908. Recovered in immediate vicinity of colony site in 1908 and in larger numbers in 1909. Not recovered in 1910. Establishment very doubtful on account of hybridization with American race.

PARASITES OF THE PUPÆ.

PIMPLA EXAMINATOR Fab.
PIMPLA INSTIGATOR Fab.
   First received in 1906 and colonized in 1906 and 1907 unsatisfactorily. Establishment doubtful. Of better promise as parasites on account of great similarity to American species.
MONODONTOMERUS ÆREUS Walk. (See Gipsy-moth Parasites.)

The gross number of each of the various species which have been colonized since the beginning of the work up to and including the season of 1910 is given in the accompanying tabulated statement:

Hymenopterous Parasites.
Schedius kuvanæ How	1,061,111
[3] *Pteromalus egregius* Först	354,300
Anastatus bifasciatus Fonsc.	177,210
Trichogramma spp	76,000
Apanteles fulvipes Hal	57,700
Apanteles lacteicolor Vier.	44,310
Monodontomerus æreus Walk	15,325
Telenomus phalænarum Nees	4,650
Meteorus versicolor Wesm.	3,113
Pimpla spp	583
Chalcis spp	338
	1,794,640

Tachinid Parasites.
Carcelia gnava Meig	15,581
[2] *Tricholyga grandis* Zett.	8,721
[2] *Zygobothria gilva* Hartig	7,502
[2] *Compsilura concinnata* Meig	6,777
[2] *Dexodes nigripes* Fall	5,040
[2] *Blepharipa scutellata* R. D.	5,109
Parexorista cheloniæ Rond.	5,026
[2] *Tachina larvarum* L	2,036

Tachinid Parasites—Con.
Parasetigena segregata Rond	1,187
[4] *Crossocosmia sericariæ* Corn	699
[2] *Pales pavida* Meig	476
Tachina japonica Towns.	471
[2] *Zenillia libatrix* Panz	161
[2] *Zygobothria nidicola* Towns	109
Masicera silvatica Fa'l	23
[2] *Eudoromyia magnicornis* Zett	5
[1] Unclassified tachinids (1906–07)	9,420
	68,343

Predatory beetles.
Calosoma sycophanta L	17,742
Carabus auratus L	478
Calosoma inquisitor L	262
Carabus arvensis Hbst	108
Carabus nemoralis Müll	100
Calosoma reticulatum Fab.	83
Carabus violaceus L	62
	18,835
Total	1,881,818

[1] Including species marked ([2]).
[2] Species which are also included under "unclassified tachinids."
[3] Does not include progeny of 114,000 individuals liberated in 1908.
[4] Including also *C. flavoscutellata*.

THE DEVELOPMENTS OF THE YEAR 1910.

At the beginning of the year 1910 the statement was made that if the parasites maintained the rate of progress which was then indicated by the results of the recent field work, the year 1916 would see the triumphant conclusion of the experiment and the automatic control of the gipsy moth through parasitism. This prophecy was also dependent upon the measurable success of the importation work which was planned for 1910.

The importations of 1910 were disappointing, and did not result in the colonization of the few parasites which have not yet been liberated in America under satisfactory conditions. Neither has the progress of the parasites in the field been quite as satisfactory as was hoped and expected.

The failure of Schedius to demonstrate as clearly as might be wished its ability to survive the winter was the first unfavorable development in 1910. Recovery of *Apanteles fulvipes*, while not expected, was hoped for, and although its nonrecovery can not be considered as surely indicative of its inability to establish itself here, it is none the less disquieting. Discovery of the error in identity which had resulted in misapprehensions concerning the status of *Tricholyga grandis* was a serious blow to expectations concerning the future of this species. Most serious of all was the nonrecovery of the important brown-tail moth parasite, *Parexorista cheloniæ*, which was considered to be thoroughly well established at the close of the season of 1909. Similarly, the failure of Monodotomerus to increase in efficiency to the extent which was expected, was viewed with apprehension, as possibly indicative of what might result with others of the imported species.

To offset these several and various reverses was the unexpectedly satisfactory increase in abundance and dispersion of Calosoma. Anastatus did better than was expected in the matter of increase in numbers and in effectiveness and slightly better in dispersion. Blepharipa was recovered, when recovery was not expected so soon following its liberation, and Compsilura was considerably more abundant, and promised more efficient assistance than had been hoped for. Among the brown-tail moth parasites, Apanteles gave evidences of a more rapid increase and wider dispersion than was expected, and Meteorus was also unexpectedly abundant over a limited area, and later showed evidence of rapid dispersion. The recovery of *Zygobothria nidicola*, after its disappearance for two or three years, was the most satisfactory and unexpected of the favorable results of the season's field work until the recovery of Pteromalus in the fall and during the winter. Although this latter is not an important parasite, its nonestablishment was practically conceded, and the circumstances surrounding its recovery are considered to be highly gratifying and significant.

It is by no means easy to draw a balance which should fairly represent the status of the work as a whole in 1910 as compared with 1909, but after long consideration it was definitely decided that the present status of the parasites was perhaps less favorable to ultimate success than was the apparent status of the work one year before. Recognition of this fact had much to do with the formulation of the policy for the continuation of the work in 1911. It is hoped that by putting forth an especial effort the small amount of lost ground may be regained, and that by 1912 it will be possible to state with assurance that the progress hoped for at the close of 1909 has been more than equaled, and that the chances are still favorable to the successful outcome of the work and to the establishment of an efficient and automatic control of the gipsy moth by the year 1916.

It should be understood that the manuscript of this bulletin was completed in the first week in January, 1911, and that no more recent developments of the situation have been considered in it, except for an incidental mention of the progress of *Apanteles lacteicolor*.

○

INDEX TO BULLETIN NO. 91, BUREAU OF ENTOMOLOGY.

	.ge.
Abraxas grossulariata, host of *Blepharidea vulgaris*..........................	91
Compsilura concinnata........................	89
Abrostola asclepiadis, host of *Zenillia libatrix*...............................	90
tripartita, host of *Carcelia excisa*...................................	89
triplasia, host of *Carcelia excisa*.....................................	89
Acer (*see also* Maple).	
food plant of gipsy moth..	124
Achætoneura fernaldi of Forbush and Fernald, status.........................	140
frenchii, parasite of gipsy moth in America......................	140
Acherontia atropos, host of *Argyrophylax atropivora*...........................	89
Blepharipa scutellata.............................	88
Acronycta aceris, host of *Compsilura concinnata*.............................	89
alni, host of *Compsilura concinnata*...............................	89
Exorista affinis......................................	89
cuspis, host of *Compsilura concinnata*............................	89
megacephala, host of *Compsilura concinnata*.......................	89
rumicis, host of *Compsilura concinnata*...........................	89
Tachina larvarum..................................	90
Zenillia fauna.......................................	92
tridens, host of *Compsilura concinnata*............................	89
Pales pavida..	92
Adopæa lineola, host of *Blepharidea vulgaris*................................	91
Ægerita webberi, fungus affecting *Aleyrodes citri*.............................	46
Agrotis candelarum, host of *Dexodes nigripes*.............................	88, 91
glareosa, host of *Echinomyia fera*..................................	89
præcox, host of *Tachina larvarum*.................................	90
sp., host of *Eudoromyia magnicornis*................................	89, 92
stigmatica, host of *Pales pavida*...................................	92
xanthographa, host of *Pales pavida*................................	92
Alabama argillacea, control by caging and permitting parasites to escape.........	19
Aleyrodes citri, quest for parasites abroad.....................................	45–46
Alfalfa, food plant of *Phytonomus murinus*....................................	46
weevil. (See *Phytonomus murinus*.)	
Amblyteles varipes, parasite of gipsy moth, recorded in literature..............	85
Ammoconia cæcimacula, host of *Parexorista cheloniæ*........................	92
Anagrus ovivorus, parasite of brown-tail moth, recorded in literature..........	87
spp., parasite of *Perkinsiella saccharicida*, introduction into Hawaiian Islands..	35
Anastatus bifasciatus, dispersion from colony center.......................	173–174
gross number colonized.................................	310
host of *Pachyneuron gifuensis*..........................	183
Tyndarichus navæ............................	183

Page.

Anastatus bifasciatus, parasite of gipsy moth in Europe, position in "sequence".. 132
       Japan, position in "sequence". 121
       introduction into United States, habits............ 75, 153, 168–176, 183
       status in United States in 1910.... 307
   superparasitized by *Schedius kuvanæ*................. 181–183
  slow dispersion as justifying the small colony.................... 94
  sp., parasite of *Apanteles fulvipes*................................. 200
Anatis 15-punctata, enemy of gipsy moth..................................... 252
Anilastus tricoloripes. (See *Limnerium tricoloripes*.)
Anisocyrta sp., parasite of gipsy moth in America......................... 138
Anomalon exile, parasite of brown-tail moth in America................ 144, 147–149
   tent caterpillar (Malacosoma)..................... 144
Antheræa mylitta, host of *Crossocosmia sericariæ*.......................... 88
  yamamai (*see also* Silkworm, Japanese).
   host of *Crossocosmia sericariæ*............................ 88
Anthonomus grandis. (*See* Weevil, cotton boll.)
  of apple, control by caging and permitting parasites to escape.... 19
Anthrenus varius, enemy of gipsy moth....................................... 252
   tussock moth (*Hemerocampa leucostigma*).......... 252
Apanteles conspersæ, comparison with *Apanteles lacteicolor*, biological differences... 285–286
  parasite of Japanese brown-tail moth (*Euproctis conspersa*)................................... 285–286
  difficilis, parasite of brown-tail moth, recorded in literature........ 86
  fiskei, host of *Mesochorus* n. sp..................................... 265
   parasite of *Parorgyia* sp................................... 265
  fulvipes, gross number colonized................................ 310
   how many individuals constitute a good colony?......... 96
   parasite of gipsy moth, importation and handling of cocoons.................... 165–166
       in Europe..................... 57
        position in "sequence" 132
        Japan, position in "sequence". 121
        Russia............... 80, 124–125
       introduction into United States, habits................. 193–202
       reared at laboratory............ 85
       recorded in literature.......... 85
       results of rearing work of 1910... 142
       status in United States in 1910.. 308
   perhaps synonymous with *Apanteles nemorum*............. 193
   possibility that it has been successfully colonized in United States.. 277–278
   prey of Corymbites...................................... 252
   secondary parasites..................................... 198–202
  liparidis (see also *Apanteles liparidis*).
    parasite of gipsy moth, recorded in literature..... 85
  glomeratus, parasite of common cabbage caterpillar (*Pontia rapæ*), first observation of exit of larvæ........... 16
    gipsy moth, recorded in literature........... 85
    Pontia rapæ, introduction into United States. 24
  hyphantriæ, host of *Mesochorus* sp.......................... 265

	Page.
Apanteles inclusus, parasite of brown-tail moth, recorded in literature	86
lacteicolor, comparison with *Apanteles conspersæ*, biological differences	285–286
gross number colonized	310
host of *Mesochorus pallipes*	263, 265, 267, 289
Monodontomerus æreus	249, 266, 267
Pteromalus egregius	266, 267, 275, 283–284
parasite of brown-tail moth, hibernating with host	262–267
in Europe, position in "sequence"	136
introduction into United States, habits	278–285
reared at laboratory	86
status in United States in 1910	309
Datana	284
Hyphantria	284
liparidis (see also *Apanteles fulvipes liparidis*).	
parasite of brown-tail moth, recorded in literature	86
nemorum, parasite of *Lasiocampa pini*	193
perhaps=*Apanteles fulvipes*	193
parasite of brown-tail moth in Europe, introduction into United States	70
gipsy moth in Europe, intruduction into United States	69
Japan, introduction into United States.	47, 73, 74
Russia	124–125
rufipes, parasite of gipsy moth	81
solitarius, parasite of brown-tail moth, recorded in literature	86
gipsy moth in Europe, position in "sequence"	132
Russia	79–82
introduction into United States, habits	189–190
reared at laboratory	85
recorded in literature	85
melanoscelus, parasite of gipsy moth, recorded in literature	85
? *ocneriæ*, parasite of gipsy moth, recorded in literature	85
sp. (*delicatus?*), host of elachertine	139
parasite of gipsy moth in America	138–139
white-marked tussock moth (*Hemerocampa leucostigma*)	138
tenebrosus, parasite of gipsy moth, recorded in literature	85
ultor, parasite of brown-tail moth, recorded in literature	86
vitripennis, parasite of brown-tail moth, recorded in literature.	86, 295–296
Apechthis brassicariæ. (See *Pimpla brassicariæ*.)	
Aphelinus mali, parasite of *Lepidosaphes ulmi*, transportation from one part of country to another part	20
Aphiochæta scalaris, reared from dead gipsy moth (*Porthetria dispar*)	90
setacea, reared from dead gipsy moth (*Porthetria dispar*)	90
Aphis, hop. (*See* Hop aphis.)	
spring grain. (See *Toxoptera graminum*.)	
Apopestes spectrum, host of *Masicera sylvatica*	92
Aporia cratægi, host of *Blepharidea vulgaris*	91
on apple and pear	133

	Page.
Apple, food plant of Anthonomus	19
Aporia cratægi	133
brown-tail moth	133
tent caterpillar (Malacosoma)	98, 104
wild haw, original food plant of *Aspidiotus perniciosus*	37
Araschinia levana, host of *Blepharidea vulgaris*	91
Compsilura concinnata	89
prorsa, host of *Blepharidea vulgaris*	91
Compsilura concinnata	89
Arbusier, food plant of brown-tail moth	134

Arbutus sp. (See Arbusier.)

Archips rosaceana, host of *Trichogramma* sp	259–260
Arctia aulica, host of *Echinomyia fera*	89
Arctia caja, host of *Carcelia excisa*	89
Compsilura concinnata	89
Exorista affinis	89
Histochæta marmorata	89
Parexorista cheloniæ	92
Tachina larvarum	90
Tricholyga grandis	88, 92
hebe, host of *Blepharidea vulgaris*	91
Carcelia excisa	89
Parexorista cheloniæ	92
quenselii, host of *Histochæta marmorata*	89
villica, host of *Carcelia excisa*	89
Histochæta marmorata	89
Parexorista cheloniæ	92
Tachina larvarum	90
Argynnis lathonia, host of *Blepharidea vulgaris*	91
Argyrophylax atropivora, parasite of gipsy moth, recorded in literature	88
recorded hosts	89
Aritranis amœnus, parasite of gipsy moth, recorded in literature	85
Arrhenotokous hymenopterous parasites, chances of successful establishment in a new country	95
Arrhenotoky in *Schedius kuvanæ*	183
Trichogramma	257, 258
Ascometia caliginosa, host of *Dexodes nigripes*	88, 91
Asecodes, parasite of *Apanteles fulvipes*	199, 200
Asphondylia lupini, control by caging and permitting parasites to escape	19
Aspidiotus perniciosus, prey of *Chilocorus similis*, attempted control by introduction of enemy	36–38
Astomaspis fulvipes (Grav.)=*Astomaspis nanus* (Grav.)	85
parasite of gipsy moth, recorded in literature	85
nanus (Grav.), *Astomaspis fulvipes* (Grav.) a synonym	85
Attacus cynthia, host of *Compsilura concinnata*	89
Pales pavida	92
lunula, host of *Pales pavida*	92
Axle grease as protective barrier against gipsy moth	124

Bagworm. (See *Thyridopteryx ephemeræformis*.)

Banana, food plant of *Omiodes blackburni*	35
Barkbeetles, prey of *Thanasimus formicarius*	36
Beetles, predaceous, enemies of gipsy moth	251–255
Biological characters for separating species indistinguishable morphologically	225–226, 257, 285–286

Page.
Birds and other predators in control of insects........................... 107–108
    enemies of cankerworms... 113
        gipsy moth.. 113
Blepharidea vulgaris, parasite of brown-tail moth in Europe, position in "sequence"................... 136
                introduction into United States..................... 304
                reared at laboratory.......... 91
    recorded hosts... 91
Blepharipa scutellata, difficulty in hibernating puparia..................... 158–159
    gross number colonized.................................. 310
    parasite of gipsy moth in Europe, position in "sequence". 132
                Russia.......................... 127
                introduction into United States, habits........................ 213–218
                reared at laboratory.............. 88
                results of rearing work in 1910... 141, 142
                status in United States in 1910.... 308
    recorded hosts... 88
Boarmia lariciaria, host of *Blepharidea vulgaris*.............................. 91
Boll weevil. (*See* Weevil, cotton boll.)
Bombus pennsylvanicus, introduction into Philippine Islands................. 45
Breeding cage. (*See* Cage.)
Brephos nothum, host of *Zenillia libatrix*...................................... 90
Brotolomia meticulosa, host of *Blepharidea vulgaris*........................... 91
Brown-tail moth and gipsy moth parasites. (*See* Parasites of gipsy and brown-tail moths.)
    caterpillars, full-fed and pupating, importation and handling....................................... 162–164
        immature, importation and handling.......... 161–162
    dipterous parasites.. 91–93
    egg masses, importation and handling.................... 160–161
    foreign tachinid parasites................................. 91
    fungous disease....................................... 135, 270, 291
    hibernation nests, importation and handling............... 161
    hymenopterous parasites.................................. 86–87
    in Europe.. 132–135
    native tachinids reared therefrom.......................... 93
    natural control by disease................................. 97–102
    parasites hibernating in webs............................. 261–295
            possible interrelations......... 267
    in Europe..................................... 132–135
    "sequence"........................... 135, 136
    of eggs, introduction, into United States, habits.. 256–261
    hibernating caterpillars, introduction into United States, habits................. 268–295
    larger caterpillars, introduction into United States, habits........................... 295–304
    pupæ, introduction into United States, habits.. 304–305
    parasitism in America..................................... 143–151
    pupæ, importation and handling.......................... 164–165
    rash at gipsy-moth parasite laboratory and elsewhere....... 65–67, 162–164, 279
    recorded hosts of foreign tachinids reared therefrom at laboratory... 91–92

	Page.
Brown-tail moth, recorded hosts of foreign tachinids recorded as parasitic thereon	92
tachinid parasites	296–304
Bumblebee. (*See* Bombus.)	
Bupalus piniarius, host of *Carcelia excisa*	89
Dexodes nigripes	88, 91
Cabbage worm, imported. (See *Pieris rapæ* and *Pontia rapæ*.)	
Cage for rearing parasites from brown-tail moth webs	269
tachinid parasites of the brown-tail moth	150–151
Cages for colonization of *Anastatus bifasciatus*	174
tachinid flies	204–207
Calliephialtes messor, parasite of codling moth, introduction into California	38–39
Callimorpha dominula, host of *Carcelia excisa*	89
Callineda testudinaria, enemy of *Perkinsiella saccharicida*, introduction into Hawaiian Islands	35
Calosoma and other predaceous beetles, importation and handling	167
disease of gipsy moth as obstacle to establishment	99
enemy of gipsy moth in Russia	80–81
inquisitor, enemy of gipsy moth, early ideas on introduction	48
introduction into United States	59, 62, 254
gross number colonized	310
native species, status as enemies of gipsy moth	48
reticulatum, gross number colonized	310
slow dispersion as justifying small colony	94
sycophanta, enemy of gipsy moth, early ideas on introduction	48
first practical handling	18
in Europe, position in "sequence" of parasites	132
introduction into United States, habits	59, 62, 253–255
status in United States in 1910	309
gross number colonized	310
respect in which similar to *Blepharipa scutellata* in relation to gipsy moth	218
Calymnia trapezina, host of *Blepharidea vulgaris*	91
Camel (*see also* Dromedary).	
host of *Rhipicephalus sanguineus*	42
Campoplex conicus, parasite of brown-tail moth, recorded in literature	86
gipsy moth, recorded in literature	85
difformis. (See *Omorgius difformis*.)	
Cankerworms, prey of birds	113
Carabus arvensis, gross number colonized	310
auratus, enemy of earwigs, practical handling	18
gross number colonized	310
nemoralis, gross number colonized	310
violaceus, gross number colonized	310
Carcelia excisa, parasite of gipsy moth, recorded in literature	88
recorded hosts	89
gnava, gross number colonized	310
parasite of gipsy moth in Europe, position in "sequence"	132
introduction into United States	231–232
reared at laboratory	88
status in United States in 1910	308
recorded hosts	88

INDEX. 319

	Page.
Casinaria tenuiventris, parasite of gipsy moth, recorded in literature	85
Catocala fraxini, host of *Tachina larvarum*	89
promissa, host of *Compsilura concinnata*	89
Ceratitis capitata, quest for natural enemies	39
Ceroplastes cirripediformis, host of *Scutellista cyanea*	31
floridensis, host of *Scutellista cyanea*	31
rubens, introduction of parasites and enemies into Hawaii	35
rusci, host of *Scutellista cyanea*	31
Chalcid flies, parasites of gipsy moth in Russia	81
Chalcidid, small, parasite of *Compsilura concinnata*	224
Chalcis callipus, parasite of gipsy moth in Japan	240
recorded in literature	86
compsiluræ, parasite issuing from brown-tail moth cocoons	145
of *Compsilura concinnata*	240
fiskei, an undesirable foreign hyperparasite	202
parasite of *Crossocosmia sericariæ* and *Tachina japonica*	240
flavipes, parasite of gipsy moth in Europe, position in "sequence"	132
introduction into United States, habits	240–245
reared at laboratory	86
status in United States in 1910	309
white-marked tussock moth (*Hemerocampa leucostigma*)	241–244
fonscolombei, parasite of sarcophagids associated with gipsy moth	241
minuta, parasite of sarcophagids associated with gipsy moth	241
obscurata, parasite of gipsy moth in Japan, position in "sequence"	121
introduction into United States, habits	240–245
reared at laboratory	86
status in United States in 1910	308
Omiodes blackburni, introduction into Hawaiian Islands	35
white-marked tussock moth (*Hemerocampa leucostigma*)	241–244
ovata, not yet reared as parasite of gipsy moth	240
paraplesia, parasite of sarcophagids associated with gipsy moth	241
parasite of *Compsilura concinnata*	224
parasites of brown-tail moth in America, results of rearing work in 1910	147–149
scirropoda, parasite of brown-tail moth, recorded in literature	87
spp., gross number colonized	310
Cherry, food plant of tent caterpillar (*Malacosoma*)	98, 104
wild, food plant of tent caterpillar (*Malacosoma*)	105
Chilocorus bivulnerus, enemy of *Diaspis pentagona*, introduction into Italy	44–45
similis (*see also* Ladybird, Chinese).	
enemy of *Aspidiotus perniciosus*, introduction into United States	37–38
Cimbex femorata, host of *Parexorista cheloniæ*	92
humeralis, host of *Compsilura concinnata*	89
Clerus formicarius. (See *Thanasimus formicarius*.)	
Clisiocampa (*see also* Malacosoma and Tent caterpillar).	
host of *Limnerium clisiocampæ*	192
Clover, red, fertilization by *Bombus pennsylvanicus*	45
Coccinella californica, transportation from one part of country to another part in control of plant-lice (Aphididæ)	22–23

	Page.
Coccinella repanda, enemy of plant-lice (Aphididæ) on sugar cane and other crops, introduction into Hawaiian Islands...................................	35
undecimpunctata, enemy of plant-lice (Aphididæ), introduction into New Zealand..	24
Coccinellid enemy of mealy bugs (Pseudococcus), introduction into United States...	46
Cochylis of grapevine, control by caging and permitting parasites to escape...	19
Codling moth (*Carpocapsa pomonella*), host of *Calliephialtes messor*, attempted control by introduction of parasites..	38–39
Coffee, food plant of *Pulvinaria psidii*...	34
Compsilura concinnata, characters of larva..	265
gross number colonized..............................	310
host of *Chalcis compsiluræ*...........................	240
Monodontomerus æreus........................	249
parasite of brown-tail moth, hibernating in webs of latter....................	263
in Europe, position in "sequence"...............	136
introduction into United States...................	296
reared at laboratory........	91
recorded in literature.....	91
results of rearing work in 1910..................	147–149
status in United States in 1910..................	308, 309
fall webworm (Hyphantria)................	224
gipsy moth in Europe, position in "sequence".....................	132
introduction into United States, habits................	218–225
reared at laboratory............	88
recorded in literature...........	88
relative abundance in Massachusetts and Russia..........	127
results of rearing work of 1910	141, 142
status in United States in 1910..	308
Pontia rapæ...............................	223
tussock moth (*Hemerocampa leucostigma*).	221–223
recorded hosts......................................	89
similarity to *Dexodes nigripes*......................	296–297
"Compsilura-like" parasite of gipsy moth..	235
Conotrachelus nenuphar. (*See* Curculio, plum.)	
Contarinia tritici, attempt at control by introduction of parasites from Europe.	23–24
burning débris in control may result in destroying beneficial parasites...	20
Corymbites, enemy of *Apanteles fulvipes*..	252
gipsy moth..	252
Cosmotriche potatoria, host of *Blepharidea vulgaris*...........................	91
Tachina larvarum...............................	90
noctuarum.............................	90
Cossus cossus, host of *Zenillia fauna*...	92
Cotton boll weevil. (*See* Weevil, cotton boll.)	

INDEX.

	Page.
Cotton caterpillar. (See *Alabama argillacea*.)	
food plant of *Hemichionaspis minor*	45
wool as protective barrier against gipsy moth	124
Craniophora ligustri, host of *Compsilura concinnata*	89
Cratægus, food plant of brown-tail moth	134
gipsy moth	82
scrubs containing nests of brown-tail moth	133
Crossocosmia flavoscutellata (?), parasite of gipsy moth, introduction into United States	234–235
in Europe, position in "sequence"	132
sericariæ and *Crossocosmia flavoscutellata*, gross number colonized	310
habits as compared with *Blepharipa scutellata*	214
host of *Chalcis fiskei*	240
parasite of gipsy moth and of Japanese silkworm, introduction into United States, habits	232–234
in Japan, position in "sequence"	121
reared at laboratory	88
recorded hosts	88
spp., parasites of gipsy moth, status in United States in 1910	308
Cryptognatha flavescens, enemy of *Aleyrodes citri*, attempted introduction into United States	46
Cryptolæmus montrouzieri, enemy of mealy bugs (*Pseudococcus*) and *Pulvinaria psidii*, introduction into United States and Hawaiian Islands	34–35
Pseudococcus, attempted introduction into Spain	35
Cryptus amœnus. (See *Aritranis amœnus*.)	
atripes. (See *Idiolispa atripes*.)	
cyanator, parasite of gipsy moth, recorded in literature	85
moschator, parasite of brown-tail moth, recorded in literature	86
Cucullia anthemidis, host of *Blepharidea vulgaris*	91
artemisiæ, host of *Ernestia consobrina*	89
asteris, host of *Blepharidea vulgaris*	91
Dexodes nigripes	88, 91
lactucæ, host of *Compsilura concinnata*	89
prenanthis, host of *Tachina larvarum*	90
scrophulariæ, host of *Carcelia excisa*	89
verbasci, host of *Blepharidea vulgaris*	91
Compsilura concinnata	89
Histochæta marmorata	89
Masicera sylvatica	92
Curculio, plum (*Conotrachelus nenuphar*), host of *Sigalphus curculionis*, artificial control by transportation of its parasite suggested	20
Cyclotophrys anser, parasite of brown-tail moth, habits	304
in Europe, position in "sequence"	136
reared at laboratory	91
Dacus, introduction of parasites into western Australia	39
oleæ (*see also* Olive fly).	
control by caging and permitting parasites to escape	19
Dasychira fascellina, host of *Tachina larvarum*	90

7362°—11——2

	Page.
Dasychira pudibunda, host of *Carcelia excisa*	89
Compsilura concinnata	89
Zenillia libatrix	90
Datana, host of *Apanteles lacteicolor*	284
Deilephila euphorbiæ, host of *Dexodes nigripes*	88, 91
Masicera sylvatica	92
Tachina larvarum	90
gallii, host of *Masicera sylvatica*	92
Tachina larvarum	90
vespertilio, host of *Masicera sylvatica*	92
Dendroctonus frontalis, attempted control by introduction of *Thanasimus formicarius* from Germany	36
Dendrolimus pini, host of *Blepharidea vulgaris*	91
Phryxe erythrostoma	89
Tachina larvarum	90
Dermestid beetles, enemies of gipsy moth	251–255
Dexodes nigripes, gross number colonized	310
parasite of brown-tail moth in Europe, position in "sequence"	136
introduction into United States, habits	296–297
reared at laboratory	91
status in United States in 1910	309
gipsy moth in Europe, position in "sequence"	132
introduction into United States	220
rarely associated with that host	235
reared at laboratory	88
recorded hosts	88, 91
Diaspis pentagona, host of *Prospaltella berlesei*, control by introduction of parasite into Italy	44
prey of *Chilocorus bivulnerus* and *Microweisea misella*, control by introduction of enemies into Italy	44–45
work with parasites	38
Dibrachys boucheanus, parasite of introduced Tachinidæ	213
secondary parasite of brown-tail moth, recorded in literature	87
gipsy moth, recorded in literature	86
host relations	200
parasite of *Apanteles fulvipes*	199, 200
Compsilura concinnata	224
Diglochis omnivora, parasite of brown-tail moth in America	144, 147–149
Europe	305
position in "sequence"	136
reared at laboratory	87
gipsy moth in America	138
Dilina tiliæ, host of *Compsilura concinnata*	89
Masicera sylvatica	92
Dilobia cæruleocephala, host of *Compsilura concinnata*	89
Dimmockia, parasite of *Apanteles fulvipes*	199
Dipterygia scabriuscula, host of *Compsilura concinnata*	89
Disease as factor in control of gipsy moth and brown-tail moth	97–102
in Russia	125
insects	108, 114

INDEX. 323

	Page.
Disease as factor in control of "pine tussock moth"	101, 103–104
tent caterpillar (Malacosoma)	98, 101, 105
white-marked tussock moth (*Hemerocampa leucostigma*)	100–101
Dog, host of *Rhipicephalus sanguineus*	42
texanus	41

Dromedary (*see also* Camel).

trypanosome disease transmitted by tabanid flies	45
Drymonia chaonia, host of *Compsilura concinnata*	89
Earwigs (Forficulidæ), prey of *Carabus auratus*, artificial control by means of its enemy	18
Staphylinus olens, artificial control by means of its enemy	18
Echinomyia fera, parasite of gipsy moth, recorded in literature	88
recorded hosts	89
præceps, parasite of brown-tail moth, recorded in literature	91
recorded hosts	92
Ecthrodelphax fairchildii, parasite of *Perkinsiella saccharicida*, introduction into Hawaiian Islands	35
Elachertine parasite of *Apanteles* sp. (*delicatus ?*)	139
Elasmus, parasite of *Apanteles fulvipes*	199

Elm (*see also* Ulmus).

food plant of gipsy moth	81
leaf-beetle (*Galerucella luteola*), host of *Tetrastichus xanthomelænæ*	62–63
work with egg-parasite, *Tetrastichus xanthomelænæ*	39–41
Emphytus cingillum, host of *Pales pavida*	92
Endromis versicolora, host of *Carcelia excisa*	89
Entedon albitarsis, parasite of *Pteromalus egregius*	263, 266, 267, 269
epigonus, parasite of *Mayetiola destructor*, introduction into United States	30
Ephyra linearia, host of *Blepharidea vulgaris*	91
Epicampocera crassiseta, parasite of gipsy moth, recorded in literature	88
recorded hosts	89
Epineuronia cespitis, host of *Blepharidea vulgaris*	91
Epiurus inquisitoriella. (See *Pimpla inquisitoriella*.)	
Erastria scitula, parasite of black scale (*Saissetia oleæ*), introduction into United States	34
Eriogaster catex, host of *Pales pavida*	92
Ernestia consobrina, parasite of gipsy moth, recorded in literature	88
recorded hosts	89
Erycia ferruginea, parasite of brown-tail moth, recorded in literature	91
recorded hosts	92
Euchloe cardamines, host of *Blepharidea vulgaris*	91
Eudoromyia magnicornis, gross number colonized	309
parasite of brown-tail moth in Europe, position in "sequence"	136
introduction into United States; habits	303–304
reared at laboratory	91
recorded in literature	88
status in United States in 1910	309
recorded hosts	89, 92

	Page.
Eulophid parasite of *Apanteles fulvipes*	200
Pteromalus egregius	202
Eupelmus bifasciatus, parasite of gipsy moth, reared at laboratory	86
recorded in literature	86
parasite of *Apanteles fulvipes*	200
Euphorocera claripennis, parasite of brown-tail moth in America	93, 145
Euphorus, parasite of ladybird adults	30
Euplexia lucipara, host of *Blepharidea vulgaris*	91
Euproctis chrysorrhœa. (See Brown-tail moth.)	
conspersa, host of *Apanteles conspersæ*	285–286
parasites in Japan	133
Eurrhypara urticæ, host of *Dexodes nigripes*	88, 91
Eurytoma abrotani Panzer=*Eurytoma appendigaster* Swed	86
parasite of gipsy moth, recorded in literature	86
appendigaster Swed., *Eurytoma abrotani* Panzer a synonym	86
Exorista affinis, parasite of gipsy moth, recorded in literature	88
recorded hosts	89
blanda, parasite of gipsy moth in America	90, 140, 142
boarmiæ, parasite of brown-tail moth in America	93, 145, 147–149
cheloniæ. (See *Parexorista cheloniæ*.)	
fernaldi, parasite of gipsy moth in America	90
pyste, parasite of gipsy moth in America	90
two undetermined species, parasites of gipsy moth in America	141
Facultative factors in control of insects	107
"Flacherie," so-called, of gipsy moth	97–102
Fruit fly. (See *Ceratitis capitata*.)	
parasites, Froggatt's journey for investigating their utility	42–44
Fungous disease of *Aleyrodes citri*	46
brown-tail moth	135, 270, 291
Funnel, integumental, of larva of *Pales pavida*	301–302
Tachina *et al*	140
tracheal, of larva of *Blepharipa scutellata*	214–216
Gastropacha quercifolia, host of *Masicera sylvatica*	92
Tachina larvarum	90
Gaurax anchora, reared from dead gipsy moth (*Porthetria dispar*)	90
Gipsy moth, additional control necessary to check increase in America	114–117
and brown-tail moth parasites. (See Parasites of gipsy and brown-tail moths.)	
caterpillars, first-stage, importation and handling	153
full-fed and pupating, importation and handling	156–159
second to fifth stages, importation and handling	154–156
conditions favoring rapid increase	112–113
dipterous parasites	87–90
egg masses, importation and handling	152–153
extent of control by parasitism abroad	117–131
hymenopterous parasites	85–86
mortality required to offset potential increase	112
native Diptera reared therefrom	90
natural control by disease	97–102
parasites in Europe, "sequence"	131
Japan, "sequence"	121
of eggs	168–188
larvæ	188–202
pupæ	236–255

INDEX. 325

	Page.
Gipsy moth, parasitism in America	136–143
Japan	120–123
Russia	123–129
Southern France	129–131
pupæ, importation and handling	159–160
rate of increase in New England	109–114
recorded hosts of foreign tachinids reared therefrom at laboratory	88
recorded as parasitic thereon	89–90
tachinid parasites	202–236
Goniarctena rufipes, host of *Histochæta marmorata*	89
Grapevine Cochylis. (*See* Cochylis.)	
Grasshoppers, hosts of sarcophagids	250
"Green bug." (See *Toxoptera graminum*.)	
Habrobracon brevicornis, hibernating in brown-tail moth nests	61, 269–270
Hadena adusta, host of *Eudoromyia magnicornis*	89, 92
secalis, host of *Parexorista cheloniæ*	92
Hæmaphysalis leporis-palustris, host of *Ixodiphagus texanus*	41
parasite of cotton-tail rabbit	41
Hæmatobia serrata, quest of parasites for introduction into Hawaiian Islands	36
Haplogonatopus vitiensis, parasite of *Perkinsiella saccharicida*, introduction into Hawaiian Islands	35
Hawthorn containing nests of brown-tail moth	133
Heliothis obsoleta, host of *Trichogramma pretiosa*	45
scutosa, host of *Dexodes nigripes*	88, 91
Hemaris fuciformis, host of *Echinomyia præceps*	92
Hemerocampa leucostigma. (*See* Tussock moth, white-marked.)	
Hemichionaspis minor on cotton, importation of *Prospaltella berlesei* for control	45
Hemiteles bicolorius, parasite of gipsy moth, recorded in literature	85
fulvipes. (See *Astomaspis fulvipes*.)	
socialis, parasite of brown-tail moth, recorded in literature	87
spp., parasites of *Apanteles fulvipes*	199, 200
utilis, parasite of *Limnerium* sp. (*fugitiva*?)	138
Hessian fly (*Mayetiola destructor*), burning stubble in control may result in destroying beneficial parasites	19
host of *Entedon epigonus*, attempted control by introduction of parasite	30
Polygnotus hiemalis, control by transportation of parasite	21
Heterocampa, control by starvation	103
studies in parasitism	103
Hippodamia convergens, enemy of plant-lice, transportation from one part of country to another part	22–23
Histochæta marmorata, parasite of gipsy moth, recorded in literature	88
recorded hosts	89
Homalotylus, parasite of ladybird larvæ	30
Hop aphis, prey of common English ladybird, artificial control by means of its enemy suggested	17
Horn fly. (See *Hæmatobia serrata*.)	
Horse chestnut, food plant of *Hemerocampa leucostigma*	101
host of *Rhipicephalus texanus*	42
Host relations of hyperparasites	201–202
tachinid parasites	203–204
Hunterellus hookeri parasite of *Rhipicephalus sanguineus*	41–42
texanus, introductions into Africa	41–42

	Page.
Hunterellus hookeri, question as to original home	42
Hybernia defoliaria, host of *Blepharidea vulgaris*	91
sp., host of *Dexodes nigripes*	88–91
Hybridization between *Tachina mella* and *Tachina larvarum*, possibility thereof.	227
Hydrocyanic-acid gas against fluted scale	24
Hyloicus pinastri, host of *Blepharidea vulgaris*	91
Carcelia excisa	89
Compsilura concinnata	89
Phryxe erythrostoma	89
Hylophila prasinana, host of *Blepharidea vulgaris*	91
Hymenopterous parasite cocoons, importation and handling	165–166
Hyperparasites, host relations	201–202
Hyphantria (*see also* Webworm, fall).	
host of *Apanteles lacteicolor*	284
Hypopteromalus, parasite of *Apanteles fulvipes*	199, 200
Icerya ægyptiaca, prey of *Novius cardinalis*	28–29
purchasi, introduction into Florida	28
prey of *Novius cardinalis*, in Cape Colony	28
Formosa	29
Hawaiian Islands	29
Italy	29
New Zealand	27
Portugal	27–28
Syria	29
United States	24–27
Ichneumon disparis, parasite of brown-tail moth, recorded in literature	86
gipsy moth in Europe	239
position in "sequence"	132
reared at laboratory	85
recorded in literature	85
fly, parasite of *Ceratitis capitata*, introduction into western Australia	39
pictus, parasite of gipsy moth, recorded in literature	85
scutellator, parasite of brown-tail moth, recorded in literature	86
Idiolispa atripes, parasite of brown-tail moth, recorded in literature	86
Insects, control by birds and other predators	107–108
disease	108, 114
parasitism	105–109
starvation	114
weather conditions	107
native, studies in parasitism	102–105
species differing in biological characters only	225–226, 257, 285–286
the three groups of factors in natural control	114
Itoplectis conquisitor. (See *Pimpla conquisitor*.)	
Ixodiphagus texanus, parasite of *Hæmaphysalis leporis-palustris*	41
Kincaid, Trevor, Russian observations on gipsy and brown-tail moths	78–82, 124–125
Ladybird, Asiatic. (*See* Ladybird, Chinese, and *Chilocorus similis*.)	
Australian. (See *Novius cardinalis*.)	
Chinese (see also *Chilocorus similis*).	
attacked by American ladybird parasites	30
common English, enemy of hop aphis, practical handling suggested	17
Ladybirds, parasites	30
Lampyrid beetles, enemies of gipsy moth	252–253
Larentia autumnalis, host of *Zenillia libatrix*	90

INDEX. 327

	Page.
"Larvipositor" of *Compsilura concinnata*	219
Lasiocampa pini, host of *Apanteles nemorum*	193
quercus, host of *Masicera sylvatica*	92
Tachina larvarum	90

Leaf crumpler, rascal. (See *Mineola indiginella*.)
   worm, cotton. (See *Alabama argillacea*.)

Lemon, food plant of *Icerya purchasi*	24
Lepidosaphes beckii, enemy of orange	28
gloveri, enemy of orange	28
ulmi, host of *Aphelinus mali*, control by transportation of its parasites	20
Lestophonus iceryæ, parasite of *Icerya purchasi*, introduction into United States	25
Lettuce, food of brown-tail moth larvæ	280, 281
Leucania albipuncta, host of *Blepharidea vulgaris*	91
lythargyria, host of *Blepharidea vulgaris*	91
obsoleta, host of *Echinomyia fera*	89
Libytha celtis, host of *Compsilura concinnata*	89
Lime-sulphur washes against San Jose scale, general use as preventing establishment of *Chilocorus bivulnerus*	38
Limnerium clisiocampæ, parasite of brown-tail moth in America	143
Clisiocampa (Malacosoma)	192
tent caterpillar (Malacosoma)	143
disparis, parasite of brown-tail moth in Europe	295
position in "sequence"	136
gipsy moth in Europe	295
position in "sequence"	132
Japan, position in "sequence"	121
introduction into United States, habits	191–192
reared at laboratory	85
parasite of gipsy moth in Russia	81
sp. (*fugitiva?*) host of *Hemiteles utilis*	138
parasite of gipsy moth in America	138
tricoloripes, parasite of gipsy moth in Europe	192
position in "sequence"	132
reared at laboratory	85

Liparis monacha. (See Nun moth and *Porthetria monacha*.)

Locust, food plant of gipsy moth	57
Lophyrus laricis, host of *Zygobothria gilva*	90
pallidus, host of *Zygobothria bimaculata*	90
gilva	90
pini, host of *Parasetigena segregata*	89
Tachina larvarum	90
Zygobothria bimaculata	90
gilva	90
rufus, host of *Zygobothria bimaculata*	90
gilva	90
socius, host of *Zygobothria bimaculata*	90
sp., host of *Dexodes nigripes*	88, 91
variegatus, host of *Zygobothria bimaculata*	90
gilva	90
virens, host of *Zygobothria bimaculata*	90

328 PARASITES OF GIPSY AND BROWN-TAIL MOTHS.

Page.

Lydella nigripes. (See *Dexodes nigripes.*)
    pinivoræ, parasite of gipsy moth, recorded in literature.............. 88, 89
Lymantria monacha. (See Nun moth and *Porthetria monacha.*)
Lysiphlebus tritici, parasite of *Toxoptera graminum*, experiments in transportation from one part of country to another part............................. 22
Macroglossa stellatarum, host of *Tachina larvarum*........................... 90
Macrothylacia rubi, host of *Compsilura concinnata*........................... 89
                    Parexorista cheloniæ............................. 92
                    Tachina larvarum............................. 90
Mænia typica, host of *Blepharidea vulgaris*................................ 91
Malacosoma (*see also* Tent caterpillar and *Clisiocampa*).
    americana, host of American *Parexorista cheloniæ*................... 299
    castrensis, host of *Carcelia excisa*................................. 89
              Tachina larvarum............................. 90
    disstria, host of American *Parexorista cheloniæ*................... 299–300
              Theronia fulvescens............................. 137
    neustria, host of *Carcelia excisa*.................................. 89
              gnava.. 88
              Tompsilura concinnata......................... 89
              Histochæta marmorata......................... 89
              Cachina larvarum............................. 90
              Zenillia libatrix............................. 90
Mamestra advena, host of *Blepharidea vulgaris*............................. 91
    brassicæ, host of *Compsilura concinnata*........................... 89
              Tachina larvarum............................. 90
    oleracea, host of *Compsilura concinnata*........................... 89
              Tricholyga grandis......................... 88, 92
    persicariæ, host of *Blepharidea vulgaris*........................... 91
              Compsilura concinnata......................... 89
    pisi, host of *Dexodes nigripes*..................................... 88, 91
              Echinomyia fera............................. 89
              Tricholyga grandis......................... 88, 92
    reticulata, host of *Blepharidea vulgaris*........................... 91
Maple (*see also* Acer).
    food plant of gipsy moth.. 80
              white-marked tussock moth (*Hemerocampa leucostigma*).... 101
Masicera sylvatica, gross number colonized................................. 310
    parasite of brown-tail moth in Europe, position in "sequence"...................... 136
                    introduction into United States. 303
                    reared at laboratory............ 91
            recorded hosts.. 92
Mayetiola destructor. (See Hessian fly.)
Mealy bugs (Pseudococcus), prey of coccinellids introduced from abroad...... 46
                    Cryptolæmus montrouzieri................. 34
Meigenia bisignata, parasite of gipsy moth, recorded in literature.............. 88, 89
Melitæa athalia, host of *Blepharidea vulgaris*................................ 91
                Erycia ferruginea............................. 92
    aurinia, host of *Erycia ferruginea*................................. 92
    didyma, host of *Tachina larvarum*.................................. 90
Melittobia acasta, parasite of introduced Tachinidæ and Sarcophagidæ....... 209–212
    parasite of tachinids, an undesirable introduction................. 202
Melopsilus porcellus, host of *Tachina larvarum*............................. 90
Mesochorus confusus, parasite of gipsy moth, recorded in literature............ 85
    dilutus, parasite of brown-tail moth, recorded in literature......... 87

INDEX. 329

	Page.
Mesochorus gracilis, parasite of gipsy moth, recorded in literature	85
n. sp., parasite of *Apanteles fiskei*	265
pallipes, parasite of *Apanteles lacteicolor*	263, 265–266, 267, 289
pectoralis, parasite of brown-tail moth, recorded in literature	87
gipsy moth, recorded in literature	85
semirufus, parasite of gipsy moth, recorded in literature	85
sp., parasite of *Apanteles hyphantriæ*	265
splendidulus, parasite of gipsy moth, recorded in literature	85
Meteorus ictericus, parasite of brown-tail moth, recorded in literature	86
japonicus, parasite of gipsy moth in Japan	121, 190–191
position in "sequence"	121
reared at laboratory	85
parasite of brown-tail moth in Europe, introduction into United States	70
gipsy moth in Russia, introduction into United States	69
pulchricornis, parasite of gipsy moth in Europe, position in "sequence"	132
introduction into United States	190
reared at laboratory	85
scutellator, parasite of gipsy moth, recorded in literature	85
versicolor, gross number colonized	310
host of *Pteromalus egregius*	266, 267
parasite of brown-tail moth in Europe, position in "sequence"	136
introduction into United States, habits	202, 264–267, 286–289, 295
reared at laboratory	86
recorded in literature	86
results of rearing work in 1910	147–149
status in United States in 1910	309
fall webworm (*Hyphantria*)	289
gipsy moth in Europe, position in "sequence"	132
introduction into United States	190
reared at laboratory	85
white-marked tussock moth (*Hemerocampa leucostigma*)	221–223, 289
Metopsilus porcellus, host of *Blepharidea vulgaris*	91
Miana literosa, host of *Dexodes nigripes*	88, 91
Microctonus, parasite of ladybird adults	30
Microgaster calceata, parasite of brown-tail moth, recorded in literature	86
gipsy moth, recorded in literature	85
consularis (Hal.)=*Microgaster connexa* Nees	86
connexa, *Microgaster consularis* a synonym	86
consularis, parasite of brown-tail moth, recorded in literature	86
fulvipes liparidis. (See *Apanteles fulvipes liparidis*.)	
tibialis, parasite of gipsy moth, recorded in literature	85
Microweisia misella, enemy of *Diaspis pentagona*, introduction into Italy	44–45
Mineola indiginella, control by permitting parasites to escape	18–19
Monedula carolina, enemy of tabanid flies, introduction into Algeria	45
Monodontomerus æreus, gross number colonized	310
hibernating in brown-tail nests	262, 266, 267
host relations	266

	Page.
Monodontomerus æreus, parasite of *Apanteles lacteicolor*	249, 267
brown-tail moth in Europe	304
position in "sequence"	136
introduction into United States, habits	245–250
reared at laboratory	87
recorded in literature	87
results of rearing work in 1910	147–149
status in United States in 1910	308, 310
Compsilura concinnata	224
gipsy moth in Europe	304
position in "sequence"	132
introduction into United States, habits	245–250
reared at laboratory	86
status in United States in 1910	308
introduced Tachinidæ	212–213
Pimpla	246, 249
tachinid and sarcophagid puparia	246
Theronia	246
white-marked tussock moth (*Hemerocampa leucostigma*)	249
Zygobothria nidicola	267
reared from brown-tail moth webs, host relations	269–270
successful colonization	276–277
unfortunately a secondary as well as a primary parasite	202
Walk.=*Torymus anephelus* Ratz	87
dentipes, parasite of brown-tail moth, recorded in literature	87
Mulberry, food plant of *Diaspis pentagona*	38
Mymarid parasite of weevil allied to *Phytonomus murinus*, introduction into United States to combat the latter	46
Natural control of insects, three groups of factors	114
Nematus ribesii, host of *Dexodes nigripes*	88, 91
Nonagria typhliæ, host of *Masicera sylvatica*	92
Notodonta trepida, host of *Argyrophylax atropivora*	89
Nun moth (see also *Porthetria monacha*).	
prey of *Calosoma sycophanta*	48
Novius cardinalis, enemy of *Icerya ægyptiaca*, introduction into Egypt	28–29
purchasi, introduction into Cape Colony	28
Formosa	29
Hawaiian Islands	29
Italy	29
New Zealand	27
Portugal	27–28
Syria	29
United States	24–27
reasons for its success	29–30

INDEX.

	Page.
Oak, food plant of gipsy moth	77, 80, 81, 124–125
Ocneria detrita, host of Tachina larvarum	90
Œonistis quadra, host of Compsilura concinnata	89
Echinomyia fera	89
Olethreutes hercyniana, host of Tachina larvarum	90

Olive fly (see also *Dacus oleæ*).

method of encouraging parasites	20
Omiodes blackburni, host of Chalcis obscurata, control by introduction of parasite	35
Omorgius difformis, parasite of brown-tail moth, recorded in literature	86
Ootetrastichus beatus, parasite of Perkinsiella saccharicida, introduction into Hawaiian Islands	35
Opiellus trimaculatus, parasite of Ceratitis capitata, attempted introduction into Western Australia	39
Orange, food plant of Aleyrodes citri	46
Icerya purchasi	24
Lepidosaphes beckii	28
gloveri	28
Orgyia antiqua, host of Carcelia excisa	89
gnava	88
ericæ, host of Pales pavida	92
Tachina larvarum	90
gonostigma, host of Tachina larvarum	90
Ortholitha cervinata, host of Dexodes nigripes	88, 91
Orthosia humilis, host of Tachina larvarum	90
pistacina, host of Parexorista cheloniæ	92
Pachyneuron gifuensis, parasite of Anastatus bifasciatus	183
Schedius kuvanæ	183
reared from gipsy-moth eggs	178
superparasitized by Pachyneuron gifuensis	183
Tyndarichus navæ	183
Pachytelia villosella, host of Exorista affinis	89
Packing and shipment of brown-tail moth egg masses	160–161
larvæ, full-fed and pupating	162–164
immature	162
pupæ	164
Calosoma and other predaceous beetles	167
gipsy-moth egg masses from Japan	152–153
larvæ, full-fed and pupating	156–159
second to fifth stages from Europe	154
pupæ	159–160
hymenopterous parasite cocoons	165–166
tachinid puparia	166–167
Pales pavida, gross number colonized	310
parasite of brown-tail moth in Europe, position in "sequence"	136
introduction into United States, habits	300–302
reared at laboratory	91
recorded in literature	91
status in United States in 1910	310
gipsy moth in Europe	235, 302
position in "sequence"	132
recorded hosts	92
Palms, food plants of Omiodes blackburni	35

332   PARASITES OF GIPSY AND BROWN-TAIL MOTHS.

	Page.
Pamphilius stellatus, host of *Parexorista cheloniæ*	92
Tachina larvarum	90
Panolis griseovariegata, host of *Echinomyia fera*	89
Pales pavida	92
Tachina larvarum	90
Papilio machaon, host of *Tachina larvarum*	90
Paranagrus optabilis, parasite of *Perkinsiella saccharicida*, introduction into Hawaii	35
perforator, parasite of *Perkinsiella saccharicida*, introduction into Hawaii	35
Parasa hilarula = *Parasa sinica*	170
sinica, hibernating in egg masses of gipsy moth	170
Parasemia plantaginis, host of *Blepharidea vulgaris*	91
Parasetigena segregata, gross number colonized	310
handling of puparia	159
parasite of gipsy moth in Europe, position in "sequence"	132
introduction into United States, habits	229–231
reared at laboratory	88
recorded in literature	88
status in United States in 1910	308
recorded hosts	89
Parasites of brown-tail moth. (*See* Brown-tail moth parasites.)	
gipsy and brown-tail moths, difficulty of naming European species.	68–69
importation	1–312
and handling	152–167
an investigation of the work	50–54
circumstances bringing about beginning of work	49–50
developments of year 1910	311–312
early ideas on introduction	47–49
establishment and dispersion	94–96
gross number of various species colonized	310
improvements in rearing methods	71–73
introduction to bulletin	13–16
narrative of progress of work	54–84
summary and conclusions	305–307
visit of junior author in Russia	125–129
Prof. Kincaid in Japan	73–7
Russia	75, 7
visits of senior author in Europe	54–

INDEX. 333

Page.

Parasites of gipsy and brown-tail moths, imported into United States, present
status............................ 307–310
known and recorded................... 84–92
localities from which material has been
received......................... 168, 169
quantity of material imported...... 167–168
moth. (*See* Gipsy moth parasites.)
injurious insects, early practical work in handling.................. 17–18
Froggatt's journey for investigation of their utility. 42–44
method of encouragement....................... 20
permitting them to escape...................... 18–20
previous work in practical handling............ 16–45
transfer from one country to another............ 23–46
transportation from one part of a given country to
another part................................. 20–23
secondary, host relations........................... 201–202
Parasitism as a factor in insect control....................... 105–109
of brown-tail moth in America................................... 143–151
Europe...................................... 132–135
gipsy moth in America... 136–143
Japan....................................... 120–123
Russia...................................... 123–129
southern France............................. 129–131
native insects, studies therein.................... 102–105
white-marked tussock moth (*Hemerocampa leucostigma*) in country
versus city... 119–120
Parexorista chelonix, American race, parasite of *Malacosoma americana* and *Malacosoma disstria*.. 299–300
biological differences between American and European
races... 257, 286, 299
gross number colonized............................. 310
parasite of brown-tail moth in Europe, position in "sequence".................... 136
introduction into United
States, habits........... 297–300
reared at laboratory......... 91
status in United States in 1910 310
probable interbreeding of American and European forms 299–300
recorded hosts... 92
Paris green against gipsy moth, formerly recommended....................... 47
Parorgyia sp., host of *Apanteles fiskei*.. 265
Parthenogenesis as a factor in establishment of hymenopterous parasites in a
new country 95
in *Melittobia acasta*... 211–212
Schedius kuvanæ.. 179, 183
Trichogramma.. 257–258
Pear, food plant of *Aporia cratægi*....................................... 133
brown-tail moth... 133
wild, containing nests of brown-tail moth................................ 133
Pediculoides ventricosus, enemy of brown-tail moth caterpillars and their parasites.. 267–268
Scutellista cyanea............................ 33
Perilampus cuprinus, an undesirable foreign hyperparasite................... 202
parasite of introduced Tachinidæ....................... 208–209

	Page.
Perilampus hyalinus, secondary parasite of fall webworm (Hyphantria), habits	208–209
Perilitus, parasite of ladybird adults	30
Perissopterus javensis, reared from gipsy-moth eggs	178
Perkinsiella saccharicida, enemy of sugar cane, introduction of parasites and enemies into Hawaiian Islands	35
Pezomachus fasciatus (Fab.)=*Pezomachus melanocephalus* (Schrk.)	85
parasite of gipsy moth, recorded in literature	85
hortensis, parasite of gipsy moth, recorded in literature	85
melanocephalus (Schrk.), *Pezomachus fasciatus* (Fab.) a synonym	85
parasites of *Apanteles fulvipes*	200
Pholera bucephala, host of *Compsilura concinnata*	89
Phora incisuralis, reared from dead gipsy moth (*Porthetria dispar*)	90
Phorocera, four undetermined species, parasites of gipsy moth in America	141
leucaniæ (?), parasite of brown-tail moth in America	93, 145
saundersii, parasite of brown-tail moth in America	93, 145
Phragmatobia fuliginosa, host of *Carcelia excisa*	89
Dexodes nigripes	88, 91
Parexorista chelonix	92
Phryxe erythrostoma, parasite of gipsy moth, recorded in literature	88
recorded hosts	89
vulgaris. (See *Blepharidea vulgaris.*)	
Phygadeuon, parasite of *Compsilura concinnata*	224
Phylloxera of grapevine, prey of *Tyroglyphus phylloxeræ*, attempted control by introduction of its enemy	24
"Physiological" species. (*See* Biological.)	
Phytonomus murinus, enemy of alfalfa, quest for parasites in Europe	46
Pieris brassicæ, host of *Blepharidea vulgaris*	91
Compsilura concinnata	89
Masicera sylvatica	92
daplidice, host of *Blepharidea vulgaris*	91
rapæ (see also *Pontia rapæ*).	
host of *Blepharidea vulgaris*	91
Compsilura concinnata	89
Pimpla brassicariæ, parasite of brown-tail moth in Europe	238
position in "sequence"	136
reared at laboratory	86
gipsy moth in Europe	238
position in "sequence"	132
reared at laboratory	85
conquisitor, not properly a host of *Theronia fulvescens*	137
parasite of brown-tail moth in America	144, 147–149
gipsy moth in America	138, 238
tent caterpillar (Malacosoma)	238
white-marked tussock moth (*Hemerocampa leucostigma*)	238
disparis, parasite of gipsy moth in Japan	238
position in "sequence"	121
reared at laboratory	85
examinator, parasite of brown-tail moth in Europe	144
position in "sequence"	136
reared at laboratory	86

INDEX. 335

	Page.
Pimpla examinator, parasite of brown-tail moth recorded in literature	86
status in United States in 1910	310
gipsy moth in Europe, position in "sequence"	132
introduction into United States	237–239
reared at laboratory	85
recorded in literature	85
host of *Monodontomerus æreus*	246, 249
inquisitoriella, host of *Pimpla instigator*	237
parasite of tussock moth (*Hemerocampa leucostigma*)	237–238
instigator, parasite of brown-tail moth in Europe	144
position in "sequence"	136
reared at laboratory	86
recorded in literature	86
status in United States in 1910	310
gipsy moth in Europe, position in "sequence"	132
introduction into United States, habits	237–239
reared at laboratory	85
recorded in literature	85
Pimpla inquisitoriella	237
pedalis, parasite of brown-tail moth in America	144, 147–149
gipsy moth in America	137–138, 237–239
pluto, parasite of gipsy moth in Japan, position in "sequence"	121
introduction into United States, habits	237–239
reared at laboratory	85
porthetriæ, parasite of gipsy moth in Japan, position in "sequence"	121
introduction into United States, habits	237–239
reared at laboratory	85
spp., gross number colonized	310
parasites of gipsy moth, status in United States in 1910	308, 310
tenuicornis, parasite of gipsy moth in America	138
Pine, food plant of "pine tussock moth"	103
Piroplasmosis, transmission by *Rhipicephalus sanguineus*	41–42
"Planidium" stage of Perilampus	208
Plant-lice, prey of *Coccinella californica*, control by transportation of enemy from one part of country to another part	22–23
repanda, control by introduction of enemy	35
undecimpunctata	24
Hippodamia convergens, control by transportation of enemy from one part of country to another part	22–23
Plum curculio. (*See* Curculio, plum.)	
Plusia festucæ, host of *Compsilura concinnata*	89
gamma, host of *Blepharidea vulgaris*	91
Compsilura concinnata	89
Dexodes nigripes	88, 91
Pales pavida	92
iota, host of *Tachina larvarum*	90
Podisus sp., enemy of *Apanteles fulvipes*	199
Pœcelocampa populi, host of *Compsilura concinnata*	89
Polygnotus hiemalis, parasite of *Mayetiola destructor*, transportation from one part of country to another part	21

Pontia rapæ (see also *Pieris rapæ*).

	Page.
control by caging and permitting parasites to escape	19
host of *Apanteles glomeratus*	24
Compsilura concinnata	223

Poplar (*see also* Populus).

black, food plant of gipsy moth	80
food plant of gipsy moth	57, 82

Populus (*see also* Poplar).

nigra, food plant of gipsy moth	124
Porthesia similis, host of *Compsilura concinnata*	89
Dexodes nigripes	88, 91
Erycia ferruginea	92

Porthetria (*Lymantria*) *monacha* (see also Nun moth).

host of *Carcelia excisa*	89
Compsilura concinnata	89
Echinomyia fera	89
Parasetigena segregata	89, 229
Tachina larvarum	90
Zygobothria bimaculata	90
Prays oleellus, control by caging and permitting parasites to escape	19
Procrustes coriaceus, host of *Blepharidea vulgaris*	91
Proctotrypid parasite of *Compsilura concinnata*	224
Prospaltella berlesei, importation into Peru for controlling *Hemichionaspis minor*	45
parasite of *Diaspis pentagona*, introduction into Italy	38, 44
lahorensis, parasite of *Aleyrodes citri*	46

Pseudococcus. (*See* Mealy bugs.)

Pseudogonatopus spp., parasite of *Perkinsiella saccharicida*, introduction into Hawaiian Islands	35
Pteromalid parasite of *Apanteles fulvipes*	200
Pteromalus chrysorrhœa D. T., *Pteromalus rotundatus* Ratz. a synonym	87
cuproideus, parasite of white-marked tussock moth (*Hemerocampa leucostigma*)	305
Pteromalus egregius Först., *Pteromalus nidulans* Thoms. a synonym	87
gross number colonized	310
host of *Entedon albitarsis*	263, 266, 267, 269
eulophid	202
parasite of *Apanteles lacteicolor*	266, 267, 283–284
brown-tail moth, colonization in United States, host relations	61, 65, 96, 262, 263, 265–267, 268–278
in Europe, position in "sequence"	136
status in United States in 1910	309
Meteorus versicolor	266, 267
unfortunately a secondary as well as a primary parasite	202
halidayanus, parasite of gipsy moth, recorded in literature	86
nidulans, parasite of brown-tail moth, reared at laboratory	87
recorded in literature	87
Thoms.=*Pteromalus egregius* Först	87
pini, parasite of gipsy moth, recorded in literature	86
processioneæ, parasite of brown-tail moth, recorded in literature	87
puparum, parasite of brown-tail moth, recorded in literature	87
rotundatus, parasite of brown-tail moth, recorded in literature	87

	Page.
Pteromalus rotundatus, Ratz.=*Pteromalus chrysorrhœa* D. T.	87
sp., near *cuproideus*, parasite of brown-tail moth	305
parasite of brown-tail moth in Europe, position in "sequence"	136
reared at laboratory	87
Pterostoma palpina, host of *Carcelia excisa*	89
Ptilotachina larvincola, parasite of gipsy moth, recorded in literature	88, 90
monacha, parasite of gipsy moth, recorded in literature	88, 90
Pulvinaria psidii, prey of *Cryptolæmus montrouzieri*, control by introduction of enemy	34–35
Pygæra anachoreta, host of *Compsilura concinnata*	89
curtula, host of *Carcelia excisa*	89
pigra, host of *Zenillia libatrix*	90
Pyrameis atalanta, host of *Compsilura concinnata*	89
Rabbit, cotton-tail, host of *Hæmaphysalis leporis-palustris*	41
Rearing cage. (*See* Cage.)	
Reproduction experiments with *Melittobia acasta*	211–212
Schedius kuvanæ	184
Trichogramma	257–258
Rhipicephalus sanguineus, host of *Hunterellus hookeri*	41–42
parasite of camel	42
dog	42
transmitter of piroplasmosis	41–42
texanus, host of *Hunterellus hookeri*	41
parasite of dog	41
horse	42
Rhizobius ventralis, enemy of black scale (*Saissetia oleæ*), introduction into California	31
Rhogas. (*See* Rogas.)	
Rhyparia purpurata, host of *Parexorista chelonix*	92
Rogas geniculator, parasite of brown-tail moth, recorded in literature	86
pulchripes, parasite of brown-tail moth, recorded in literature	86
testaceus, parasite of brown-tail moth, recorded in literature	86
Rose, food plant of gipsy moth	82
Saissetia oleæ. (*See* Scale, black.)	
Salix (*see also* Willow).	
food plant of gipsy moth	124
Sarcophaga sp., reared from dead gipsy moth (*Porthetria dispar*)	90
Sarcophagid puparia from gipsy moth pupæ, results of rearing work of 1910	141
Sarcophagids, hosts of *Monodontomerus æreus*	246
parasites of grasshoppers	250
probable parasites of "pine tussock moth"	251
that feed upon gipsy moth pupæ; are they parasites or scavengers?	250–251
Saturnia pavonia, host of *Exorista affinis*	89
Masicera sylvatica	92
Tricholyga grandis	88, 92
pyri, host of *Carcelia excisa*	89
Exorista affinis	89
Masicera sylvatica	92
Tachina larvarum	90
Tricholyga grandis	88, 92
spini, host of *Masicera sylvatica*	92

Scale, black (*Saissetia oleæ*), host of *Tomocera californica*, control by transportation of parasite suggested.......... 21
    international work with enemies.......... 31–34
    prey of *Rhizobius ventralis*, control by introduction of latter.......... 31
 coffee, host of *Scutellista cyanea*.......... 31
 cottony cushion. (See *Icerya purchasi*.)
 fluted. (See *Icerya purchasi*.)
 long. (See *Lepidosaphes gloveri*.)
 oyster-shell. (See *Lepidosaphes ulmi*.)
 purple. (See *Lepidosaphes beckii*.)
 San Jose. (See *Aspidiotus perniciosus*.)
 wax. (*See* Ceroplastes.)
 West Indian peach. (See *Diaspis pentagona*.)
Schedius kuvanæ, gross number colonized.......... 310
 host of *Pachyneuron gifuensis*.......... 183
   Tyndarichus navæ.......... 183
 parasite of gipsy moth in Japan, position in "sequence"..... 121
  introduction into United States, habits.......... 75, 176–188
   reared at laboratory.......... 86
   status in United States in 1910.......... 307
 superparasitized by *Schedius kuvanæ*.......... 181–183
 superparasitizing *Anastatus bifasciatus*.......... 181–183
Scutellista cyanea, parasite of coffee scale, black scale (*Saissetia oleæ*), and Ceroplastes spp., introduction into United States and Italy.... 31
 prey of *Pediculoides ventricosus*.......... 33
"Sequence" of parasites.......... 106
 of brown-tail moth in Europe.......... 135, 136
  gipsy moth in Europe.......... 131
   Japan.......... 121
Sericaria mori, host of *Crossocosmia sericariæ*.......... 88
Sigalphus curculionis, parasite of plum curculio, transportation from one part of country to other parts suggested.......... 20
Silkworm, common. (See *Sericaria mori*.)
 Japanese (see also *Antheræa yamamai*).
  "uji" parasite.......... 232–234
Smerinthus ocellatus, host of *Zenillia fauna*.......... 92
 populi, host of *Compsilura concinnata*.......... 89
Species of insects differing in biological characters only........ 225–226, 257, 285–286
Sphinx ligustri, host of *Carcelia excisa*.......... 89
  Masicera sylvatica.......... 92
  Tricholyga grandis.......... 88, 92
Spilosoma lubricipeda, host of *Compsilura concinnata*.......... 89
  Parexorista chelonix.......... 92
 menthastri, host of *Compsilura concinnata*.......... 89
Staphylinid beetle, enemy of *Ceratitis capitata*, introduction into Western Australia.......... 39
Staphylinus olens, enemy of earwigs, practical handling.......... 18
Starvation as factor in control of Heterocampa.......... 103
  insects.......... 114
Stauropus fagi, host of *Compsilura concinnata*.......... 89
  Zygobothria gilva.......... 90

INDEX. 339

	Page.
Stilpnotia salicis, host of *Carcelia excisa*	89
gnava	88
Compsilura concinnata	89
Parexorista cheloniæ	92
Tachina larvarum	90
Sugar-cane borer (*Rhabdocnemis obscurus*), quest of parasites for introduction into Hawaiian Islands	36
food plant of *Perkinsiella saccharicida*	35
plant-lice	35
leafhopper. (See *Perkinsiella saccharicida*.)	
Sycamore, food plant of white-marked tussock moth (*Hemerocampa leucostigma*)	101
Syntomosphyrum esurus, parasite of brown-tail moth in America	139, 144, 147–149
gipsy moth in America	139
parasite of ladybird larvæ	30
Tabanid flies, prey of *Monedula carolina*	45
transmitters of trypanosomiasis of dromedaries	45
Tachina, biological character separating the species *mella* and *larvarum*	257
japonica, gross number colonized	310
host of *Chalcis fiskei*	240
parasite of gipsy moth in Japan, position in "sequence"	121
introduction into United States	227–228
reared at laboratory	88
status in United States in 1910	308
recorded host	88
larvarum, gross number colonized	310
parasite of brown-tail moth in Europe, position in "sequence"	136
introduction into United States	296
reared at laboratory	91
status in United States in 1910	308, 309
gipsy moth in Europe, position in "sequence"	132
introduction into United States, habits	225–227
reared at laboratory	88
recorded in literature	88
status in United States in 1910	308
recorded hosts	90
latifrons, parasite of brown-tail moth, recorded in literature	91
recorded hosts	92
"Tachina-like" parasite of gipsy moth, results of rearing work of 1910	142
Tachina mella and *Tachina larvarum*, biological differences	286
parasite of brown-tail moth in America	93, 145, 147–149
gipsy moth in America	90, 139–140
white-marked tussock moth (*Hemerocampa leucostigma*)	221–223
noctuarum, parasite of gipsy moth, recorded in literature	88
recorded hosts	90
parasite of gipsy moth, relative abundance in Massachusetts and Russia	127
Tachinidæ, hyperparasites	207–213
Tachinid flies, rearing and colonization, large cages versus small cages	204–207
parasites of the brown-tail moth	296–304
gipsy moth	202–236

	Page.
Tachinid puparia, importation and handling	166–167
undetermined, parasite of brown-tail moth in America	145–146, 147–149
Tachinids, host relations, physiological and physical restrictions thereto	202–204
hosts of Melittobia	202
Monodontomerus æreus	246
miscellaneous, parasites of gipsy moth, results of rearing work of 1910.	141
parasites of gipsy moth in Russia	81, 125
undetermined species, reared from gipsy moth in America	141
Tæniocampa stabilis, host of *Compsilura concinnata*	89
Tapinostola elymi, host of *Dexodes nigripes*	88, 91
Telenomus phalænarum, gross number colonized	310
parasite of brown-tail moth in Europe, position in "sequence"	136
introduction into United States	64, 260–261
reared at laboratory	87
recorded in literature	87
status in United States in 1910	309
Tent caterpillar (*see also* Clisiocampa and Malacosoma).	
control by disease	98, 101, 105
host of *Anomalon exile*	144
Limnerium clisiocampæ	143
Pimpla conquisitor	238
parasitism	102, 104–105
forest. (See *Malacosoma disstria*.)	
Tephroclystia virgaureata, host of *Dexodes nigripes*	88, 91
Tetrastichus xanthomelænæ, parasite of elm leaf-beetle (*Galerucella luteola*), introduction into United States	39–41, 62–63
Thalpochæres pannonica, host of *Carcelia excisa*	89
Thalpochares cocciphaga, parasite of black scale (*Saissetia oleæ*), introduction into United States	34
Thamnonona wavaria, host of *Blepharidea vulgaris*	91
Thanasimus formicarius, enemy of barkbeetles, introduction into United States	36
Thaumetopœa pinivora, host of *Dexodes nigripes*	88, 91
pityocampa, host of *Blepharidea vulgaris*	91
Compsilura concinnata	89
Tricholyga grandis	88, 92
processionea, host of *Blepharidea vulgaris*	91
Carcelia excisa	89
Compsilura concinnata	89
Epicampocera crassiseta	89
Pales pavida	92
Zenillia libatrix	90
Thelyotokous hymenopterous parasites, chances of successful establishment in a new country	95
Thelyotoky in Trichogramma	257, 258
Theronia atalantæ, parasite of brown-tail moth in Europe	304
position in "sequence"	136
reared at laboratory	86
recorded in literature	86

INDEX.   341

	Page.
Theronia atalantæ, parasite of gipsy moth in Europe	236
position in "sequence"	132
reared at laboratory	85
recorded in literature	85
fulvescens, not properly a parasite of *Pimpla conquisitor*	137
parasite of brown-tail moth in America	144, 147–149, 304
gipsy moth in America	137, 141, 142, 236–237
host of *Monodontomerus æreus*	246
japonica, parasite of gipsy moth in Japan	236
position in "sequence"	121
melanocephala, not known to be parasite of gipsy moth.	137
Thyridopteryx ephemeræformis, control by caging and permitting parasites to escape	19
Ticks, work with parasites	41–42
Timandra amata, host of *Compsilura concinnata*	89
Tobacco, food plant of *Heliothis obsoleta*	45
Tomocera californica, parasite of *Saissetia oleæ*, transportation from one part of country to another part suggested	21
Torymus anephelus Ratz.=*Monodontomerus æreus* Walk	87
Toxocampa pastinum, host of *Blepharidea vulgaris*	91
Toxoptera graminum, host of *Lysiphlebus tritici*, experiments in control by transportation of its parasite	22
Trachea atriplicis, host of *Compsilura concinnata*	89
Tray, "tanglefooted," for brown-tail caterpillars	280
Trichiocampus viminalis, host of *Compsilura concinnata*	89
Trichogramma, biological characters separating *pretiosa* from European *pretiosa*-like form	257
pretiosa-like form, parasite of brown-tail moth in Europe, position in sequence	136
introduction into United States, habits	256–260
pretiosa, parasite of brown-tail moth in America, habits	143, 256–260
Heliothis obsoleta, shipment to Sumatra	45
sp., parasite of *Archips rosaceana*	259–260
brown-tail moth in Europe, position in "sequence"	136
spp., gross number colonized	310
parasites of brown-tail moth, status in United States in 1910	309
sp. I, parasite of brown-tail moth, reared at laboratory	87
II, parasite of brown-tail moth, reared at laboratory	87
Tricholyga grandis, gross number colonized	310
parasite of brown-tail moth in Europe	296
position in "sequence"	136
reared at laboratory	91
gipsy moth in Europe, position in "sequence"	132
introduction into United States, habits	228–229
reared at laboratory	88
status in United States in 1910	308
recorded hosts	88, 92

	Page.
Trogoderma tarsale, enemy of gipsy moth	252
white-marked tussock moth (*Hemerocampa leucostigma*)	252
Trogus flavitorius [sic] *lutorius* (Fab.)?, parasite of gipsy moth, recorded in literature	85
Trypanosomiasis of dromedaries, transmission by tabanid flies	45
Tussock moth, pine, control by disease	101
parasitism	103–104
probable host of sarcophagids	251
white-marked (*Hemerocampa leucostigma*), host of *Anthrenus varius*	252
Apanteles sp. (*delicatus?*)	138
Compsilura concinnata	221–223
Meteorus versicolor	221–223, 289
Monodontomerus æreus	249
Pimpla conquisitor	238
Pimpla inquisitoriella	237
Pimpla instigator	237
Pteromalus cuproideus	305
Tachina mella	221–223
Trogoderma tarsale	252
parasitism	102
in country versus city	119–120
Tyndarichus navæ, parasite of *Anastatus bifasciatus*	183
Schedius kuvanæ	183
reared from gipsy moth eggs	153, 171, 178
superparasitized by *Tyndarichus navæ*	183
superparasitizing *Pachyneuron gifuensis*	183
Tyroglyphus phylloxeræ, enemy of grapevine Phylloxera, introduction into Europe	24
"Uji" parasite of silkworm in Japan. (See *Crossocosmia sericariæ*.)	
Ulmus (*see also* Elm).	
food plant of gipsy moth	124
Vanessa antiopa, host of *Blepharidea vulgaris*	91
Blepharipa scutellata	88
Compsilura concinnata	89
Tachina larvarum	90
io, host of *Argyrophylax atropivora*	89
Blepharidea vulgaris	91
Compsilura concinnata	89
Dexodes nigripes	88, 91

INDEX. 343

	Page.
Vanessa io, host of *Erycia ferruginea*	92
Tachina larvarum	90
Tricholyga grandis	88, 92
polychloros, host of *Dexodes nigripes*	88, 91
Tachina larvarum	90
urticæ, host of *Blepharidea vulgaris*	91
Compsilura concinnata	89
Dexodes nigripes	88, 91
Tachina larvarum	90
xanthomelas, host of *Blepharidea vulgaris*	91
Compsilura concinnata	89
Varichæta aldrichi, parasite of fall webworm (Hyphantria), adaptations in habits to host	203
Vedalia cardinalis. (See *Novius cardinalis*.)	
Verania cardoni, enemy of *Aleyrodes citri*, attempted introduction into United States	46
frenata, enemy of *Perkinsiella saccharicida*, introduction into Hawaiian Islands	35
lineola, enemy of *Perkinsiella saccharicida*, introduction into Hawaiian Islands	35
Weather conditions as affecting rate of increase of the gipsy moth	110
factors in control of insects	107
Webworm, fall (Hyphantria), host of *Compsilura concinnata*	224
Meteorus versicolor	289
Varichæta aldrichi	203
Perilampus hyalinus a secondary parasite	208–209
studies in parasitism	104
Weevil, cotton boll (*Anthonomus grandis*), control by transportation of parasites	21–22
method of encouraging parasites	20
Wheat midge. (See *Contarinia tritici*.)	
White fly of orange. (See *Aleyrodes citri*.)	
Willow (*see also* Salix).	
food plant of gipsy moth	80
"Wilt" disease of gipsy moth	97–102
Yponomeuta evonymella, host of *Tachina larvarum*	90
Zenillia libatrix	90
padella, host of *Compsilura concinnata*	89
Zenillia libatrix	90
Zenillia fauna, parasite of brown-tail moth, recorded in literature	91
recorded hosts	92
libatrix, gross number colonized	310
parasite of brown-tail moth in Europe, position in "sequence"	136
introduction into United States, habits	302–303
reared at laboratory	91
recorded in literature	91
gipsy moth, recorded in literature	88
recorded hosts	90
Zygæna achilleæ, host of *Blepharidea vulgaris*	91
filipendulæ (?), host of *Blepharidea vulgaris*	91
Tachina latifrons	92
janthina, host of *Blepharidea vulgaris*	91

	Page.
Zygobothria bimaculata, parasite of gipsy moth, recorded in literature	88
recorded hosts	90
gilva, gross number colonized	310
parasite of gipsy moth in Europe, position in "sequence"	132
reared at laboratory	88
recorded in literature	88
status in United States in 1910	308
recorded hosts	90
nidicola, gross number colonized	310
host of *Monodontomerus æreus*	267
parasite of brown-tail moth, colonization in United States, habits	262, 264–267, 289–295
in Europe, position in "sequence"	136
reared at laboratory	91
results of rearing work in 1910	147–149
status in United States in 1910	309
gipsy moth, introduction into United States, habits	232

ERRATA TO BULLETIN NO. 91, BUREAU OF ENTOMOLOGY.

Page 6, line 15 from bottom, indent *Limnerium (Anilastus) tricoloripes* Vier. 2 ems more.
Page 9, line 3, for *The Calosoma beetles* read *Calosoma sycophanta.*
Page 9, line 18, for *The gipsy moth* read *Different stages of the gipsy moth.*
Page 9, line 19, for *The brown-tail moth* read *Different stages of the brown-tail moth*
Page 10, line 18 from bottom, for *webs* read *web.*
Page 24, line 20, after *Muls* insert period.
Page 35, line 7, for *Dactylopius* read *Pseudococcus.*
Page 41, line 2 from bottom, for *trypanosome* read *protozoan.*
Page 42, line 6, for *trypanosomiasis* read *piroplasmosis.*
Page 88, line 11 from bottom, for *polychlorus* read *polychloros.*
Page 89, last line, for *Haloicus* read *Hyloicus.*
Page 90, line 13 from bottom, for *Lymantria* read *Porthetria.*
Page 91, left-hand column, omit *Digonichæta setipennis* Fall., *Digonichæta spinipennis* Meig., *Nemorilla sp.*, and *Nemorilla notabilis* Meig., and close up the column.
Page 91, omit last three lines.
Page 92, omit lines 2, 3, 11, 12, and 13.
Page 93, lines 4, 5, and 6, omit *Blepharipeza leucophrys* Wied., *Sturmia discalis* Coq., and *Exorista griseomicans* V. de Wulp.
Page 93, line 4, left-hand column, insert *Phorocera saundersii* Will.
Page 93, line 6, left-hand column, insert *Exorista boarmiæ* Coq.
Page 93, right-hand column, move *Tachina mella* Walk. from line 6 to line 5.
Page 95, line 14 from bottom, for *pathogenetically* read *parthenogenetically.*
Page 116, line 10 from bottom, for *III* read 111.
Page 145, line 2, for *Chalcis sp.* read *Chalcis compsiluræ Crawf.*
Page 145, lines 5 to 9, inclusive, read as follows: has been examined and compared with other species of the genus by Mr. J. C. Crawford, who has found it to be distinct and has described it under the name *Chalcis compsiluræ.*
Page 145, line 14 from bottom for "*Native Parasite of chrysorrhœa*" read (?) *Phorocera leucaniæ* Coq.
Plate VI (facing p. 156), line 2, for *below,* at read *Japanese variety, lower.*
Plate VI (facing p. 156), line 3, omit *enlarged* and *somewhat reduced.*
Plate VI (facing p. 156), line 4, omit *somewhat reduced.*
Plate VI (facing p. 156), line 5, after *center* insert *All slightly reduced.*
Plate VII (facing p. 160), line 2, for *another winter nest* read *cocoon in leaf.*
Plate VII (facing p. 160), line 3, omit *somewhat reduced.*
Plate VII (facing p. 160), line 4, for *the eggs hatching* read *torn open, showing eggs.*
Plate VII (facing p. 160), line 5, after *leaf* insert *All somewhat reduced.*
Page 170, line 5 from bottom, after *Staud* insert period.
Page 183, line 2 from bottom, for *thelyotoky* read *arrhenotoky.*
Page 193, line 2, *Apanteles fulvipes* Hal. should be printed in 10-point small capitals.
Page 198, line 11, the words *secondary parasites attacking apanteles* should begin with capitals.
Page 199, figure 34, the upper *b* in cut should be *d.*
Page 199, figure 34, the lower *c* in cut should be *a.*
Page 213, line 12, for *are* read *is.*
Page 229, line 5 from bottom, for *Liparis* read *Porthetria.*
Page 237, line 12 from bottom, for *Hoplectis* read *Itoplectis.*
Page 237, line 17 from bottom, after *Say* omit period.
Page 237, line 18 from bottom, for *Hoplectis* read *Itoplectis.*
Page 258, line 8, for *arrhenotokous* read *thelyotokous.*
Page 258, line 9, for *thelyotokous* read *arrhenotokous.*
Page 258, line 15, for *arrhenotokous* read *thelyotokous.*
Page 267, Table XI, column 2, line 6, after *Mesochorus pallipes,* for [2] read [1].
Page 267, Table XI, column 2, line 8, after *æreus,* for [2] read [1].
Page 285, line 18 from bottom, for *conspersæ* read *conspersa.*
Page 309, line 7, for *Caosoma* read *Calosoma.*
Page 310, left-hand column, line 5, between *Anastatus* and *bifasciatus* insert space.
Page 310, right-hand column, line 11, for *silvatica* read *sylvatica.*

HISTORY OF ECOLOGY
An Arno Press Collection

Abbe, Cleveland. **A First Report on the Relations Between Climates and Crops.** 1905

Adams, Charles C. **Guide to the Study of Animal Ecology.** 1913

American Plant Ecology, 1897-1917. 1977

Browne, Charles A[lbert]. **A Source Book of Agricultural Chemistry.** 1944

Buffon, [Georges-Louis Leclerc]. **Selections from Natural History, General and Particular, 1780-1785.** Two volumes. 1977

Chapman, Royal N. **Animal Ecology.** 1931

Clements, Frederic E[dward], John E. Weaver and Herbert C. Hanson. **Plant Competition.** 1929

Clements, Frederic Edward. **Research Methods in Ecology.** 1905

Conard, Henry S. **The Background of Plant Ecology.** 1951

Derham, W[illiam]. **Physico-Theology.** 1716

Drude, Oscar. **Handbuch der Pflanzengeographie.** 1890

Early Marine Ecology. 1977

Ecological Investigations of Stephen Alfred Forbes. 1977

Ecological Phytogeography in the Nineteenth Century. 1977

Ecological Studies on Insect Parasitism. 1977

Espinas, Alfred [Victor]. **Des Sociétés Animales.** 1878

Fernow, B[ernhard] E., M. W. Harrington, Cleveland Abbe and George E. Curtis. **Forest Influences.** 1893

Forbes, Edw[ard] and Robert Godwin-Austen. **The Natural History of the European Seas.** 1859

Forbush, Edward H[owe] and Charles H. Fernald. **The Gypsy Moth.** 1896

Forel, F[rançois] A[lphonse]. **La Faune Profonde Des Lacs Suisses.** 1884

Forel, F[rançois] A[lphonse]. **Handbuch der Seenkunde.** 1901

Henfrey, Arthur. **The Vegetation of Europe, Its Conditions and Causes.** 1852

Herrick, Francis Hobart. **Natural History of the American Lobster.** 1911

History of American Ecology. 1977

Howard, L[eland] O[ssian] and W[illiam] F. Fiske. **The Importation into the United States of the Parasites of the Gipsy Moth and the Brown-Tail Moth.** 1911

Humboldt, Al[exander von] and A[imé] Bonpland. **Essai sur la Géographie des Plantes.** 1807

Johnstone, James. **Conditions of Life in the Sea.** 1908

Judd, Sylvester D. **Birds of a Maryland Farm.** 1902

Kofoid, C[harles] A. **The Plankton of the Illinois River, 1894-1899.** 1903

Leeuwenhoek, Antony van. **The Select Works of Antony van Leeuwenhoek.** 1798-99/1807

Limnology in Wisconsin. 1977

Linnaeus, Carl. **Miscellaneous Tracts Relating to Natural History, Husbandry and Physick.** 1762

Linnaeus, Carl. **Select Dissertations from the Amoenitates Academicae.** 1781

Meyen, F[ranz] J[ulius] F. **Outlines of the Geography of Plants.** 1846

Mills, Harlow B. **A Century of Biological Research.** 1958

Müller, Hermann. **The Fertilisation of Flowers.** 1883

Murray, John. **Selections from *Report on the Scientific Results of the Voyage of H.M.S. Challenger During the Years 1872-76*.** 1895

Murray, John and Laurence Pullar. **Bathymetrical Survey of the Scottish Fresh-Water Lochs. Volume one.** 1910

Packard, A[lpheus] S. **The Cave Fauna of North America.** 1888

Pearl, Raymond. **The Biology of Population Growth.** 1925

Phytopathological Classics of the Eighteenth Century. 1977

Phytopathological Classics of the Nineteenth Century. 1977

Pound, Roscoe and Frederic E. Clements. **The Phytogeography of Nebraska.** 1900

Raunkiaer, Christen. **The Life Forms of Plants and Statistical Plant Geography.** 1934

Ray, John. **The Wisdom of God Manifested in the Works of the Creation.** 1717

Réaumur, René Antoine Ferchault de. **The Natural History of Ants.** 1926

Semper, Karl. **Animal Life As Affected by the Natural Conditions of Existence.** 1881

Shelford, Victor E. **Animal Communities in Temperate America.** 1937

Warming Eug[enius]. **Oecology of Plants.** 1909

Watson, Hewett Cottrell. **Selections from *Cybele Britannica*.** 1847/1859

Whetzel, Herbert Hice. **An Outline of the History of Phytopathology.** 1918

Whittaker, Robert H. **Classification of Natural Communities.** 1962